Henry William Fuller

On Diseases of the Lungs and Air-Passages

Their Pathology, Physical Diagnosis, Symptoms and Treatment

Henry William Fuller

On Diseases of the Lungs and Air-Passages
Their Pathology, Physical Diagnosis, Symptoms and Treatment

ISBN/EAN: 9783337812553

Printed in Europe, USA, Canada, Australia, Japan

Cover: Foto ©berggeist007 / pixelio.de

More available books at **www.hansebooks.com**

ON DISEASES

OF THE

LUNGS AND AIR-PASSAGES:

THEIR

PATHOLOGY, PHYSICAL DIAGNOSIS, SYMPTOMS, AND TREATMENT.

BY

HENRY WILLIAM FULLER, M.D., CANTAB.,

FELLOW OF THE ROYAL COLLEGE OF PHYSICIANS, LONDON;
PHYSICIAN TO ST. GEORGE'S HOSPITAL;
ETC. ETC.

From the Second and Revised London Edition.

PHILADELPHIA:
HENRY C. LEA.
1867.

PREFACE.

There are few students who do not experience misgivings on commencing the study of diseases of the chest. The subject embraces such a wide field of research, and is at once so difficult and complicated, that they are quite at a loss as to the best means of approaching it. Unfortunately their dislike of "dry detail," and their anxious longings after "practical" knowledge, induce them too often to attempt a method which is quite impracticable. Instead of making themselves familiar with the mechanism in which the physical signs of disease originate, and connecting each one separately with its corresponding change in the physical condition of the heart or lungs, they begin by attempting to group together certain physical signs and general symptoms as indicative of particular forms of disease. Before they are familiar with the very elements of auscultation or have formed a clear conception of the nature and value of any of its signs, they endeavor to estimate the precise importance attaching to the presence of particular signs, and thus to diagnose existing mischief. No wonder that they so often find themselves bewildered, and are ready to discard the stethoscope as, to them at least, a hopeless mystery. Auscultation has a language of its own, a language clear and forcible when properly understood, but otherwise obscure, mysterious, and unintelligible. This language it is almost impossible to acquire without a previous knowledge of its alphabet; and if, by extraordinary diligence and perseverance, the more zealous have learned to interpret it correctly, the less fortunate and less persevering have notoriously failed in mastering its difficulties.

My object has been to lessen these difficulties, and to render attainable by men of ordinary capacities and ordinary opportunities a science which is indispensable to every medical practitioner. I have endeavored to begin at the beginning, to assume nothing, and to explain every auscultatory sign by reference to the morbid condition and consequent altered mechanism in which each takes its origin. My wish has been to inculcate the necessity for regarding each physical sign, not as indicative of a certain disease, but rather as the natural consequence of a certain physical alteration in the tissues, the source and true interpretation of which must be determined by concomitant circumstances.

I have endeavored to use the simplest language, so as to obviate the formidable difficulty presented by the confused and varied phraseology made use of by many writers on the subject, to give a definite meaning to each term which is employed, and to present a classification of the various sounds which shall be intelligible even to a novice at auscultation. How far I have succeeded it will be for others to determine; but I shall have done good service if I have even cleared the way for other laborers in the same field.

My explanation of the mechanism, and true significance of ægophony, and of sundry other abnormal sounds, is at variance with that which is generally received; but in this, and in every other instance, I have stated the grounds on which my opinion rests, and it will be for future observers and future experimentalists to estimate their real value.

Under the head of "Consumption" I have brought together

a variety of facts which serve to elucidate several points on which erroneous ideas are commonly entertained, and have discussed and expressed my views as to others about which there is considerable difference of opinion.

I have purposely avoided incumbering my pages with the details of remedies and modes of treatment which experience has proved to be undeserving of confidence, and have contented myself with describing the particular methods of treatment which have appeared to me most generally successful, and based upon the soundest physiological grounds.

Throughout the work I have availed myself of the labors of my predecessors, and have profited largely by their investigations. Whenever I have seen reason to dissent from their conclusions, I have not hesitated to avow it, and have stated the grounds of my opinion. In no other way is it possible to obtain a full discussion of the principles on which our science is founded, or to arrive at a satisfactory conclusion respecting their merits.

I need scarcely say how gratifying it would be to me to learn that the views which I have propounded have received the approval of my professional brethren; but as my sole object is to elicit the truth, I trust that those gentlemen who meet with facts which in any way elucidate the questions at issue will kindly favor me with the result of their experience.

13 Manchester Square, London, W.;
1862.

CONTENTS.

PART I.

THE PRINCIPLES OF PHYSICAL DIAGNOSIS, AND THEIR APPLICATION TO THE INVESTIGATION OF DISEASES OF THE LUNGS.

For further particulars, see Index.

PART II.

THE PATHOLOGY, DIAGNOSIS, SYMPTOMS, AND TREATMENT OF DISEASES OF THE LUNGS.

CHAPTER I.

CHAPTER II.

CHAPTER III.

CHAPTER IV.

CHAPTER V.

FOR FURTHER PARTICULARS, SEE INDEX.

Plate I.

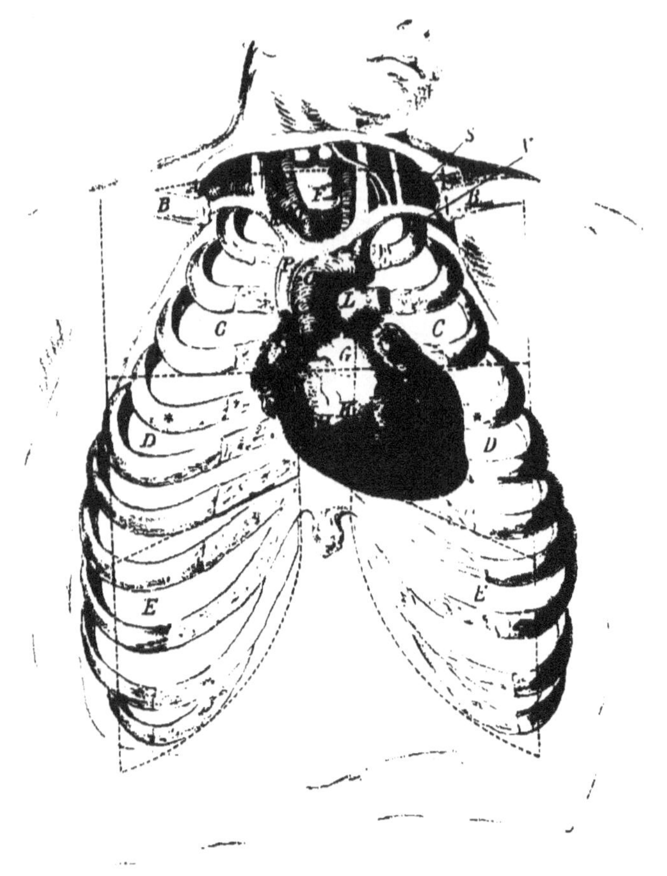

[illegible] A. [illegible] region. B. [illegible] C. [illegible] D. [illegible] E. [illegible] F. [illegible] G. [illegible] H. [illegible] + Nodules. J. [illegible] K. [illegible] L. [illegible] M. [illegible] N. [illegible] O. [illegible] P. [illegible] Q. [illegible] R. [illegible] vein. S. [illegible] V. [illegible]

Plate II.

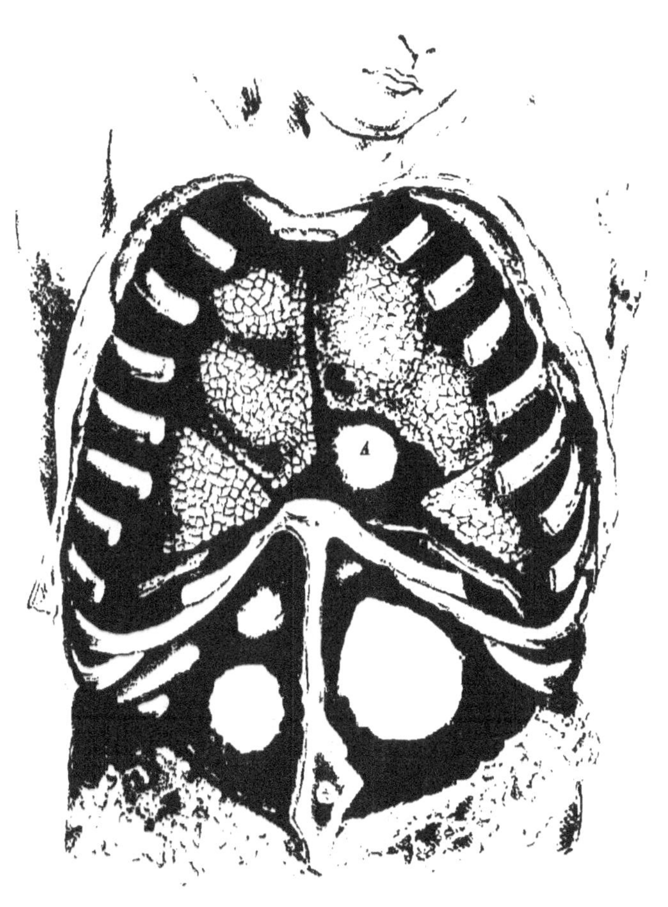

DISEASES

OF THE

LUNGS AND AIR-PASSAGES.

PART I.

CHAPTER I.

PRINCIPLES OF PHYSICAL DIAGNOSIS—TOPOGRAPHY OF THE WALLS OF THE CHEST—CONTENTS OF THE VARIOUS REGIONS.

"Felix, qui potuit rerum cognoscere causas."

WHEN a student's attention is first directed to the physical diagnosis of diseases of the chest, he rarely forms a just estimate of its importance. He is apt either to underrate or else over-estimate its true value. In either case he starts with erroneous impressions, which, in one way or another, must ultimately cause vexation and disappointment. It may be well, therefore, to point out the true scope and object of the study on which we are about to enter, in order that beginners may be encouraged to persevere in mastering its difficulties.

The lungs, the heart, and the large vessels are placed within a bony framework, where the eye cannot penetrate and the sense of touch is of little avail. Thus the senses usually employed in the investigation of disease are rendered almost useless, and as the *general* symptoms of diseases of the chest are uncertain and fallacious, the practitioner who knows not how to conduct a physical examination of the chest, must treat a whole class of dangerous maladies in ignorance of the true nature of his patient's disorder. Here it is that the stethoscopist steps in. His aim is to discover the physical condition of the heart and lungs; to detect the early inroads of disease; to chronicle its onward progress, or haply note its gradual subsidence. His first object is to unveil the nature of the changes going on within the chest, with a view to the adoption of remedial measures calculated to alleviate his patient's suffering; his next to ascertain the precise extent of those changes, in order that he may be able either to speak hopefully as to the issue or to warn the patient of his danger.

It is right, however, to approach the subject with the full and clear understanding that physical signs will not of themselves enable him to accomplish the object he has in view. As the constitutional symptoms which accompany chest diseases are often perplexing and fallacious, so the physical signs, when taken by themselves, are apt to lead to grievous error. The former are so uncertain and variable in their character that constant reference to the physical signs is required to confirm or negative their indications; whilst the latter are mere exponents of physical changes in the organs within the chest, and need all the light which can be thrown upon them by a careful consideration of the general symptoms before they can be trusted as interpreters of disease or guides to its rational treatment. An exclusive reliance upon the one class of symptoms is as mischievous as a blind confidence in the other, and both prove fruitful sources of mistaken diagnosis and consequent improper treatment.

With a view to a clear understanding of the subject, I purpose beginning by an exposition of the principles on which the physical diagnosis of chest diseases is founded. It will then be seen how naturally, how necessarily the physical signs arise out of physical changes which have occurred, and how readily, after tracing morbid changes to the production of characteristic physical signs, the mind is able to invert the picture, and refer morbid sounds to their corresponding physical changes.

Physical diagnosis, then, rests on the fact that disease produces anatomical changes in the organs it affects, that such changes give rise to corresponding alterations in the physical properties of those organs, and that, knowing the physical properties of each part in a state of health, we are enabled to judge of the morbid changes which have occurred by the character of the alteration which has taken place in its physical properties. Of these we judge, 1st, by the careful inspection or ocular examination of the chest; 2dly, by palpation or manual examination; 3dly, by mensuration; 4thly, by percussion; 5thly, by auscultation; 6thly, by succussion.

The first of these methods of examination demands a previous knowledge of the form, size, and capacity of a healthy chest. the nature and position of its contents, and its action during ordinary and forced respiration, as also of the various changes it may undergo in one or all of these particulars from influences external to the lungs, heart, and great vessels.

The three last require for their elucidation an intimate acquaintance with the laws which govern the production and transmission of sound, and, before they can be made available for the diagnosis of disease, a careful study of the sounds emitted under various circumstances by different portions of a healthy chest.

With the view of localizing the physical signs, and thus rendering description as accurate as possible, it is expedient to map out the surface of the chest into regions admitting of easy recognition. The natural divisions of anterior, posterior, and lateral are not in this instance sufficiently minute, as each of the spaces thus marked out embraces an extent of surface which renders accuracy of description impossible. It therefore becomes necessary to adopt some other and smaller boundaries; and the following, which approximate closely to those proposed by many authors, will be found well defined, and sufficiently small for practical purposes:

Anteriorly on either side—1st, supra-clavicular; 2d, clavicular; 3d, infra-clavicular; 4th, mammary; 5th, infra-mammary.
" in the centre—1st, supra-sternal; 2d, superior sternal; 3d, inferior sternal.
Laterally 1st, axillary; 2d, infra-axillary.
Posteriorly on either side—1st, superior scapular; 2d, inferior scapular; 3d, inferior dorsal.
" in the centre—inter-scapular.

Of these regions it will be observed that the inter-scapular region and the three sternal regions are single, whilst all the rest are double, existing equally on either side of the chest.

The *supra-clavicular* space has for its upper boundary a line drawn from the claviculo-scapular articulation to the upper rings of the trachea; below it has the clavicle, and inside the trachea.

The *clavicular* is that which lies beneath the inner three-fifths of the clavicle, and has the bone for its anterior boundary.

The *infra-clavicular* has for its limits the clavicle above, the inferior margin of the third rib below, the sternum inside. and outside a line falling vertically from the junction of the outer and middle third of the clavicle.

The *mammary* is bounded above by the third rib; below, by the inferior margin of the sixth rib; inside, by the edge of the sternum; outside by a vertical line continuous with the outer boundary of the infra-clavicular region.

The *infra-mammary* is that portion of the anterior surface of the chest which lies below the mammary. Above it has the sixth rib; below, a curved line corresponding to the edges of the false ribs; inside, the inferior portion of the sternum; outside, a continuation of the outer boundary of the mammary region.

The *supra-sternal* is the hollowed space which lies immediately above the notch of the sternum, and is bounded on either side by the sterno-mastoid muscle.

The *superior-sternal* is the space bounded anteriorly by that portion of the sternum which lies above the lower margin of the third rib.

The *inferior-sternal* corresponds to that portion of the sternum which lies below the lower margin of the third rib.

The *axillary* has for its limits the axilla above; below, a line carried backwards from the lower boundary of the mammary region to the inferior angle of the scapula; in front, the outer margin of the infra-clavicular and mammary regions; and behind, the external edge of the scapula.

The *infra-axillary* has the axillary region for its upper, and the edges of the false ribs for its lower boundary; in front it is limited by the infra-mammary, and behind by the inferior dorsal region.

The *superior and inferior scapula* together occupy the space from the second to the seventh rib, and are respectively identical in their outlines with the upper and lower fossa of the scapula.

The *inter-scapular* is the space situated between the inner margins of the scapulæ and the spines of the dorsal vertebræ from the second to the sixth.

The *inferior dorsal* lies below the inferior scapula and the inter-scapular regions, which form its superior boundary; below it extends to the twelfth rib; inside to the spine; outside to a line falling vertically from the inferior angle of the scapula.

To these, then, or to somewhat similar divisions, constant reference should be made in describing the position of any alteration observed either in the walls of the chest or in the organs contained within them; and therefore at the very outset of his investigation the student should make himself familiar with their limits.

When once this preliminary knowledge has been acquired, his next object should be to study the natural position of the thoracic organs, and ascertain precisely what portions of them correspond to these artificial external divisions. Without such a knowledge he cannot judge what signs he may expect to find there in a state of health, nor can he distinguish the accession of disease.

The following statement of the contents of the several regions will be found sufficiently minute for practical purposes:

The *supra-clavicular* region on either side contains the very apex of the lung, with portions of the subclavian and carotid arteries, and of the subclavian and jugular veins.

The *clavicular* is occupied by the parenchyma of the lung. On the right side, at its outer extremity, lies the subclavian artery; at the sterno-clavicular articulation, the arteria innominata; and on the left side, deeply seated, are the carotid and subclavian arteries.

The *infra-clavicular* has on either side the superior lobe of the lung and the main bronchus: the right bronchus lies behind, the left a little below the second costal cartilage; on the right side is the superior cava and a small portion of the right auricle of the heart; on the left is a portion of the pulmonary artery, and in some instances a portion of the subclavian artery. The aorta and the pulmonary artery lie immediately behind the second sterno-costal articulation—the one on the right, the other on the left side of the sternum. On the left side the lower boundary of this region very nearly corresponds to the base of the heart.

The *mammary* region differs greatly in its contents on the two sides. On the right, the lung is found in front extending down to the sixth rib, where its thin, sharp border very nearly corresponds to the lower boundary of this region. The fissure which separates the upper and middle lobes commences about the fourth cartilage; that which separates the middle and lower lobes, about the fifth interspace. The right wing of the diaphragm, though not attached higher than the seventh rib, is usually pushed up by the liver as high as the fourth interspace, the lower surface of the lung being concave to receive it. A portion of the right auricle and the superior angle of the right ventricle lie close to the sternum, between the third and the fifth ribs.

On the left side, the lung lies in front down to about the fourth sterno-costal articulation; thence its anterior border passes downwards and outwards (leaving an open space for the heart), until it reaches the fifth rib, when it curves forwards and downwards to opposite the sixth rib or interspace, and thence passes outwards almost horizontally. The open space where the heart, so to speak, comes in contact with the chest is usually about an inch and a half or two square inches in extent. The left auricle and ventricle, and a portion of the right ventricle towards the apex, are found in the left mammary region. The apex lies immediately above the sixth costal cartilage. The fissure which separates the two lobes of the left lung usually commences about the fifth interspace.

The *infra-mammary* on the right side contains the liver, with the lung protruding in front on full inspiration; on the left, the anterior border of the spleen, and the stomach, with a portion of the left lobe of the liver, lying in front of it. towards the mesian line. The stomach usually rises to a level with the sixth rib.

The *supra-sternal* is occupied chiefly by the trachea: but on the right side is the arteria innominata. The arch of the aorta sometimes reaches to its lower border, and may be felt pulsating there, on deep pressure.

In the *superior sternal*, the lung lies in front, and immediately behind it are the ascending and transverse portions of the arch of the aorta, and the pulmonary artery from its origin to its bifurcation. The pulmonary valves are situated close to the left edge of the sternum, on a level with the lower margin of the third rib: the aortic about half an inch lower down, more deeply seated, and midway between the mesian line and the left edge of the sternum. The trachea bifurcates on a level with the second ribs.

The *inferior sternal* has lung in front on the right side throughout its whole extent, and on the left down to about the fourth sterno-costal articulation. Below this, on the left side, lies the chief part of the right ventricle of the heart, and a small portion of the left ventricle. The mitral valve is situated close to the left edge of the sternum, on a level with the fourth sterno-costal articulation; and the tricuspid lies nearer the mesian line, and more superficially. Inferiorly is the attachment of the heart to the diaphragm, and below this again a small portion of the liver, and sometimes of the stomach.

In the *axillary*, the subjacent parts are the upper lobes of the lungs, with the main bronchi deeply seated.

The *infra-axillary* contains the lung, with its lower margin sloping down from before backwards. On the right side is the liver; on the left the stomach and spleen.

The *superior and inferior scapular* have beneath them the parenchyma of the lung.

In the *inter-scapular*, towards the centre, opposite the third dorsal vertebra, will be found the bifurcation of the trachea, the main bronchi, and the bronchial glands; on either side parenchyma of the lungs; on the left the œsophagus; and from the upper part of the fourth dorsal vertebra downwards, the descending aorta.

In the *inferior dorsal*, the lung lies superficially as low as the eleventh rib. On the right side, the liver extends downwards from the level of the eleventh rib; and on the left lie the stomach and intestines. Along the left side of the spine runs the descending aorta; and on both sides, close to the spine, is found a small portion of the kidney.

CHAPTER II.

INSPECTION OF THE CHEST—HOW TO BE CONDUCTED; WHAT TO BE OBSERVED.

A CAREFUL ocular examination of the chest will often furnish important information respecting the condition of the thoracic viscera. It shows us not only the shape of the chest and any irregularities, congenital or acquired, presented by any portion of its surface, but it makes us acquainted with the motions of the thoracic walls, the character of the respiratory efforts, the frequency with which they are repeated, their evenness or irregularity, the ease or difficulty with which they are performed, and the part played by the various muscles of respiration. It enables us to note the rhythm[1] of the respiration, to estimate the exact time occupied by the acts of inspiration and expiration respectively, and to compare the movements of the two sides of the chest in regard to their symmetry, regularity, and force. These are all important points in the diagnosis of disease, as will be seen when their practical bearing is explained.

[1] By the term rhythm is meant the mode of progression and evolution of the sounds, which may be gradual and equable from first to last, or irregular, jerking, or interrupted.

But if we wish to make inspection subservient to the diagnosis of thoracic disease, it behooves us, first of all, to examine carefully the form, size, and movements of a *healthy* chest. Without this preliminary study it will be impossible to detect any physical alteration, still more to estimate its importance.

In conducting an examination of the chest whether by inspection or by any other means, it is of the utmost importance that the patient's position should be free and unrestrained, and that, whether he be lying down, sitting, or standing, the plane on which he rests should be perfectly even, and his limbs in the same position on either side. Inattention to these points may lead to irregularity more or less discernible between the two sides of the thorax, when the one is carefully compared with the other, and may give rise to differences more or less marked, when corresponding portions of the chest are contrasted by the aid of mensuration, percussion, and auscultation.

The patient, then, being placed in an easy posture (sitting is the easiest and most convenient), with the surface of the chest exposed, in a good light opposite the observer, and the arms similarly disposed on either side, the thorax should be examined in front, behind, sideways, and from above downwards. Its form, size, and movements should be noted during ordinary and forced respiration, every deviation from the healthy standard marked, and every region on the one side compared with its corresponding region on the other.

The apparent size of the thorax in relation to the body varies greatly in different individuals enjoying equally good health, and no certain rule, therefore, can be laid down respecting it. It is sufficient for our purpose at present to know that a large broad chest does not (as is popularly supposed) denote a healthy chest, nor a small or somewhat narrow chest an unhealthy.[1]

The form of a well-proportioned chest being well known to anatomists, it is needless to enter into a description of it here. Suffice it to say, that in a normal state its two sides are symmetrical in every part, and the intercostal spaces more or less distinctly visible, according as the individual is more or less fat.

Considerable alterations, however, in the form of the chest are compatible with a healthy condition of the thoracic viscera, and it therefore becomes necessary to make ourselves acquainted with the frequency of their occurrence, in order that we may be enabled to estimate correctly the indications they severally afford.

[1] See "Med.-Chir. Trans.," vol. xxix, p. 172–4.

It appears from the investigations of M. Woollez[1] that a *perfectly* symmetrical chest is of rare occurrence, existing in scarcely more than one out of every five of the adult male population. This may be owing to causes either of physiological or pathological origin; the former unconnected with thoracic disease, being the result of congenital malformation, or of tight lacing, or else produced by the habits or pursuits of the individual; the latter the effects of existing or former pectoral disease. In either case the defects may be *general*, affecting the whole form or shape of the thorax; or *partial*, giving rise to local alterations, without interfering with its general shape. But whether the defect be general or local, it is obvious that mere want of symmetry in the chest cannot be depended on as an indication of visceral disease. Its chief value is that it arrests attention and suggests inquiry as to the existence of disease at a former period, as also into the presence of characteristic signs denoting subjacent structural alteration.

When the form of the chest is altered, it may be in one or other of the following ways: The entire thorax, or any part of it, may either bulge or fall in to an unnatural extent; there may be abnormal elevation, or lowering of the whole or any part of it, or it may be more or less distorted.

The *bulging* may be either general or local; the walls of the chest on both sides may be unusually prominent; or the prominence may be confined to one side, or it may be even more circumscribed, existing only in one or two small localities. But whatever its form, its cause, supposing it to arise from disease, is always the outward pressure of some force existing within the chest. Therefore, if it is general, it takes its origin in some disease which produces such pressure over the whole of either or both sides; as, for instance, general vesicular emphysema, pneumothorax, empyema, and other rarer forms of mischief, whilst, if it is partial only, it is caused by an expanding force of an equally limited nature; such, for instance, as partial emphysema, circumscribed empyema, aneurismal and other intrathoracic tumors, or enlargement of the heart, liver, or spleen. Simple hydrothorax has been said to be incapable of occasioning bulging, whether general or local; but I am satisfied that the assertion is incorrect; for although it holds good as a general rule, I have met with several well-marked instances of bulging chest, which, from the effect of treatment, I cannot doubt to have resulted from long-standing and copious serous effusion into the pleural cavity; and in one case in which bulging was

[1] "Recherches sur l'Inspection et la Mensuration de la Poitrine." Paris, 1838.

well marked, the trochar afforded unmistakable evidence of mere serous effusion.

Retraction or falling in of the chest may, in like manner, be either general or local. In both cases it results from diminution in the size of the lung, consequent either on pressure from without or on change in the lung-substance. The pulmonary tissue, being deprived of elasticity, no longer offers its normal resistance, and the chest-walls fall in under atmospheric pressure. If the mischief, be what it may, pervades the whole of one side of the chest, the retraction or falling in of the chest-walls on that side will be general; if it be limited to one or two spots, the depression will be circumscribed, occurring only at those spots. So it happens that general depression or retraction of the chest is most frequently found in cases where, at a former period, there has been extensive pleuritic effusion, and the lung is carnified and reduced to a small inelastic mass by the continued pressure of the fluid; but it is also observed to a less extent among consumptive patients, as also in cases where the lung-substance is collapsed and atrophied by inaction, consequent on occlusion of the bronchi, whether produced by the pressure of an aneurismal sac, a cancerous deposit, enlarged bronchial glands, or any other means. Instances of partial falling in of the chest, are constantly met with in the infra-clavicular regions when tuberculous cavities exist in the apices of the lungs and occur in other parts of the thorax, according to the position of the lung-substance which has undergone contraction, whether from the infiltration of tubercle or any other matter, or from the formation of pulmonary abscess.

Elevation and lowering of the shoulders and chest-walls, or certain portions of them, are of common occurrence. The former is a frequent accompaniment of general emphysema, and is also seen not uncommonly on one side of the chest, when there is chronic pleuritic effusion on the other: the latter may be observed when the chest has fallen in as a result of chronic pleurisy.[1] Without strict investigation, however, it would be wrong to attach much importance to such phenomena, for they often exist without visceral disease, and to such an extent in some instances, that in a case recorded by Dr. Walshe, elevation of the shoulder on the one side, occurred on the same side as retraction of the parietes from chronic pleurisy.[2]

Distortion of the chest, when extensive, is usually the result

[1] See "Hospital Case-book," xxxviii; case of George Godfrey, admitted into Hope ward, Jan. 16th, 1859.

[2] Walshe, "On Diseases of the Lungs and Heart," p. 14.

of congenital malformation or of accidental circumstances unconnected with thoracic disease. It may take place, however, to a certain extent as a consequence of disease within the thoracic cavity. Thus chronic pleurisy, when followed by compression and carnification of the lung, and consequent falling in of the chest-walls, not unfrequently causes lateral curvature of the spine, the convexity being towards the sound side,[1] as also twisting of the ribs to such an extent in certain cases that their upper edges become external. The sternum, ribs, and clavicles may yield in any direction under the pressure of aneurismal and other tumors, and the clavicles may bend downwards and inwards in cases of extreme disorganization of the apices of the lungs.

Such are the chief modifications which may occur in the form of the chest; and if the causes on which they may depend are constantly borne in mind, there will seldom be much difficulty in forming a correct estimate of any irregularity observed during the progress of visceral disease.

Valuable, however, as are the indications derived from this source, they are less so than those afforded by the movements of the chest. In a healthy person inspiration is effected partly by the elevation and expansion of the ribs, partly by the descent of the diaphragm; and the eye watching the naked chest may readily perceive the part taken by the thoracic and abdominal muscles respectively. If the descent of the diaphragm is checked, the thoracic muscles have to perform double duty, and "thoracic" respiration ensues, characterized by "heaving" of the chest. If, on the other hand, anything occurs to prevent a free expansion of the chest, the diaphragm is brought more actively into play, and diaphragmatic or abdominal breathing is the result. Thus it often happens that even a cursory survey of the respiratory movements suffice to throw important light on the seat of mischief.

But the movements of the chest, if carefully examined, may be made to furnish still more accurate information. For this purpose they may be divided into *general* and *partial;* the first being those in which the entire thorax is concerned; the last those arising from the motion of the ribs on one another, and therefore confined to particular regions.

The *general* motions during inspiration are of two kinds, viz., expansion and elevation; and during expiration the converse of these, viz., retraction and depression. During tranquil respira-

[1] Walshe "On Diseases of the Lungs and Heart," p. 14.

tion, more especially in men, these costal movements are exceedingly slight, the breathing being carried on chiefly by the diaphragm; but during forced respiration, and in women who by lacing restrict the action of the diaphragm and the lower ribs, they are much more palpable and may be seen without difficulty over the entire chest. The expansion and elevation movements are intimately associated, and in health are closely proportioned to one another; but their actual extent is found to increase in a direct ratio to the mobility of the chest-walls and the height of the individual, whilst the frequency of their repetition is regulated by the force, duration, and completeness of the pulmonary respiratory movements, which vary with the requirements of the system. They usually occur with greater rapidity towards the middle of the act of respiration than at its commencement or towards its close; but from first to last they take place evenly, without any jerking irregularity of rhythm. Inspiration is followed immediately by expiration; at the close of that act there is usually a pause, and then inspiration begins again. The inspiratory and expiratory efforts are of nearly equal length; and if the duration of an entire respiratory act be represented by 12, the duration of inspiration may be estimated at 6, that of expiration at 5, and that of the succeeding pause at 1; so that the period of thoracic movement is to that of rest as 11 to 1.

The *partial* costal motions arising from the movement of the ribs on one another are not very readily appreciated by the eye; indeed, during tranquil respiration they are barely perceptible, but in forced inspiration the ribs will be seen to diverge from one another, the divergence being greatest in the lower interspaces. During expiration they converge towards each other.

Dr. Sibson maintains that the five upper ribs converge during inspiration, and moreover that their convergence is "great."[1] Any one, however, by placing a finger in the upper intercostal spaces of a thin but healthy person during forced respiration, may satisfy himself that such is not usually the case. During the act of expiration the finger will be compressed, and will be relieved of all pressure during inspiration. The real fact appears to be, that each of the ribs alluded to not only rises during inspiration more than the rib above it, but moves outwards and forwards to a greater extent, so that when the elevation movement takes place without the expansion movement—as it does in certain forms of disease which interfere with the due expansion of the lung—these upper ribs do "move nearer to each

[1] "Med.-Chir. Trans.," vol. xxxi, p. 360.

other," as stated by Dr. Sibson; whereas, when the expansion movement is fully performed and is carried out in due proportion to the elevation movement, the ribs, though differently placed in relation to each other, cannot be said to approximate.

But, besides these costal motions, there are other movements connected with respiration which are deserving of close attention. As the lungs expand and the diaphragm descends, the subjacent viscera are forced down, and the abdomen consequently protrudes. In tranquil respiration this abdominal expansion movement commences before the costal, and varies in an inverse proportion to it; so that in men, in whom the costal movement is slight, the abdominal is considerable; whilst in women who lace tightly the reverse holds good. In both sexes, however, forced inspiration gives rise to thoracic movement, which is out of all proportion greater than the abdominal, and commences before it even in men.

As age advances and the costal cartilages become rigid, thoracic expansion necessarily diminishes, the breathing becomes more diaphragmatic, the movement of the sternum exceeds that of the ribs, and its lower extremity advances more than the upper. On the other hand, in infancy and childhood, the costal cartilages being extremely flexible, admit of greater movement of the ribs; the upper part of the sternum advances more than the lower, and thoracic expansion is greater and abdominal breathing less than in the adult. Sometimes, however, if the abdomen is large and the respiration hurried and forcible, the air does not find entrance to the lungs quick enough, and in quantity sufficient to fill the space created by the descent of the diaphragm; in which case the lower end of the sternum and the adjoining cartilages will be observed to fall in at the beginning of respiration, and to advance again as the chest becomes more fully inflated. In females, on the contrary, owing to the restraint of stays and tight lacing, the action of the diaphragm and lower ribs is interfered with, and thoracic expansion is exaggerated.

Such, then, are the movements of the healthy chest and the modifications to which they are exposed from causes independent of visceral disease.

The mechanism of respiration, however, is liable to be interfered with by disease; and if the alterations so produced are carefully studied they may be made to furnish most important information respecting the changes going on within the thorax. Sometimes the expansion and elevation of the chest-walls may be simultaneously and abnormally increased or diminished; sometimes the relationship naturally existing between the elevation and expansion movements may be disturbed, the former of these actions being greatly augmented, whilst the other is lessened or almost wholly suspended; sometimes the relative proportion of the thoracic and abdominal movements may be altogether changed, the one being abnormally diminished whilst the other is increased in a corresponding degree; and sometimes, again, the respiratory rhythm may be interfered with, and may become jerking, or the expiratory movement may become of longer duration than the inspiratory. All of these changes may be easily recognized by the eye, and each possesses its proper signification.

An increase in the movements of expansion and elevation may take place simultaneously on *both* sides of the chest under three different conditions: 1st. When the descent of the diaphragm is checked by mechanical causes existing within the abdominal cavity, such as pressure of the gravid uterus, dropsical effusion, flatulent distension of the stomach and bowels, and tumors of various kinds. 2dly. When the action of the diaphragm is checked by the existence of abdominal pain, whether due to inflammation or other causes. 3dly. When increased muscular exertion is used to overcome some obstruction situated low in the chest, as in spasmodic asthma.

Increased movement may be produced on *one* side of the chest by anything which impairs the function of the other, such, for instance, as pleuritic effusion, pneumonia, or tubercular infiltration, pressure on the main bronchus, and other similar circumstances; and, in like manner, increased action in a part of one side may be occasioned by deficient action in the remainder of that side.

Diminution of the natural movements of the chest may be observed under three conditions:

1st. When disease of the lungs or pleura exists of such a nature as to offer a mechanical impediment to the entrance of air, as in tubercular, pneumonic, or other infiltration of the lung-substance; in pleurisy and pneumothorax, and in pressure upon and consequent obstruction to the main bronchi. 2dly. When, in consequence of pain in the chest, there is voluntary fixing of the chest-walls, and respiration is carried on chiefly by the diaphragm, as in inflammation of the costal and upper pulmonary pleura; in pneumonia, pleurodynia, and intercostal rheumatism. 3dly. When some paralysis of the respiratory muscles exists, as a result of cerebro-spinal mischief. A diminution of thoracic movement, however, is also observed when an obstacle exists to the entrance of air in the upper part of the air-passages, as in croup, cynanche tonsillaris, laryngeal disease, and œdema and spasm of the glottis, as also in cases of pressure of an enlarged thyroid gland or of pharyngeal, aneurismal, or other tumors. In such cases, more especially when the amount of obstruction is great, not only do the thoracic walls not expand, but, owing to the forcible descent of the diaphragm, they are actually drawn in during inspiration, whilst the abdominal walls advance in a corresponding degree. If the obstruction affects the bronchi on one side only of the chest, the phenomena referable to such obstruction will occur *on that side only*, whilst the amount of thoracic movement on the other side will be increased.

2dly. *The relationship naturally subsisting between the elevation and expansion movements may be disturbed*, the former existing in full or overfull activity, while the latter is almost suppressed. This is seen either partially or generally, according to the extent of mischief, in all cases where the pulmonary tissue is consolidated and rendered impermeable to air, as in tubercular, pneumonic or other infiltration; in pleurisy, and in instances in which the lung is compressed by intra-thoracic tumors. Volition may cause extraordinary efforts to expand the chest, and may effect its elevation; but as the lungs do not admit of inflation, no expansive movement follows. So also, when the chest is already fully expanded, as in cases of extreme vesicular emphysema, volition may drag the chest-walls upwards, but cannot effect any further expansion. Indeed, in some such instances, the chest may even fall in at the base during full inspiration.

3dly. *The relative proportion of the thoracic and abdominal movements may be altogether changed; the one being abnormally diminished, whilst the other is increased in a corresponding degree.* Thus it is found that whatever causes a diminution of thoracic movement gives rise to a corresponding increase of diaphragmatic action, and whatever checks the action of the diaphragm produces increased thoracic movement, with heaving or thoracic respiration. It is not necessary to recapitulate the causes which lead to an increase or decrease of thoracic movement, as these have been referred to in connection with the causes of the other varieties of breathing; but it may be well to add, that any disease or injury of the spinal cord below the point where the phrenic nerve is given off will give rise to increased action of the diaphragm.

4thly. *The respiratory rhythm may be interfered with, the expiratory movement becoming of longer duration than the inspiratory.* This occurs when obstruction exists in any part of the air-passages to the free egress of the air. It receives its most constant and best-marked illustration in cases of vesicular emphysema, where the elasticity and resiliency of the lung are destroyed, or when aneurismal, cancerous, or other intra-thoracic tumors are pressing upon the trachea or main bronchi; but it is also met with in many cases of tuberculous deposit, and forms a valuable accessory sign of its existence. In epilepsy, hysteria, chorea, and many other affections, the rhythm of the breathing becomes jerking and uneven, without any disease of the thoracic organs.

But other important facts, beyond mere changes in the rhythm of respiration, are discoverable by ocular inspection. The eye takes cognizance of dilatation and distension of the superficial veins of the chest and neck, and of pulsation on any part of the thoracic walls. Dilatation of the veins indicates deepseated obstruction to the venous circulation, and suggests the necessity of examining into the condition of the heart and great vessels, and, in the event of no cause of obstructed circulation being there discovered, of searching for some pulmonary, or other intra-thoracic disease, capable of producing the obstruction. Nothing is more common than to find the superficial veins distended in connection with an extensive deposit of tubercle or cancer, and not unfrequently, when the deposit is confined to one side of the chest, the veins of that side alone are dilated. It should be added, however, that the limitation of the distended vessels to one side is not diagnostic of pulmonary disease; for it is observed in connection with aneurism, cancer, and other varieties of intra-thoracic tumors. So again in respect to pulsation on the surface of the chest. Not only does ocular inspection detect the existence of increased cardiac pulsation, and of pulsation due to aortic aneurism, but it often informs us of a wave-like pulsation at the second intercostal space, attributable indirectly to tubercular disease of the lungs; for when the lung has been excavated by vomicæ, and has subsequently contracted, the left auricle or the pulmonary artery may come into contact with the anterior surface of the chest, and give rise to pulsation in the situation above mentioned.[1]

[1] For further information on the subject of inspection of the chest, see Part III, Chapter I.

CHAPTER III.

MANUAL EXAMINATION OF THE CHEST, OR PALPATION—HOW TO BE PERFORMED; INDICATIONS DERIVABLE FROM IT.

THE sense of touch may often be made available in the diagnosis of thoracic disease. It acquaints us with the actual condition of the chest-walls, and the nature of their movements, whether general or partial; it informs us of the character and amount of vibration communicated under certain circumstances to the parietes, and thus makes us aware of certain physical alterations in the lungs and pleura whereby that vibratory action is modified; and it indicates the presence or absence of fluctuation in the intercostal spaces, and thus gives notice of the existence or non-existence of fluid in the cavity of the pleura.

In order to arrive at satisfactory results from its employment, it is necessary to observe the various precautions already pointed out as conducive to accurate inspection. Beyond this it need only be mentioned that the hand or the fingers should be applied to the chest gently and evenly; that the two sides of the thorax should be examined simultaneously, the one with the right hand, the other with the left; and that each corresponding portion on the two sides should be thus examined and compared.

In many instances the unassisted eye is incompetent to determine whether the local expansion of the chest bears its due proportion to the elevation movement, as also to estimate the motion of the ribs in respect to one another. Both these important practical points may be determined by the aidof palpation. In the upper part of the chest the hand readily detects the absence of local expansion, and thus, by obtaining evidence of some physical condition which prevents the due inflation of the lung, becomes a valuable auxiliary in the diagnosis of consolidation which so often exists in the early stage of consumption. In the lower part of the chest the thumb or the index-finger, placed in one of the intercostal spaces, not only detects the absence of local expansion, but feels convergence of the ribs taking place coincidently with the continuance of the elevation movement, and thus furnishes additional presumptive proof as to the impermeability of the pulmonary tissue.

Again, touch is the only sense whereby it is possible to appreciate the character and amount of the vibration communicated by the voice to the chest-walls—a fact of much practical importance. Thus, if the hand be laid lightly upon the surface of the chest whilst a healthy person is speaking, a delicate, tremulous vibration will be felt, varying in a direct ratio to the loudness and coarseness of the voice, and the lowness of its pitch. As a general rule, therefore, this "vocal fremitus" is more pronounced in adults than in children, and in males than in females, and is stronger during the utterance of certain sounds than it is during the utterance of others. It is more strongly felt in thin than in fat persons; whilst in children and females, more especially if stout, and having shrill weak voices, it is often altogether absent. On the right side of the chest it is usually more marked than on the left, especially in the infra-clavicular region,[1] whilst in the space where the heart is uncovered by the lungs it is altogether deficient, on account of there being no proper medium for its transmission by the chest-walls.

In disease the vocal fremitus may be either increased or diminished. In all cases, however, a careful comparison of the two sides of the chest, due regard being had to the marked difference naturally existing between them, is necessary to determine the presence of disease. Moderate consolidation of the pulmonary tissue, whatever its cause, intensifies the vibration, by increasing the reflecting power of the bronchial tubes, preventing the diffusion of the vibrations, and forming a better medium for their transmission to the chest-walls. Thus tuberculous deposits, when of moderate extent, partial pneumonic infiltration, more especially if in the vicinity of the larger bronchi, and œdema in its early stage, increase the fremitus; whilst more extreme tuberculization of the lungs and large intra-thoracic tumors are apt to deaden or altogether destroy the vibration, partly by pressing upon or blocking up the bronchial tubes, and thus interfering with the columns of air through which the vibrations are propagated, and partly by presenting a solid inelastic mass which the voice is incapable of throwing into vibration.

Again, the fremitus is annihilated when the lung is removed

[1] My friend Dr. Herbert Davies (on "Diseases of the Lungs and Heart," p. 112) asserts that, "although the natural vocal vibration is stated by some writers to be more marked on the right than the left side of the chest, the difference may be practically disregarded, and both sides may be taken to present in health the same amount of vocal fremitus." This certainly does not accord with my experience, which enables me to state that, out of 300 persons of both sexes, and average health and prospects of longevity, who have come before me for examination for life assurance, and whom I have examined specially with a view to this inquiry, above three-fourths (234) have had vocal fremitus more marked on the right than on the left infra-clavicular region, the spot, above all others, where the existence or non-existence of such inequality possesses the greatest importance in relation to the presence of tubercular disease.

from the chest-walls by the intervention of air or liquid in the pleural cavity—a fact which sometimes enables us to discriminate between pleuritic effusion and pneumonic or other consolidation of the lung, in which, as already stated, the vocal fremitus, as felt by the hand, is ordinarily increased, rather than diminished or destroyed.

When the removal of the lung from the walls of the chest is due to the presence of solid matter, the fremitus may or may not be destroyed, according as the foreign matter is more or less fitted by its nature to serve as a conducting medium, and is more or less remote from the larger bronchi, where the vibrations are most intense. If the solid matter be large in amount and inelastic, it will altogether stop the vibration; if less in amount, it may only deaden it; whilst if it exist in a still smaller quantity, and be contiguous to the larger bronchi, it may even raise it above the usual standard by increasing the reflecting power of the bronchi, concentrating the waves of sound, and intensifying the vibration.

The increased size of the bronchial tubes, with the adjacent consolidation often met with in chronic bronchitis, usually tends to increase the fremitus; whilst the condition of the lung which exists in vesicular emphysema produces an opposite result, by interfering with the homogeneity of the pulmonary tissue, and thus impairing its conducting power. In certain instances, however, where inflammation supervenes in an emphysematous lung and gives rise to pneumonic infiltration and consolidation, the vibration is increased rather than diminished.

The voice, however, is not the only source of vibration in the chest-walls; everything which throws the air contained within the lungs into a state of vibration may occasion fremitus of the parietes. The only conditions essential to its production are, that the vibratory motion be sufficiently strong, and the lungs in a state to form a good medium for its transmission. The act of coughing, therefore, may give rise to it, and so may certain rhonchi; but tussive fremitus and rhonchal fremitus, unlike the vocal, possess little value as indications of disease.

Palpation, however, is often able to furnish us with valuable indications of thoracic disease, besides those derivable from the vocal fremitus. It may be employed to detect the friction caused by the rubbing together of the two surfaces of the pleural membrane, or of the pericardium roughened by plastic exudation; to recognize a peculiar thrill occasioned by certain forms of valvular disease of the heart; and to guide us to a more certain knowledge of the condition of the heart itself, by informing us of the character of its impulse and of the extent of the area over which it is felt. Sometimes, when the intercostal spaces are prominent, it may assist in determining the presence of fluid in the pleural cavity; for in such cases fluctuation may be felt when the fingers are applied after the manner adopted for the detection of fluid in an abscess. More generally, however, it is necessary to percuss one portion of an intercostal space with one hand whilst a finger of the other hand is applied to another portion; and sometimes, as in cases of hydro-pneumothorax, it is even necessary to have recourse to succussion, when the splashing of the fluid will be felt. It must be remembered, however, that when the lung is adherent to the walls of the chest and a large superficial cavity exists containing a considerable quantity of fluid, certain rhonchi will give rise to fluctuation or splashing sufficiently marked to be perceptible even by the fingers, so that the mere discovery of fluctuation in an intercostal space is not of itself sufficient to justify an opinion as to the presence of fluid in the pleural cavity. The existence of fluid in a pulmonary cavity is by no means a common cause of fluctuation; but I have met with two instances in which it was well-marked when the patients were placed in a semi-erect posture, leaning somewhat forward.

CHAPTER IV.

MEASUREMENT OF THE CHEST—HOW TO BE PERFORMED; ITS INDICATIONAL VALUE.

Amongst other expedients for throwing light upon the condition of the chest is mensuration. By its aid, the comparative dimensions of the two sides, and the relative positions of their different parts, may be ascertained, and the extent of movement, the precise amount of the thoracic expansion and retraction, accurately determined.

The simplest mode of measuring the circular dimensions of the chest is by means of a piece of graduated tape, or a thin whalebone measure, passed round the thorax from the mesian line anteriorly, to the spine behind. Practically, however, considerable uncertainty attaches to this mode of mensuration, from the fact that it is almost impossible to determine accurately the precise point of the measure which corresponds to the spine, and equally difficult to carry the measure round exactly corresponding portions of the two sides. With the view of obviating these difficulties, it has been suggested by Dr. Hare that two pieces of tape, each similarly graduated, should be joined together, and padded on their inner surface, close to the line of junction. The saddle thus formed, when placed over the spine, readily adjusts itself to the spinous processes, and becomes fixed sufficiently for the purpose of mensuration. Each side of the chest is thus provided with its own graduated tape, which is more readily managed than the single tape, and greater accuracy is consequently insured.

The circumference of the chest varies so greatly in different individuals enjoying equally good health, that little would be gained by the determination of its mean size.[1] Moreover, such a discovery would afford no clue to the expansibility of the thorax, and therefore practically would be of little value. The really important point for investigation is, the increase or diminution of the circular dimensions of either side, as shown *by a comparison of one side with the other.* An increase will exist under the conditions already specified as contributing to the production of morbid expansion of the thorax;[2] whilst a diminution will accompany those morbid states which give rise to retraction or falling in of the parietes.[3] The only point to be borne in mind, beyond the necessity for extreme caution in conducting the mensuration, is, that in adults the two sides of the chest are usually unequal in size, the right, on the average, being larger than the left by about the third of an inch. This irregularity does not exist in early youth, is less marked in females than in male adults, and does not obtain in left-handed persons, clearly showing that it is due to the greater exercise to which the right side is subjected.

When it is desired to measure the expansion movements of the chest during inspiration, the double tape already described will enable us to do so with tolerable accuracy. Applied to the chest closely, but not so tightly as to interfere with respiration, it indicates and enables us to read off the amount of thoracic expansion. It shows that on a level with the sixth costal cartilage the expansion accompanying tranquil inspiration in health does not average more than a quarter of an inch in a male adult, with a chest thirty-four inches in circumference; that in forced inspiration it is increased to from one and a half to three inches, and that the difference in the circular dimensions of the chest after forced inspiration and forced expiration varies from

[1] The average circular dimension of the chest in an adult, at the level of the sixth cartilage, is about thirty-three inches.

[2] *Ante*, p. 25.

[3] *Ante*, p. 26.

two and a half to five and a half inches, the average being about three inches. These several amounts are contributed equally, or nearly equally, by the two sides of the chest, a slight excess only existing on the right side; so that, although no inference can be drawn from variations in the total expansion of the thorax, admeasurement may assist in the detection of disease, by determining the amount of difference between the expansibility of the two sides.

Another and a very convenient mode of measuring the expansion of the chest is by means of an instrument contrived by Dr. Quain, and named by him the stethometer. It consists of a case like a watch-case, on the upper surface of which is a dial furnished with an index. This case contains a single movement, by means of which a silken cord, which passes through an aperture on one side of the case, can act on the index. When the instrument is held firm on the spine with one hand, and the cord is carried round the chest to the sternum, and fixed there by pressure of the fingers of the other hand, the degree of thoracic expansion which accompanies inspiration will be indicated by the movement of the index on the dial, consequent on the traction exerted on the cord. The expansion of one side of the chest having been thus ascertained, the cord can be carried round the opposite side, and thus any difference subsisting between the mobility of the two sides at any given point can be determined without difficulty. This instrument, like the tape, exhibits any difference which exists in the general or circular expansion of the chest on the two sides; but it fails to indicate the deficiency in the antero-posterior movement, and to discover the point at which the deficiency occurs.

Another mode of measuring the chest is by the employment of callipers. The extremity of one of the limbs of the instrument is fixed on the back, whilst the other is placed on the infra-clavicular or other region the expansion of which it is desired to ascertain; and when this has been done on one side of the chest, the same operation is performed on the other, care being taken to place the extremities of the instrument on corresponding portions of the chest on the two sides. The instrument is furnished with a graduated scale, and thus the antero-posterior expansion movement of the chest may be easily and accurately ascertained. The callipers differ from the tape in not showing the circular or general expansion of the chest, but they exhibit its antero-posterior movement, and can be readily applied to the discovery of any deficiency in the antero-posterior expansion of the infra-clavicular region, or of any other given portion of the chest. In this respect they are superior to the tape and the stethometer, and in the readiness of their application they are preferable to every other form of instrument.[1]

Another, though practically a less applicable mode of measurement, is by means of an instrument proposed by Dr. Sibson. He has named it the "chest-measurer." In principle it is a calliper with a movable branch, to which is attached a rack, having at its extremity a small graduated dial, and an index, which serves to indicate any movement of respiration. "It can readily be applied to any part of the body, and by successive applications of it over the chest and abdomen all the movements of respiration can be ascertained with minute accuracy. It indicates the rhythm of respiration, showing whether the respiration is equal to, longer or shorter than expiration." Such is the description given of it by Dr. Sibson—a description, however, which requires to be modified in several important particulars. The chest-measurer indicates only the antero-posterior movement of the chest; it does not show its circular or general expansion. Hence, if trusted to exclusively, it may lead to grievous error; for when the costal cartilages are ossified, the forward movement of the parietes of the chest is by no means in the ratio of its general expansion; and in such a case, the treacherous index would indicate an undue deficiency of thoracic expansion. Again, during labored inspiration, when the action of the diaphragm is excessive, and the bulk of air inspired is smaller than the space left by the movement of the muscle, the lower part of the sternum and the adjoining cartilages may be actually drawn in, whilst a fair amount of general expansion takes place nevertheless. Here the indications of the chest-measurer would be completely at fault. Moreover, unless great caution is observed, the very delicacy of the instrument may occasion inaccuracy; for, as pointed out by Dr. Walshe, the torsion movement of the ribs will cause an apparent increase or decrease of the forward motion, according as the movable rack of the instrument is fixed near their lower or upper edge. Add to these sources of uncertainty the extreme nicety required in the application of the instrument whenever accuracy of results is needed,[2] and the length of time necessarily consumed in making such an application, and it will be obvious that, for many purposes at least, the chest-measurer is not available by the bedside of the patient. But for localizing the excess or deficiency of antero-posterior motion, and ascer-

[1] The most convenient, and, in every respect, the best form of this instrument, is that devised by Dr. G. Nelson Edwards, and described by him in the "Med. Times and Gazette" for December 27th, 1856.

[2] In proof of this fact, see "Med.-Chir. Trans.," vol. xxxi, note at p. 365.

taining its precise amount, the instrument, like the ordinary callipers, is extremely valuable, as it thus becomes an efficient aid in the diagnosis of consolidation, with its consequent imperfect expansion, in the early stage of phthisis.

Dr. Scott Alison has suggested another form of instrument, which he has styled the "hydrostatic pneumatoscope." It consists of a small cup containing water, connected with a graduated glass tube. The cup is covered with a piece of thin India-rubber membrane, which confines the water, and yet admits of motion being communicated to it from without, so that the slightest pressure on the India-rubber membrane causes the water to rise in the tube, and so to indicate the degree of pressure. The cup and tube are fixed on a stand, and when the instrument is to be used for the examination of the chest, "the stand is placed on a table, and the cup is made to touch the patient's chest when he has expired fully. An act of inspiration is now to be slowly and fully made. The liquid rises in the tube, and the number of degrees travelled over indicate the amount of thoracic elevation and advancement. When one side of the chest has been examined, the other is to be proceeded with."[1] "This instrument, applied over the subclavicular region of a patient affected with deficiency of inspiration, in the first stage of phthisis, will indicate a deficient movement." Such is the description given by Dr. Alison; but the instrument is so liable to be broken, and, if accurate results are to be obtained, requires so much delicacy and care in its application, that for practical purposes it is not equal to the "chest-measurer," or to the simpler yet equally efficacious callipers.

Spirometry, which has received its fullest development from Dr. Hutchinson, is another mode of estimating the expansibility of the thorax. It hinges on the fact that the lungs in health contain a certain volume of air which varies in a certain ratio with the height, age, and weight of the individual, and that anything which interferes with their permeability or their action will alter the volume of air they can be made to receive to the exact amount of that interference. The spirometer or instrument employed by Dr. Hutchinson for measuring what he terms the "vital capacity of the chest," or, in other words, the largest volume of air which the chest can be made to contain, consists of a cylinder closed at its upper extremity, and suspended in a reservoir of water by means of two cords fixed to opposite sides of the cylinder. Each cord passes over a pulley, and has a weight attached to its extremity; and the two weights are

[1] Alison, "On the Chest," p. 343.

together sufficient to counterbalance the weight of the cylinder. When the instrument is ready for use, the cylinder is nearly but not quite full of water. A pipe, which forms a continuation of the tube through which the patient has to breathe, passes under the lower extremity of the cylinder, and rises within it above the level of the water. As the patient forces the air through this tube, each cubic inch of air which he expires displaces a corresponding amount of water, and raises the cylinder to a proportionate extent; and the exact amount of air discharged from the lungs in any given expiration is indicated by a graduated scale affixed.[1] By means of the "spirometer" Dr. Hutchinson has ascertained that in a person five feet in height the mean volume of air expelled from the chest by the deepest possible expiration succeeding the fullest possible inspiration is 174 cubic inches,[2] and that eight additional cubic inches of air are given off at 60° Fahr. for every additional inch of stature. Exceptions to this law are occasionally met with, whilst age, weight, and other circumstances modify its influence in a determinate manner; but the law is deduced from experiments upon more than 3000 individuals, and in most instances approximates closely to the truth.[3]

This fact being established, however, it becomes necessary to ascertain whether any and what amount of deficiency below this healthy standard can be relied upon as indicating the existence of disease. The advocates of spirometry assert that a deficiency of from ten to fifteen cubic inches per hundred—though, possibly, arising from physiological peculiarity—is always suspicious, whilst any further deficiency is undoubtedly morbid; hence, they would infer that spirometry affords sufficiently accurate data for a positive diagnosis.

Unfortunately, experience does not warrant our assenting to this proposition. The exceptions to the law Dr. Hutchinson has laid down are so numerous as to forbid our trusting to any conclusions derived from its supposed uniformity of operation. It appears that an individual's breathing capacity,[4] which is measured by the volume of air he is able to expel at one effort from the chest, is apt to vary widely from Dr. Hutchinson's "standard of health." In one person it may exceed, in another fall

[1] A modification of the spirometer, constructed on the principle of an ordinary gas-meter, has been suggested by Dr. Edward Smith. A full description of it will be found in the "Medical Circular."

[2] "Med.-Chir. Trans.," vol. xxix, pp. 157–8.

[3] This is confirmed by the recent observations of Dr. Balfour. See "Med.-Chir. Trans.," vol. xliii, p. 263.

[4] "Med.-Chir. Trans.," vol. xxix, p. 143.

below the general standard. In the former case an individual with disease in his lungs may be able to expel a volume of air fully equal to the average standard of men of his height, and may thus be marked by spirometry as healthy; whilst in the latter, with a perfectly healthy chest, he may figure in the list as unhealthy. In the course of my examinations at insurance offices, as also in private practice, I have met with several well-marked examples of both of these conditions. The fact appears to be that the *individual*, and not simply the *general* standard, is requisite as a foundation for any positive conclusions as to the existence or non-existence of disease: the man must be compared with his former self, and not with the average of other men. When once the individual healthy standard has been ascertained, spirometry is a test of infinite value; but without such a standard, the practitioner will do well to regard its indications as mere hints for his guidance, to be carefully tested by other means of diagnosis.

Unlike the instruments already described, the spirometer affords no clue to the nature or situation of existing disease. Its indications are simply those of perfect or imperfect expansion of the lungs: it does not reveal the causes of such imperfect expansion, neither does it show whether such causes exist in the lungs themselves, in other parts of the thorax, in the abdomen, or in the nervous centres, which, if acted on by disease, may interfere with the expansion of the chest on the principle of diminished nervous energy and consequent impaired muscular action.

Hence it would appear that, for ascertaining the relative size of the two sides of the chest and the exact amount of their general expansion, mensuration by the tape or by the stethometer is most effectual; whilst the callipers and the chest-measurer form the most efficient instruments we possess for ascertaining the precise amount of antero-posterior movement of the ribs and localizing the earliest inroads of disease. The spirometer is chiefly valuable as enabling us to estimate the condition of the thorax rapidly though roughly, and, by means of observations repeated at intervals, to ascertain whether any recent mischief has been set up; or whether disease, known to have existed at a former period, has been making progress, or is happily arrested. Each mode of mensuration has its peculiar advantages, and each, therefore, will be made use of by the judicious practitioner, according to the circumstances of the case.

CHAPTER V.

PERCUSSION.

We have now to make ourselves acquainted with percussion, the first of those methods of examination which enable us to ascertain on what depends any alteration discovered in the form, the size, or the movements of the chest.

Everybody knows that different substances emit different sounds when struck—that a solid, inelastic body yields a dull sound, and a hollow body, with thin, firm, elastic walls, a full-toned clear sound. Further, it is notorious that the resonance of a hollow body is modified by the nature of its contents—that an empty cask, for example, is very resonant when struck, and emits a full clear sound; that filled loosely with wood-shavings or other light substance it is still resonant, but less so than before, and yields a note less full and clear; and that filled with fluid, or with tallow or any other solid matter, its resonance is lost, it is dull on percussion, and the sound it yields is short, abrupt, and dead. The same holds good in regard to the cavity of the human chest. As the different portions of the thoracic walls and of the structures which lie beneath them vary greatly in their texture, so even in health each portion of the chest emits its characteristic sound, and offers a peculiar sensation of resistance to the finger; whilst in disease alterations are produced in these respects, according to the nature, position, and extent of the physical changes which occur in the chest-walls and subjacent parts. Such are the simple yet important facts on which the employment of percussion is founded, and which, strange to say, escaped observation until Avenbrugger published his researches on the subject.[1] Then it became evident how valuable a method of examination had been neglected; and at the present time percussion ranks among the most important means at our disposal for exploring the different cavities of the body.

The immediate object of percussion as a means of diagnosis is the determination of the density of subjacent parts; and this is inferred partly from the degree of resistance offered to the finger, and partly from the sound elicited by the percussion stroke. It is obvious, therefore, that as preliminary to the use of percussion

[1] Published at Vienna, A.D. 1761. Avenbrugger's work, however, appears to have attracted little attention until fifty years afterwards, when Corvisart translated it into French, and introduced the practice of percussion into the French hospitals.

in the diagnosis of diseases of the chest, we should make ourselves acquainted with the relative position of the thoracic viscera, their several boundaries, and their physical characters, and should endeavor to attain to perfect knowledge on three most important points, viz.:

1st. *The resistance offered and the sounds emitted on percussion by the different portions of a healthy chest*, in order that we may be able to recognize any alteration in their character occasioned by disease.

2dly. *The conditions which govern the production of sound and regulate the degree of resistance offered by a body under percussion*, so that we may be competent to judge of any physical change in the subjacent textures by the signs elicited on percussing the chest-walls.

3dly. *The various forms of thoracic disease and their usual seat*, with the view of being able to form an opinion as to what disorder has produced any given change, from the position such change is found to occupy in the chest.

Let us then endeavor, in the first instance, to arrive at a clear understanding on the circumstances which govern the production of sound, and regulate the degree of resistance offered to the finger by a body under percussion.

The *sensation of resistance* is regulated almost wholly by the elasticity of the vibrating body: the more solid and inelastic the mass, the greater and more unyielding the resistance.

The *sound* is dependent upon three circumstances: 1st, the force of the percussion stroke; 2dly, the nature and bulk of the vibrating body; 3dly, the nature of the conducting medium.

The *force of the stroke* sets the molecules of the vibrating body in motion, and thus regulates the loudness and intensity of the sound, the vibrations being stronger, and the sound louder and more intense, in proportion to the force of the stroke.

The *nature and bulk of the vibrating body* lead to differences in the *quality*, *duration*, *clearness*, *fulness of tone*, and *pitch* of the resulting sound.

The *quality* of the sound varies with the form of the sonorous waves, and consequently with the size, form, and composition of the vibrating body.

The *duration* of the sound is dependent on the frequency with which the vibrations are repeated, and therefore is determined, *cæteris paribus*, by the molecular elasticity of the vibrating body; the vibrations being free and unobstructed, and the sound consequently well sustained when its elasticity is great; shorter and more abrupt when its elasticity is less. The prolonged ringing sound produced by striking a gong and the short abrupt sound which results from striking the thigh, are examples of the fact.

The *clearness* of the sound varies according as the vibrations take place more or less freely, and interfere with or destroy each other to a greater or less extent—conditions which are regulated by the elasticity and homogeneity of the vibrating body. If it be inelastic, so that vibrations can hardly be excited, or, if excited, do not penetrate beyond the surface, and are instantly destroyed, it matters not what the nature of its texture, for the result is necessarily a short, abrupt, shallow sound—a sort of dead tap—which, for convenience sake, stethoscopists term dulness.. If it be elastic but non-homogeneous the vibrations will interfere with and destroy each other, and the sound will be, more or less, dull or muffled in consequence; whereas if it be elastic and homogeneous in structure, the vibrations will be free and non-interferent, and the resulting sound will be clear. Therefore, as all clear sounds are caused by vibrations more or less free and non-interferent, and as air furnishes the medium which admits of the most free and least interferent of all vibrations, so all clear sounds emitted by the chest convey more or less an impression of hollowness; or, in other words, of the presence of air; whilst all dull sounds, in which the vibrations are necessarily short and abrupt, convey an impression of solidity.

The *fulness* of the sound, or in other words its volume, is determined by the length of the sonorous wave and thus, *cæteris paribus*, by the size of the vibrating body, the sound being *full* when the vibrating mass is large, and *shallow*[1] when it is small—a fact which may be exemplified by the difference existing between the sound emitted by the feeblest vibrations of a large bell, and that produced by even the strongest and most intense vibrations of a small bell. And as the force of an ordinary percussion stroke could not excite any great extent of vibration except in matter possessed of considerable molecular elasticity—a property which belongs to air alone of all the contents of the chest—it follows that in percussion of the chest-walls fulness of tone implies the presence of a large amount of air beneath, and shallowness of tone its comparative absence.

Finally, the *pitch* of the sound is determined by the number

[1] We have no word to convey our meaning adequately, but the term "shallow" is more expressive and less objectionable than any other. We constantly speak of a full-toned instrument, and as frequently of an instrument having a poor, shallow tone. The term "empty," which is often employed, not only seems to imply the presence of a cavity or hollow space, which often does not exist. but conveys an impression which, in many cases, is not produced by the sound in question.

of equal vibrations excited in a given time—the pitch being high when the vibrations are numerous, and low when they are few in number.

Thus, then, the sounds elicited by percussion may be clear or dull, full-toned or shallow-toned, high-pitched or low-pitched; or, resulting from a combination of these conditions, they may be "clear and full-toned," "clear, but shallow-toned," "dull, but full-toned," "dull and shallow-toned." Placed in a tabular form they stand thus:

Clear-toned, and its opposite, *dull-toned*.
Full-toned, and its opposite, *shallow-toned*.
High-pitched, and its opposite, *low-pitched*.

Or combinations of these sounds; thus—

<table>
<tr><td>Clear and full toned.</td><td colspan="2">Normal, as over-healthy lung, constituting what is termed "good pulmonary resonance," a fair term, and may be retained.
Abnormal, as in pneumothorax, constituting what has been styled "tympanitic resonance,—a fair term, clearly expressing the character of the sound, and not suggestive of any theory as to its origin; objectionable only as encumbering our phraseology.</td></tr>
<tr><td>Clear, but shallow-toned.</td><td colspan="2">Never normal, except over the trachea and larger bronchi. The clearness of the sound is greater, whilst its depth or fulness is less than that even of pulmonary resonance, and far less than that of tympanitic resonance. Varieties of this sound have been termed "tubular," "tracheal," "bronchial," "cavernous," "amphoric," and "cracked-pot." The four first-named terms are extremely objectionable, as implying a theory as to the origin of the sound, and though the last-named, "amphoric," with its variety, "cracked-pot," is less objectionable, it has no pathological signification, and is therefore useless, and to be avoided. The expression "abnormally clear, but shallow-toned," represents the true character of this resonance.</td></tr>
<tr><td>Dull, but full-toned.</td><td>Never normal, except when the chest-walls are fat, flabby, or œdematous, and the clear, full sound of healthy pulmonary resonance is rendered dull and muffled in consequence.</td><td rowspan="2">These sounds constitute "dulness" of greater or less degree—a convenient and not objectionable term, and therefore to be retained. The term "wooden," as applied to a variety of this sound, has no clinical significance, and is to be avoided as a useless encumbrance to our phraseology.</td></tr>
<tr><td>Dull and shallow-toned.</td><td>Never normal, except in the scapular and other regions, in which a thick mass of fat or muscle is interposed between the skin and the chest-walls.</td></tr>
</table>

A clear and full-toned sound results from the free vibrations of a large and more or less homogeneous mass possessed of considerable molecular elasticity. Its normal type is the sound emitted by a well-developed chest, the walls of which are thin and elastic, but slightly covered with fat or muscle. From this it varies in every possible degree, until in pneumothorax, where air exists in the cavity of the pleura, it reaches its extreme point of abnormal development, and somewhat resembling the sound of a drum, has been termed, not inaptly, "tympanitic."

A clear but shallow-toned sound arises from the vibrations of a small homogeneous mass possessed of great molecular elasticity. In the chest it results from the strong vibrations of a small quantity of air, and in a normal condition of the thoracic viscera can be elicited only over the trachea or a large bronchus. Hence this type of resonance has been termed "*tubular*," "*tracheal*," or "*bronchial*." In disease it may be met with in any part of the chest where there exist small empty superficial cavities bounded by thin tense walls capable of reflecting sound, and hence, sometimes, it has been termed "*cavernous*," and at others, when it is supposed more nearly to resemble the sound emitted on striking a jar (amphora), it has been styled "*amphoric*" resonance. These terms, however (tubular, tracheal, bronchial, cavernous, and amphoric), especially the first four when applied to diseased conditions, are extremely objectionable, as suggesting an explanation of the sound which it is not in the power of any one to verify, and which may be clinically incorrect. In most cases this clear but shallow resonance causes an impression of emptiness. A good example of it may be obtained by filliping the inflated cheeks.

A dull but full-toned sound results from the vibrations of a large mass, the vibrations of which are damped or muffled. It is the tone of the muffled bell or muffled drum. In the chest it may be produced by fatness or flabbiness of the thoracic parietes, by thickening of the pleural membrane, or by the presence of a thin layer of fluid in the pleural cavity; or again, by the existence of a thin superficial layer of solidified lung or other solid matter lying over healthy air-containing lung. In either case the transmission of vibrations from the lung beneath is impeded, and the sound, though of a full-toned character, is in consequence weakened or muffled.

A dull and shallow sound is that variety of sound which is emitted by a body not possessed of much molecular elasticity or vibratile power; it is the sound which in percussion of the chest is elicited in the scapular regions, and is usually termed "dulness." It varies in degree from slight to absolute or perfect dulness, according to the vibratile power of the part percussed. Slight dulness commences directly the vibrations are shorter, and the sound, therefore, shallower than normal pulmonary resonance. Perfect dulness and extreme shallowness are attributable to the same condition, and have the same significance, and may be typified by the sound produced by percussing the thigh or a mass of putty. Their cause is the almost instant cessation of vibration, and the consequent shortness and abruptness of the sound.

The pitch of the sound being regulated by the number of equal vibrations excited in a given time, is dependent on a variety of circumstances. In percussion of the chest variations of pitch are not of much diagnostic value; but occasionally we may be assisted in the diagnosis of tubercles by noting a difference in the pitch of the percussion sound over corresponding portions of the two lungs. Shallow tones are generally of higher pitch than full tones, and as the vibrations are shorter, the pitch of the sound is higher over solidified lung than over healthy pulmonary tissue.

Lastly, all sound is influenced by *the nature of the conducting medium.* If its elasticity be great, and its composition perfectly uniform and homogeneous, the sonorous waves are transmitted clearly and readily to the ear; if it be elastic, but non-homogeneous, the waves of sound repeatedly change their medium, and interfere with or destroy each other every time they do so, and thus they either fail to reach the ear, or else reach it much diminished in intensity and clearness. If, again, it possess but little molecular elasticity, the sonorous vibrations are soon obstructed and little or no sound is heard. In percussion of the chest, where the sound results chiefly from the vibration of air contained in the lungs, the chest-walls and the external air together form the conducting medium to the observer. The atmosphere, of course, forms an excellent conductor, but the fat and the muscles which cover the chest are not possessed of much molecular elasticity, and, therefore, are very imperfect conductors. Consequently, if the thoracic parietes are thick, the sound emitted by the chest is necessarily dull and muffled when it reaches the ear, however healthy the condition of the lung beneath.

These, then, are the facts on which all percussion is founded; and we shall see their bearing exemplified in the percussion of a healthy chest, as well as in those altered conditions of the thorax and of the thoracic organs produced by various forms of disease.

When a well-developed, healthy chest is struck lightly by the ends of the fingers, it yields a clear and somewhat full-toned sound. This resonance, which is often called "a good clear sound," does not arise from the tissue of the lung, which, when deprived of air by compression, is inelastic, and incapable of sonorous vibration, but is caused by the vibration of the elastic chest-walls, and of the air contained in the pulmonary cells and bronchial tubes beneath. It derives its precise character partly from the force and direction of the stroke, partly from the physical condition of that portion of the chest-walls which receives the stroke, partly from the density of the subjacent matter. The primary impulse by which the sonorous vibrations are set in motion varies, *cæteris paribus*, with the force of the stroke, and produces a corresponding loudness of sound; so that a weak stroke gives rise to feeble vibration and a weak sound, whilst a forcible stroke occasions a loud sound. But in order that clear sonorous vibrations be produced, other conditions are indispensable. Not only must the stroke be sufficiently forcible—the parts struck must possess a certain degree of elastic tension; for, if the natural elasticity of the chest-walls be destroyed by ossification and stiffening of the cartilages, if flaccidity exists, or if the integuments be thickened by periosteal swelling, by great muscular development, or by œdema, fat, or other cause, the sense of resistance will be increased, the vibrations and the sound will be proportionably diminished or deadened, and more or less "dulness" on percussion will result. The subjacent matter also exercises an important influence; for the larger the amount of air present, and the greater the elasticity, and the more homogeneous or uniform the composition of the media through which the vibrations of that air are propagated, the more freely will they be transmitted to the surface, and the clearer, fuller, and more prolonged will be the resulting sound.[1]

Now, the natural spongy tissue of the lung when inflated with air presents a uniform elastic mass, well calculated to admit of free and unresisted vibration. Consequently, when an impulse is given to the thin elastic chest-walls, and is propagated to such a mass, the sense of resistance to the finger is slight, the vibrations are free, and the resulting sound is proportionately clear, full, and prolonged. When, on the other hand, the chest-walls are thick or inelastic, or the subjacent structures more solid, the sense of resistance is greater, and the vibrations are not transmitted to any depth in the thorax, but are short, abrupt, and quickly returned to the ear, producing more or less dulness of sound, and conveying an impression of solidity or hardness. When, again, the subjacent matters are liquid, or are solids of a soft, inelastic nature, the sense of resistance is still greater,

[1] See "Encyclopædia Metropolitana," article "Sound," by Sir John Herschel.

the vibrations are almost instantly destroyed, and the resulting sound is a sort of short, dead tap, exemplifying in the highest degree what stethoscopists term "dulness."

The sensation of resistance offered by the finger is closely proportioned to the elasticity of the part which receives the stroke; and, as the elasticity of the chest is in great measure attributable to the air therein contained, it is usually found that the degree of resistance varies inversely as the clearness of the percussion sound. The upper part of the sternum and the sternal extremity of the clavicle form exceptions to this rule, in consequence of there being a large amount of air beneath, while the walls at the same time are bony, and incapable of yielding under percussion; but, except in the instances referred to, the resistance is slight when the sound is clear, and considerable when the sound is dull. In proof of this, the peculiar elastic vibratory sensation imparted to the fingers on percussing the infra-clavicular regions in a healthy chest may be contrasted with the dead, inelastic feel conveyed by percussion of the thigh or of the scapular regions. Indeed, it may be laid down as a general rule that whatever deadens vibration diminishes the clearness of the sound, and increases the sense of resistance to the finger.

Thus it will be seen how anything which affects the elasticity, the thinness, and the tension of the chest-walls, or which alters the quantity of the contained air, removes the lungs from the parietes, or modifies the physical condition of the lung-substance, must tend to alter the character of the sound resulting from percussion, and no less so the sense of resistance offered to the finger. And it will be readily understood that those chests, and those portions of a healthy chest, which are most elastic, least covered with fat and muscle, and which have beneath them the spongy air-filled tissue of the lungs, possess the greatest vibratile properties, emit the clearest and fullest-toned sound on percussion, and, with the solitary exceptions before alluded to, offer least resistance to the finger. This being borne in mind, let us briefly inquire into the results of percussion on the various regions of a healthy chest, as it will then be seen how closely the character of the sounds and the sense of resistance to the finger accord with the condition of the thoracic parietes and with the anatomical character of the subjacent structures.

The *supra-clavicular* regions, which contain the triangular apices of the lungs, emit a tolerably clear, though shallow sound on percussion. They vary, however, in this respect according to the degree of inflation of the lungs and the extent to which the apices of the lungs rise above the clavicle. In some instances, especially where there is great muscular development, or much fatty deposit, the outer part of this region emits a dull sound, and is very inelastic and resistant.

In the *clavicular regions* the resonance is very clear towards the sternum, owing to the proximity of the trachea, but the resistance is considerable, owing to the bony nature of the chest-walls; the resonance is still clear, but more distinctly pulmonary[1] in its character about the centre of the bone; much less clear, and, in some instances, almost dull towards its humeral extremity.

The *infra-clavicular regions* afford a good type of pulmonary resonance. The sound they yield is clear and full, the resistance slight, the elasticity manifest to the finger.

The *right mammary region* gives a clear, full sound over nearly its whole extent on ordinary gentle percussion; but, when the stroke is firm and forcible, it brings out slight dulness below the fourth interspace, consequent on the presence of the liver behind the shelving border of the lung. Between the third and fifth ribs, close to the sternum, the sound is sometimes deadened by the presence of the right auricle and the superior angle of the right ventricle of the heart.

The *left mammary region* also emits a clear, full resonance, except in those portions lying below the fourth sterno-costal articulation, where the presence of the heart deadens the sound and increases the resistance. In women a clear sound can be obtained in the *mammary* regions only by means of firm pressure on the breasts.

In the *infra-mammary*, on the right side of the chest, the resistance increases, and the sound becomes gradually duller, until at the lower margin of the chest it is perfectly dull except on deep inspiration, when the lung forces the liver down, pushes in front of it, and gives rise for a time to partial resonance on gentle percussion. Dulness, however, is still perceived on increasing the force of the stroke, and thus bringing out the shallowness which results from the presence of the liver behind. The lower part of this region on the left side is usually dull, in consequence of the presence of the spleen and the left lobe of the liver; but sometimes when the stomach is distended with flatus, the resonance becomes exceedingly clear and full-toned, or, in other words, tympanitic below the sixth rib.

The *supra-sternal* yields an unusually clear but shallow-toned resonance, owing to the proximity of the trachea.

[1] The resonance yielded by those portions of the chest which have beneath them the spongy air-filled tissue of the lung has been termed "pulmonary" resonance.

In the *superior sternal*, down as low as the second rib, the resonance is unusually clear in thin persons, owing to the presence of the trachea and its bifurcations; but in certain persons a large quantity of fat accumulates at the upper part of the mediastinum, obscures the tracheal sound, and causes dulness on percussion. From the bony nature of its anterior boundary, this region is resistant under the finger.

In the *inferior sternal* the presence of the heart and great vessels, together with the left lobe of the liver, occasions dulness on percussion, which is modified in some degree during inspiration by the overlapping of a portion of the lung, and sometimes to a still greater extent by the resonance which results from flatulent distension of the stomach.

The *axillary regions* are extremely resonant at their upper part, from having beneath them the parenchyma of the lung and the main bronchi. Below the fourth interspace on the right side a clear sound is elicited on gentle percussion, and a duller sound by a more forcible stroke, which brings out the dulness occasioned by the liver; whilst on a level with and below the sixth rib complete dulness, whether with gentle or forcible percussion, results from the presence of the liver.

In the *infra-axillary* region, on the right side, there is complete dulness on percussion; on the left side pulmonary resonance may be elicited by gentle percussion, but is modified by the presence of the stomach and spleen.

In the *inferior and superior clavicular* regions the sounds are dull, and the resistance great, in consequence of the thick muscular tissue which fills the superior and inferior scapular fossæ.

The *interscapular* is more resonant, and offers less resistance than the other scapular regions, though it does not yield a sound so clear or full as good pulmonary resonance, in consequence of the intervention of soft muscular tissue. With the view of eliciting a clear sound from this region it is necessary to make the patient cross his arms in front, incline his head forwards, and bend his back, so as to put the muscles upon the stretch, and make them as tense and thin as possible.

The *inferior dorsal* regions emit a clear sound on gentle percussion, but a dull sound on forcible percussion, especially on the right side, owing to the presence of the liver. At the lower part of the left side the spleen and the stomach and intestines modify, in their respective ways, the character of the percussion-sound and the sense of resistance to the finger.

Having thus traced out the principle on which percussion is founded, and seen that its effects do practically accord with what theory would have led us to expect, our attention must next be directed to the differences in the sound emitted, and the resistance offered by the chest in different individuals, irrespectively of actual disease. In childhood and in youth the cartilages are more elastic, and the chest-walls more susceptible of vibration, and the resonance, therefore, is clearer, and the resistance less than in middle age; whilst, in middle age, for the same reasons, the chest, though less clearly resonant and less yielding than in youth, is more so than in old age, when the cartilages are ossified and stiffened. In males the resonance of the chest is usually less clear than in females, partly on account of the greater development of the pectoral muscles, and partly in consequence of the greater ossification of the cartilages. In thin persons, again, the resonance of the chest, for obvious reasons, is clearer than in stout. Some chests, too, are occasionally met with in which, quite irrespectively of visceral disease, or of any external peculiarities of the parietes, the resonance is much clearer, or else less clear than in the average of healthy chests. That the dulness is not due, as is often stated, to deficiency in the power of expanding the lung, is obvious, from the fact that the persons in whom it exists are not necessarily short-breathed; and I am rather inclined to attribute this peculiarity to a difference in the relative size of the respiratory and circulatory apparatus, the lungs being relatively small, and of necessity, therefore, more fully inflated when the resonance is clear; large, and therefore less fully inflated, when the resonance is dull. It is almost impossible, from the nature of the case, to adduce direct proof of this position; but I am the more disposed to consider it correct, from the fact that, after forced inspiration, whilst the lungs are fully expanded, and the chest filled with air, the differences previously observed almost entirely disappear.

We will now pass on to the modes of employing percussion, and to the various considerations relative thereto. When percussion was first introduced as a means of diagnosis, it was practised directly or immediately on the chest. This "immediate" percussion—the only method of percussion employed by Avenbrugger and Laennec—is performed by striking the chest either with the palmar surface of the fingers held fully and firmly extended, or with the two or three first, or the four fingers of the right hand, held closely to each other, and so bent as that their points may be brought down perpendicularly and simultaneously on its surface. If the blow be sharp and quick, and the chest covered with a towel, shirt, or some thin dress, or other covering kept tightly stretched by the left hand, a sound is produced which varies with the condition of the subjacent textures, and furnishes valuable information.

But there are many objections to the general employment of "immediate percussion." In the hands of the unskilful or inexperienced operator it causes pain and suffering to the patient; it cannot be employed satisfactorily in examining the intercostal spaces; and when the integuments are anasarcous, emphysematous, loaded with fat, or flaccid, it fails in eliciting any resonance, and therefore becomes absolutely useless. These difficulties are in great measure overcome by interposing some solid substance between the chest and the percussion agent; hence, "mediate percussion," the invention of M. Piorry, has superseded the older method of "immediate percussion." M. Piorry employed as his pleximeter[1] or percussion-plate a piece of ivory or wood about a line in thickness, and an inch and a half in diameter, provided with two projections or handles, placed at right angles to its surface, and almost at opposite points of its circumference. These enable the operator to hold the plate firmly and evenly on the chest with the left hand, whilst he percusses with the right. Various modifications of this pleximeter have been proposed, made of wood, leather, metal, and other substances; but amongst them, the only two deserving of notice are: 1st, a flat piece of ordinary India-rubber, as suggested by M. Louis; and 2dly, a pleximeter made of vulcanite. Unlike M. Piorry's wood or ivory plate, the ordinary India-rubber can be adjusted to the chest-walls with tolerable accuracy even in the thinnest persons, and its elasticity is such, that it breaks the force of the percussion-stroke, and saves pain. At the same time, however, it deadens the sound—a circumstance of little consequence, when the patient can bear forcible percussion, inasmuch as our inference must be drawn from a comparison of different parts, rather than from the actual sound produced, but of material importance when the patient's chest is tender, the percussion-stroke necessarily very gentle, and the sound elicited therefore weak. In some of these cases, when the India-rubber is used, it is difficult to catch and recognize slight differences of tone which are manifest when another medium is employed. This, however, does not hold good in respect to the pleximeter made of vulcanite, which is in every respect well adapted for the purpose, and is an excellent conductor of sound.

In the method of percussion now generally had recourse to, the use of artificial pleximeters is discarded, and the index and middle fingers of the left hand are made to take their place.

[1] So named from πλῆξις, percussion, and μέτρον, a measure.

They offer the advantages of being constantly at hand, capable of ready application in the intercostal spaces, and of perfect adjustment to the various inequalities of the chest. They may be applied singly or together, parallel to the ribs, or at various angles with them, and not unfrequently, more especially in thin persons, it is desirable to vary the direction in which they are applied. Most persons apply their palmar surface to the chest, and this forms the most convenient and most efficient pleximeter; but some persons prefer the dorsal surface, whilst others apply the palmar surface when they wish to ascertain the resistance of the chest-walls, and the dorsal when their object is simply to elicit sound. One great advantage of this digital plexmeter is, that the sound made by striking it is not so loud as that caused by striking most of the artificial pleximeters, and does not to the same extent interfere with that which is dependent on the condition of the chest.

As percussing agent, no instrument hitherto contrived proves equal to the fingers, with their tips brought to precisely the same level. When gentle percussion is practised, the index-finger may be used alone; and when a more forcible stroke is required, the index and middle fingers supported by the ball of the thumb. Sometimes the first three fingers may be used with advantage, instead of two; and sometimes again, in rough examinations of an extensive surface, when the four fingers of the left hand are employed as a pleximeter, the four fingers of the right hand may be used for percussing. In this latter case they should be held firmly extended, and made to fall horizontally, instead of perpendicularly, the palmar surface being lightly tapped against the dorsal surface of the fingers of the left hand. Thus, with the fingers of the left hand as a pleximeter, and those of the right hand as a hammer, the physician is ready armed for the operation of percussion, and can have recourse to its various modifications, according to the requirements of the case. In the determination, however, of the method to be adopted, much must depend upon the part to be examined, and upon the plan he is in the habit of employing.

Some physicians, though employing the fingers of the left hand as a pleximeter, make use of various percussing agents in lieu of the fingers of the right hand. Thus hammers and other instruments are to be found in shops, calculated in the opinion of their inventors to supersede the hammer provided for us by nature. Practically, however, they fall short of the instrument they are intended to replace; for not only do they deprive the operator of the evidence derivable from the sense of resistance in the parts percussed, but they are apt to cause alarm to timid patients. Sometimes, however, I have derived assistance from the use of a small hammer, the end of which has a piece of India-rubber affixed to it. It is the invention, I believe, of Dr. Winterlich, of Würzburgh, and possesses the advantage of producing a remarkably clear tone, and one which can be relied upon, even in the hands of an inexperienced operator. I never employ it to the exclusion of my fingers, but sometimes have recourse to it in confirmation of the evidence they afford; and it has happened on more than one occasion that in the early stage of phthisis I have been enabled thus to satisfy myself of the existence of slight dulness on percussion, and of a variation in the pitch of the percussion-note on the two sides, of which I had previously entertained some doubts.

It need hardly be stated that the practice of percussion demands considerable manual dexterity, and that the correctness of its indications depends in great measure upon the mode in which it is performed. It may be well, therefore, to direct attention to certain precautions which are necessary to insure satisfactory results.

In the first place, it is essential that the position of the patient should be rigidly attended to, and that, whether he be lying down, sitting, or standing, his body should rest on the same plane and his linbs be similarly disposed on either side, for the slightest irregularity in that respect gives rise to differences in the sounds elicited from the two sides of the chest. Sometimes the state of the patient is such as to render a recumbent posture necessary; but when this is not the case, the sitting or standing posture should be selected, for when the patient is lying down in bed not only are the sounds deadened by the effect of the bed-clothes, but the physician is often forced, for the purpose of examination, to place himself in an awkward and constrained position, by no means favorable to accuracy of observation. While the anterior surface of the chest is being examined, the patient should be made to hold himself upright and allow his arms to hang loosely on either side; to cross his arms in front and bend slightly forward whilst the back is being examined, and to raise and clasp his hands above his head whilst the lateral regions are under examination.

Secondly. Attention should be directed to the condition of the thoracic parietes, inasmuch as, if "immediate" percussion be practised, the flaccidity of a mass of relaxed muscle would interfere with vibration and deaden sound, whereas that very muscular relaxation would admit of a closer approximation of the finger or pleximeter to the chest-walls, and so would tend to render the sound clear if "mediate" percussion were employed.

Hence, before percussion is had recourse to, the muscular tissue covering the thoracic walls should be put upon the stretch and rendered tense in the one case, but allowed to remain in a state of relaxation or flaccidity in the other.

Thirdly. The two sides of the chest should be percussed at precisely the same stage of the respiratory act; for the expansion of the lung during full inspiration not only gives rise to a decided increase in the superficial extent of surface over which good pulmonary resonance can be elicited, but, by diminishing the density of the lung-substance, renders the percussion-sound clearer and the sense of resistance less. Moreover, the expanded lung pushes down the liver and spleen and presses in front of the heart, giving rise to more or less resonance on percussion, where there would otherwise exist well-marked dulness. The reverse obtains after full expiration; and as the difference is very great in all these respects between the chest after a full *in*spiration and a deep *ex*piration, it behooves us in all cases of delicacy to exercise great caution in selecting the same moment or the same period of respiration for percussing the two sides. Perhaps the best method of accomplishing our object is by desiring the patient to expire deeply, or else take a deep inspiration and then hold his breath; we may then make sure, for a time, at least, of having both lungs in the same state.

In disease, this caution is especially needful; for if tubercles or any disease which interferes with pulmonary expansion exist in one lung and not in the other, the clearness of the resonance and the area over which it is heard will not increase on inspiration, as they ought to do, on the diseased side, and the difference between the two sides thus rendered manifest will furnish valuable information. On the other hand the dulness resulting from the presence of tubercles or of any disease producing pulmonary consolidation, becomes particularly apparent when the air has been expelled from the lung by a full *ex*piration; and thus differences existing between the two sides may often be detected in delicate cases when they have previously escaped observation. So, again, when any disease exists on one side only of the chest, of a nature to prevent the expulsion of the air and the collapse of the lungs during expiration, there will not be a proper reduction in the clearness of the resonance nor of the area over which it is heard on that side after a full expiration; and this very point may sometimes enable us to determine the presence of some aneurismal or other intra-thoracic tumor pressing upon and obstructing the main bronchus.

Fourthly. Care should be taken to compare corresponding portions of the two sides of the chest; and as in percussion, just as in inspection and mensuration, our inferences as to the condition of the lungs should be deduced from the comparative rather than from the absolute sound emitted by different parts of the chest, it is essential that the same conditions in every respect should be observed in percussing the two sides. The best practical rule is to apply the finger or pleximeter with equal firmness to both sides of the chest in succession, to examine corresponding portions of the chest, and to percuss with equal force on both sides. Due allowance must of course be made for the position of the heart, liver, and spleen, and for the alterations in their position which necessarily accompany the act of respiration.

Fifthly. In all cases, but more especially in doubtful cases, it is desirable to repeat the observation several times at different stages of the respiratory act and while the patient is in various postures. By taking these precautions, a careful observer will rarely fail to arrive at a correct conclusion; for any uncertainty which may have existed at one examination, in one posture, and at one period of the respiratory act, will be cleared up by an examination at another.

Sixthly. The greatest care should be taken, not only to apply the finger or the pleximeter to precisely corresponding parts on the two sides, but to apply it firmly on the spot to be examined, so that it may be closely and uniformly in contact with the chest-walls. No extraneous condition exercises so great an influence in modifying the sound, and nothing, therefore, demands more constant and more jealous attention in instituting a comparison between any two corresponding portions of the chest. If, for instance, in the examination of a stout person, the finger or pleximeter be applied lightly at one time and firmly at another, a dull sound will be produced in the one case, from the muscles and integuments alone being influenced by the force of the percussion-stroke, whereas in the other a clear, full sound will be emitted in consequence of the vibrations excited in the air contained within the lung. The rule to be ever borne in mind is, that the finger or pleximeter must be applied with equal force on both sides and always with sufficient firmness of pressure to render it, so to speak, a part of the organ about to be percussed.

Seventhly. The act of percussion should be performed by a movement of the wrist alone, the forearm and the arm being held motionless. By this means the pain or uneasiness occasioned by the weight of a blow in which the arm takes part is avoided and percussion can be practised with care and nicety, the force of the blow regulated, a precisely similar blow therefore given to different parts of the chest, and a uniformity obtained in the character of the sound elicited by successive blows at the same spot.

Eighthly. Throughout the examination, the percussing fingers should be kept at the same angle, in respect to the chest-walls, and should be made to fall upon the pleximeter simultaneously. Percussion, practised with the percussing fingers, held at varying angles with the chest, is a fertile source of doubt and perplexity to the inexperienced operator, who thus elicits a new and different sound by each act of percussion, thus rendering the operation valueless, and likely to lead to an erroneous diagnosis.

Ninthly. The object of percussion being to determine the position and the density of the thoracic viscera, the force of the stroke should be proportioned to the depth at which the part to be examined is seated,—gentle when it is superficial, more forcible when it is deeper seated. In this way, the density of the structures at various depths within the chest may be determined, the clearness or dulness of the sound varying greatly in many instances with the force of the stroke. Thus, the sound elicited from the lower part of the right axillary and mammary regions will be clear or dull according as the stroke is gentle or strong; or, in other words, according as the superficial lung-substance is alone percussed, or the subjacent liver is made to feel the force of the impulse. It is obvious, therefore, that in every instance both gentle and forcible percussion should be practised, as by such means alone is it possible to arrive at a correct conclusion respecting the real condition of the structures at various depths within the chest.

Tenthly. Care should be taken that the percussion-stroke be not strong enough to occasion pain to the patient. All fear on this score may be obviated by never giving a "heavy" blow. A sharp, quick rap, or, in other words, a blow lightly given, is best calculated to impart the requisite impulse to the chest-walls; and the speedy withdrawal of the fingers removes all impediment to free vibration. This, therefore, is the mode in which percussion should be performed, the only exception being in certain cases of disease, hereafter to be described, in which it is desired to elicit a sound known as the "cracked-pot" sound, the "*bruit du pot fêlé*" of Laennec. In these cases, to which I shall presently allude, the blow should be firmer, heavier, and more sustained, as favoring those conditions on which the production of the sound depends.

Having thus explained the theory of percussion, described the method of performing it, and detailed the character of the signs which it elicits from a healthy chest, it only remains to point out the alterations those signs may undergo, and how far such alterations can be made available in the diagnosis of disease.

From what has been already stated, it will be obvious that anything which interferes with the density of the matters which lie beneath the part struck will usually change or modify their molecular elasticity, and will thus occasion either an increase or a decrease in the clearness, fulness and duration of the percussion-sound, and in the sense of resistance to the finger. Thus, it is found that in proportion as the subjacent textures are rendered more dense than natural, and, being so, are situated more or less superficially in the chest, so does the percussion-sound pass through every gradation, from normal clearness to absolute dulness, the duration of the sound becoming at the same time proportionably shorter, and the sense of resistance greater. On the other hand, when the subjacent structures are of less than their normal density, precisely the reverse obtains: the clearness of the percussion-sound is abnormally raised, its duration prolonged, and the resistance to the finger lessened. This holds good in all cases in which the quantity of air is not so extreme as to put the chest-walls upon the stretch, and thus, by excessive tension, to interfere with their vibration. It is not necessary, at the present moment, to go into the particulars of every cause which may give rise to one or other of these variations: it is sufficient that their true import be recognized, and that it be fully understood that the precise nature of the change which has occurred can only be determined inferentially, by a due consideration of the position and extent of the mischief, the concomitant symptoms, and other circumstances. This much, however, it may be well to point out, that whether the cause of the increased density be in the chest-walls, the lung, the bronchi, the mediastina, or the pleural cavities;—whether it be that the chest-walls are loaded with fat or muscle, or infiltrated with serum; the lung imperfectly inflated, as an effect of spasmodic asthma; or of any obstructive disease in the upper air-passages,[1] or else congested or inflamed, infiltrated with tuberculous matter, or otherwise solidified; the bronchi enlarged or thickened, or else clogged with muco-purulent or other secretion; the mediastina loaded with fat or other materials; the pleural cavities filled with liquid or solid matter,—the result is in all cases the same, namely, a diminution in the vibration and the production of more or less dulness on percussion, and decreased duration of sound, with increased resistance to the finger. On the other hand, provided

[1] This holds good unless emphysema exists, in which case the increased density of the lung-tissue is thereby masked.

the tension of the thoracic walls be not so great as to prevent free vibration, whatever diminishes density has a direct tendency to favor free vibration, to increase the duration and clearness of the percussion-sound, and to lessen the resistance to the finger, whether the cause of diminished density be in the lung itself, as in emphysema, or in the pleural cavity, as in pneumothorax, or be found in the thinness, tension, and elasticity of the chest-walls.

It must always be remembered, however, that the percussion-sound varies in different individuals. Thus in some instances it is unusually clear, and in others unusually dull, over the entire chest, without the existence of any thoracic disease, and without its being possible to discover any extraneous cause for the peculiarity. It is only when dulness or unusual resonance on percussion is found on one side of the chest, and not on the other, or at any portion of one side as compared with the corresponding portion on the other, that any certainty can be felt as to the existence of disease.

Further, it must be borne in mind that in certain cases the *cause* of dulness or of increased clearness of resonance may not be stationary or immovable, nor the position of the sign therefore fixed or unchangeable. For instance, when the dulness is dependent on effusion within the pleura, or the resonance on hydro-pneumothorax, the fluid in either case will gravitate to that part of the pleural cavity which is made most dependent by the position of the patient, whilst the air, in the latter case, will as certainly rise and occupy the part which is placed uppermost. Thus, when the area over which dulness or increased clearness of resonance exists is found to shift according as the patient lies on one side or on the other, reclines backwards or leans forwards, the fact of its so shifting affords valuable information respecting the nature of the mischief, and often enables us to discriminate between the dulness resulting from solidification of the lung and that which is caused by liquid effusion into the pleura. But mere unchangeableness in the position of dulness is not a *certain* criterion as to the existence of pulmonary solidification; for where adhesions have taken place between the two layers of the pleura, the movement of the fluid is restrained or prevented, and the boundaries of the dull sound are necessarily fixed and unvarying.

Another modification in the relationship subsisting between the clearness and duration of the percussion-sound, and of the resistance offered to the finger, is sometimes met with in cases of phthisis, viz., increased clearness and duration of the sound coexisting with increased resistance to the finger. Practically, the occurrence of such an unusual complication is not of much importance, as the mischief which causes it can hardly fail to be discovered by other means. Its origin, however, should be noted and understood. The conditions necessary for its production are such as exist naturally in the upper part of the superior sternal region, viz., resistant chest-walls, and an unusually large volume of air beneath. When they are found elsewhere on the chest, they are generally dependent on the formation of a superficial cavity in the lung, with a thin, tense, indurated wall adhering to the parietes of the thorax, and increasing its resistance.

M. Piorry and others have described a great variety of percussion-sounds, to each of which they have endeavored to attach a different significance; and those persons who are curious in nice distinctions may consult the works of those authors. But repeated observations has led me to the conviction that many of the sounds described exist only in the imagination of their discoverers, and that others which are met with occasionally cannot be connected with any peculiar physical change, and therefore have no pathological significance. *Practically*, then, the changes of sound to which it is necessary to direct attention are very few in number. Even these may be referred to one or other of the classes of sound described at the beginning of this chapter; and I am satisfied that in most cases it is better thus to describe them than to introduce terms which, however appropriate in certain instances, possess no real significance, and have the great disadvantage of needlessly encumbering our phraseology, and confusing a study already sufficiently complicated and obscure.

Of these changes the first to be mentioned is a modification of the full, clear-toned resonance elicited over healthy lung. It is abnormally clear and full-toned, of unusual duration, and accompanied by a condition of the chest-walls which renders the sense of resistance to the finger slight, and gives the impression of tension and elasticity. It is termed tympanitic resonance, and, as its name imports, resembles the sound obtained by striking a drum. It requires for its production a large space filled with air and bounded by tense elastic walls capable of reflecting sonorous vibrations. Hence it is heard of maximum intensity in cases of pneumothorax. If, however, the tension of the chest-walls be extreme, not only is their power of vibration diminished, but the included air undergoes compression, the vibratory oscillation of its particles, upon which the sound depends, is interfered with, and the tympanitic quality is lessened or destroyed. So that, although a tympanitic tone is usually indicative of air in the pleural cavity, the mere fact of its not being present in full perfection does not necessarily denote the non-existence of pneumothorax.

There are two, and only two, other conditions under which this abnormally clear and full-toned resonance is met with occasionally in the chest, namely, during the existence of emphysema with bulging of the thoracic parietes, as also, though more rarely, when there exists a very large tuberculous excavation in the lungs, with tense elastic walls.[1] In neither case, however, is the term "tympanitic" so appropriate as in most cases of pneumothorax, for although the sound is often clearer, more full-toned, and of longer duration than in healthy pulmonary resonance, it is seldom so truly tympanitic as in the cases referred to, and the chest-walls are more resistant.

A peculiar sound sometimes accompanies tympanitic resonance, which has been termed metallic clanging, or metallic resonance. A good imitation of it may be produced by covering the ear with the palm of one hand and lightly tapping on the back of it with the other. It is a clear ringing sound of a metallic character such as may be produced by a sharp blow on an empty cask. It may be generally obtained by sharp and somewhat forcible percussion in those cases of pneumothorax which are characterized by much bulging of the chest and consequent excessive tension of the parts, and not unfrequently may be produced over large empty thoracic cavities with tense walls. The conditions which impart this ringing character to the percussion-sound are the force and sharpness of the stroke, and the tension of the walls of the air-containing cavity. The phenomenon is not of much importance in a diagnostic point of view, but is interesting as showing how completely the causes which modify the production of sound under ordinary circumstances, coincide with those which regulate its emission by the chest.

The next change is also a modification of the ordinary clear sound of pulmonary resonance. The sound is clearer but more shallow than that of healthy pulmonary resonance, and usually is of a higher pitch. The trachea and a large bronchus yield this sound in perfection, and hence it has been termed "tubular," "tracheal," or "bronchial." It is the natural sound emitted in the supra-sternal region, and must be regarded as morbid only when it exists in regions which do not usually yield this peculiar sort of resonance. Small empty superficial cavities

[1] This is a rare occurrence, but I have met with three well-marked examples of it.

with tense walls furnish excellent examples of it, and hence it has been termed "cavernous," but it may be also met with when any of the superficial bronchi are dilated; when a thin layer of partially condensed lung intervenes between the chest-walls and a large bronchus, and in the inner part of the infra-clavicular regions when the lung adherent to the upper and anterior part of the chest is pushed upwards by pleuritic effusion, and is more or less condensed by pressure. In these latter instances, inexperienced and careless auscultators are apt to mistake the nature of the disease, and to regard the unusual clearness of the percussion-sound, and the increased resonance of the voice which accompanies it, as indicating the existence of a vomica. However, as the sound in question may arise from a variety of causes, it is far better to describe it as an abnormally clear, but shallow resonance, than to employ either of the terms "tubular," "tracheal," "bronchial," or "cavernous," each of which suggests a theory as to its mode of origin, the correctness of which it is impossible to verify.

A modification of this clear, but shallow-toned percussion-sound, resulting from the size of the cavity which emits it, is known as the "amphoric" sound, and derives its title from its close resemblance to the sound obtained by percussion of an empty jar (amphora). Unlike the tympanitic sound above described, which gives an impression of fulness, this amphoric note is suggestive of shallowness or emptiness. It is produced in perfection if, when the mouth is closed, the cheeks are inflated, but not too tensely, and then filliped with the finger. Its most common source is a large superficial cavity with thin tense walls, and hence it is most commonly indicative of consumption, but a sound of a more or less amphoric character may occur under certain other circumstances, to which I shall presently have occasion to allude.

Sometimes in connection with this amphoric resonance, but occasionally without it, there is heard a peculiar sound somewhat resembling that produced by striking a cracked earthenware or metal jar, or other vessel. This is the *bruit du pot felé* of Laennec, and is commonly called the "cracked-pot sound." It was formerly attributed to the forcible collision of air and liquid; but the fallacy of this opinion is made evident by three facts, viz.: Firstly, that the sound is producible in cavities in which no gurgling or bubbling sounds can be heard, and in which, therefore, it is probable no fluid exists; secondly, that it cannot be produced unless there be a free communication between the cavity and the external air, whereas, if it depended solely on the collision of air and liquid, the sound ought to be produced as well when the cavity is closed as when its communication with the external air is uninterrupted; and thirdly, that it is met with occasionally when no abnormal cavity exists, and when the physical and general symptoms, confirmed by the subsequent career of the patient, contraindicate the existence of any serious disease. Therefore another explanation must be sought; and the schoolboy's trick of forming a hollow with the palms of the two hands and striking the back of one of them against the knee, furnishes a good illustration of its mode of production. If when the hands are thus struck, a portion of the contained air be forcibly expelled by the blow, a peculiar sound will be produced analogous in its character to the cracked-pot sound; whereas, if no air be suffered to escape, the sound produced will want the peculiar cracked-pot character. In short, when this sound is met with in the chest, it is occasioned by the sudden expulsion of air and its forcible contact with the sides of the passages through which it is driven. It appears in fact to be a compound sound made up of the ordinary percussion-sound of the part which, in these cases, is always a loud, yet short, hollow, metallic sound, and of a peculiar hissing noise resulting from the sudden and forcible expulsion of air from the lung beneath, through passages in most cases constricted or otherwise altered in character. The more elastic and yielding the chest-walls, the heavier and more sustained the percussion-stroke, and the more free and uninterrupted the communication with the bronchi and upper air-passages, and through them with the external air, the larger, *cæteris paribus*, will be the quantity of air expelled by each act of percussion, the less impediment will there be to its escape, the more rapid will be its current, the more forcible its contact with the sides of the air-tubes, and the louder therefore the cracked-pot sound. Hence, whenever our object is to elicit this sound, the patient should be desired to hold his mouth open so as to remove all possible obstruction to the free escape of the air, and to turn his mouth towards us, so that we may the more readily catch the sound;[1] whilst we on our part should not place the finger which receives the stroke in an intercostal space, in order that the full force of the stroke may be felt by the lung beneath, but should employ somewhat slow and heavy percussion, so as to maintain the direction of the impulse and prevent the resilience of the chest-walls. If we take these precautions, and a large superficial cavity exists, and

[1] This precaution, though long adopted by many auscultators, was never, I believe, insisted on in print until Dr. Cotton enforced it in a paper on the cracked-pot sound, published in the "Lancet" for 1857.

its walls be hard and tense, its communication with the external air through the bronchi free, and the thoracic walls beneath which it lies elastic, we shall never fail to elicit the distinctive character of the sound. But if the cavity be deepseated in the chest, and consequently protected against the force of the percussion-stroke; if its walls be dense, soft, relaxed, or inelastic, so that very little impulse can be propagated to its contents, and little or no air, therefore, expelled by the act of percussion; if its communication with the bronchi be obstructed so that no portion of the contained air can be expelled, and the remainder cannot be greatly agitated; or if the mouth and nostrils be closed so that the escape of air is thereby prevented, then the conditions essential to the production of the cracked-pot sound are absent, and however skilful the operator, he will assuredly fail to elicit its peculiar character.

What, then, is the diagnostic value of this cracked-pot sound? In most works on auscultation it is said to denote the presence of a cavity, and by the majority of the profession it is still believed to do so. But I have often known those who thus interpret it misled as to the character of the disease from which the patient was suffering; and I do not hesitate to affirm, as the result of long and careful observation, that it does not necessarily indicate the existence of a cavity. Dr. Hughes Bennett goes so far as to assert[1] that a cracked-pot resonance may be elicited in various diseases of the chest, and even when the chest is perfectly sound. This statement, however, is thoroughly opposed to common experience, and must have its foundation in some use of the term "cracked pot," which is not usually recognized in practice; for, if care be taken in applying the finger or pleximeter firmly to the chest-walls, so as to avoid the production of the jarring sound which results from the forcible expulsion of air from beneath the finger when carelessly applied, a cracked-pot sound is rarely producible. Nevertheless, it does sometimes occur under circumstances not commonly supposed to give rise to it. In no instance have I heard its distinctive character more marked than in the case of a young man, admitted into the York Ward of St. George's Hospital in the autumn of 1855, suffering from acute pneumonia of the upper lobe of the right lung. So strongly was it developed in that case, that several gentlemen who examined the patient insisted on the presence of tuberculous excavation, and yet the patient recovered perfectly in a fortnight, healthy breathing was re-established, and the cracked-pot resonance ceased. In children, too, who are suffer-

[1] "Edinb. Monthly Journal," January, 1856.

ing from bronchitis a jarring sound, resembling a cracked-pot resonance, may be sometimes elicited. My belief, therefore, is, that this sound simply indicates the sudden and forcible expulsion of a quantity of air from a space or spaces possessed of a certain degree of resilient elasticity, and through passages having a free communication with the external air, but, probably, somewhat flattened, contracted, or otherwise altered. Tuberculous or other cavities in the condition of those above alluded to are undoubtedly its most common source, and therefore it usually indicates their existence; but as already stated, it may arise under certain conditions in pneumonia, and when, as in children, the elasticity of the chest-walls is great, it may be developed to some extent during an attack of bronchitis, even when no solidification of the lung exists.

I would maintain, then, that wherever its presence is detected, its true significance must be carefully gauged before any diagnostic value is attached to it. Care must be taken to ascertain in every instance that it is not caused by the imperfect application of the finger or the pleximeter to the chest-walls and the expulsion of the air from beneath it, for nothing proves a more fertile source of error to inexperienced operators than the sound which is thus produced. When this source of fallacy is guarded against, we may generally arrive at a correct conclusion as to the cause of the phenomenon by a careful examination as to the presence or absence of abnormal resonance or dulness on percussion, and the existence or non-existence of general and auscultatory signs of pulmonary excavation; but when there is pneumonic consolidation of the upper lobe, the diagnosis is at all times extremely difficult. On the whole, therefore, it must be concluded that this cracked-pot sound, though valuable as an accessory sign of tubercular excavation, is not a symptom to be implicitly relied upon, and in no case can be regarded as of itself sufficient to warrant a positive opinion, unless corroborated by other symptoms, general as well as physical. When strongly developed, it justifies the gravest suspicion as to the presence of some abnormal excavation in the lung, and, if persistent, may almost be regarded as pathognomonic of a vomica. It must be remembered, however, that it is liable to cease from obstruction of the bronchus leading to the cavity, by the filling of the vomica with fluid secretion, and also by the collapse or cicatrization of the vomica, and therefore that its temporary or even permanent cessation must not be regarded as evidence of the non-existence of a cavity.

The last change worthy of mention is a modification of the ordinary dull sound of percussion; and, from the fact that it closely resembles the peculiar sound obtained by mediate percussion of a common table, it has been invested with the title of the "wooden" sound. Essentially it is a dull sound, but it gives the impression of "hardness." It is rarely, if ever, produced by fluid in the pleural cavity, but is often developed in great intensity when adhesion has recently taken place between the two layers of the pleura by means of interposed lymph. Indeed, it has been asserted that when this sound is strongly marked, it may be regarded as almost pathognomonic of this form of mischief. Such, however, is not the case. It may exist to an equal extent in certain cases of pneumonic hepatization, as also in extensive tuberculous or other infiltration of the lung, and all that can be fairly stated of it is, that in its highly developed character it is a more common accompaniment of pleuritic adhesion than of pulmonary consolidation. In no instance is it observed more frequently than when tuberculous disease of the upper part of the lung, whether accompanied or unaccompanied by the formation of vomicæ, has given rise to pleuritic inflammation and to adhesion of the lung to the anterior surface of the chest. However, as it is not necessarily connected with any particular form of pulmonary disease, and therefore has no special signification, the term is little more than a useless incumbrance to our phraseology.

There are certain apparent exceptions to the laws already enunciated as regulating the emission of sound by the chest under percussion, which are not commonly recognized or understood, and which, nevertheless, deserve special attention, as calculated to give rise to serious errors of diagnosis. They occur in instances in which the lung is supposed to present the conditions already described as contributing to the production of dulness on percussion. Thus it sometimes happens that a clear, high-pitched sound, of a somewhat metallic character, may be elicited over lung-tissue more or less solidified, and that an abnormal resonance may be met with in the infra-clavicular regions when the pleural cavity is three parts full of fluid, and the lung is thereby partially compressed. Reference is made to this fact in most modern works on percussion, but no satisfactory explanation has been hitherto offered of it. The resonance alluded to has been variously described by authors as of a tracheal, tubular, metallic, empty, hollow amphoric, or tympanitic character, and, in fact, it varies greatly with the varying condition of the lung; but the point to be remembered is, that the sound is neither the dull sound of solidification, nor the peculiar resonance of healthy lung, but is clearer, and usually more shallow-toned. It may differ little from healthy pulmonary resonance, so that comparative percussion of the healthy lung may be necessary to establish its abnormal character; but from this point the sound may range through every degree of intensity, until in many cases it is decidedly amphoric, and now and then very closely resembles a cracked-pot sound. The precise amount of change, however, does not serve to establish any difference in the pathological condition of the parts out of which it arises, and therefore it appears to me better to speak of the sound simply as an abnormally clear, but shallow resonance, rather than to attempt to define by words the precise character of a sound which varies from day to day, and, in many cases, does not admit of accurate definition.

Well, then, the occasional existence of this abnormal resonance under the conditions above referred to, does not admit of doubt, and it is essential to determine, if possible, the causes to which it is attributable, and the mode of distinguishing it from the resonance which occurs under an apparently opposite condition of the subjacent parts.

With a view to a full understanding of the subject, I propose to bring together some of the more striking and suggestive facts connected with its various bearings.

Firstly, then, during the progress of tubercular disease of the lungs, it is observed not unfrequently that the part of the chest at which the least respiratory murmur is audible is that at which there is the clearest resonance under percussion. The existence of emphysema, which may be suggested as one method of accounting for this abnormal resonance, fails utterly in many cases in affording an explanation; for I have repeatedly noted cases in the wards of St. George's Hospital, in which this resonance existed during life, and in which no trace of emphysema was found after death. And when, in connection with this fact, it is remembered that the abnormal resonance is not always observed as a consequence of tuberculous consolidation, and that, when it does occur, it is not a persistent condition, but, with the progress of the disease, and the increase of solidification, is replaced by dulness more or less complete, the conclusion seems inevitable that its source is to be looked for not in simple pulmonary consolidation, but in some peculiar condition of the lung which is merely an accidental accompaniment of consolidation, and is apt, under certain circumstances, to arise during its progress.

So, too, in regard to inflammation of the lungs. Abnormal resonance of a clear, high-pitched, ringing, metallic character is sometimes elicited by percussion over a lung which is undergoing pneumonic consolidation, and occurs, therefore, under circumstances which, *à priori*, would lead to the expectation of partial or complete dulness on percussion.[1] The fact that such resonance does not *always* occur during the progress of pneumonia, and ceases as soon as complete condensation of the lung-tissue has occurred, seems to prove, as in the last case, that the cause of the abnormal resonance is not mere condensation of the pulmonary tissue, but is to be sought in some condition into which the lung is apt to pass during the process of consolidation.

The same holds good in regard to pleurisy, when there is an abundant secretion into the pleural cavity, and the lung is adherent to the anterior and upper part of the chest. As the lung becomes partially condensed by the fluid, the upper part of the side on which the effusion exists is found to become unusually resonant on percussion, the resonance, as in the other cases, being abnormally clear, but of a shallow character.[2] Experiment, however, proves that lung-tissue entirely deprived of air by compression yields a perfectly dull sound on percussion, and the conclusion, therefore, follows, that an explanation of the resonance must be sought in some other cause than mere pulmonary consolidation.

Viewing the above facts separately, it might appear that although a certain degree of pulmonary consolidation leads to partial dulness on percussion, and complete solidification of the lung to perfect dulness, yet that there exists a stage between these two extremes in which the percussion-sound, instead of being dull, becomes abnormally clear and resonant. But such an hypothesis is utterly repugnant to common sense, and to all that is known respecting the emission of sound by a body under percussion, and close investigation at the bedside of the patient, serves thoroughly to establish its fallacy; for it shows, as already stated, that abnormal resonance does not always occur in connection with the progress of pulmonary consolidation, and that when it does occur, it is often of very short duration, and is met with under every variety of condition in regard to the presence or absence of respiration, occurring sometimes when

[1] This has been often noted, though never properly explained. See a paper by Dr. Hudson, in the "Dublin Journal," vol. vii; a paper by Dr. Graves, in the same journal; a memoir by Dr. Roger, in the "Archives Générales de Médecine," vol. xxix; Dr. Stokes, "On Diseases of the Chest," p. 332; Dr. Walshe, loc. cit., p. 76; a report by Dr. Addison, in the "Guy's Hospital Reports," vol. iv, 1846; a paper by Dr. Markham, in the "Edinb. Monthly Journal," for June, 1853; and another in the same journal for August, 1853.

[2] Skoda, Hudson, Walshe, Davies, and many observers in this country, have noted this fact; a marked example of it, in the person of Thomas Ringrose, is at present (May 22d, 1862) under my care, in the King's Ward of St. George's Hospital.

little or no respiratory murmur is audible, and at others when the respiration is loud, harsh, and hollow.

What, then, can be the source of this singular phenomenon? Several hypotheses have been offered in explanation, but none of them afford the slightest solution of the mystery. Some persons,[1] who have observed this abnormal resonance chiefly in the infra-clavicular regions, have suggested that when the vesicular structure of the lung is compressed or solidified, its sound-conducting power is increased, and that it will then admit of the transmission of the sounds elicited by percussion from the trachea and larger bronchi, just as though they had been directly percussed. But the fallacy of this argument is obvious from the fact that the abnormal resonance is by no means confined to the inner portion of the infra-clavicular regions, and to the vicinity of the larger bronchi, and is not more marked on the left side than on the right, as, from the greater proximity of the trachea, it ought to be; and further, that when complete solidification has occurred, this reputed sound-conducting power ceases, and the resonance is replaced by dulness.[2]

Others,[3] who have met with this resonance in the lower part of the chest, have endeavored to account for it by the transmission through solidified lung of sound excited by the act of percussion in the stomach and intestines. But although in Dr. Addison's case, and in certain cases reported by other observers, this suggestion may be supposed to explain the phenomenon, still, its occasional occurrence in other instances over the middle portions of the lung, whilst a stratum of perfectly solidified lung, emitting a dull sound on percussion, lies immediately below the part at which the resonance is heard, suffices to prove that the phenomenon will not admit in all cases of such an explanation.

Others,[4] again, have expressed their belief in the existence of air or gaseous exhalations in the pleura over the solidified lung; and in some rare instances, when the resonance exists in the

[1] Dr. C. J. B. Williams and his followers. See his "Lectures on Diseases of the Chest."

[2] I have never met with a case in which this abnormal resonance was more strongly marked than in a patient admitted under my care into the York Ward of St. George's Hospital, suffering from acute pneumonia of the upper lobe of the left lung. Within thirteen days the right infra-clavicular region had passed through a stage of extraordinary increased resonance into complete dulness on percussion, and from that again into the normal condition. At the same time the respiration, which was accompanied at first by crepitations, became markedly hollow in character, and gradually reassumed its normal character. The expectoration was very characteristic throughout.

[3] Dr Hudson, "Dublin Journal," vol. vii; Dr. Addison, "Guy's Hospital Reports," vol. iv, 1849.

[4] Drs. Stokes, Graves, and Walshe.

upper part of the chest, it may be difficult to *prove* that the presence of air may not possibly be its cause. But the result of careful post-mortem investigation is greatly against the agency of such a cause, and, in those instances in which the resonance is observed only at the base, or about the middle of the lungs, the phenomenon cannot be so explained, unless, indeed, it be supposed that the air is confined in a circumscribed sac, formed by the presence of old adhesions of the pleura—an hypothesis which is at variance with reasonable expectation, and is often contradicted by the shifting in the limits of the resonance during life, and by the results of an inspection of the parts after death.

Others, again, as Skoda and his followers, who are conscious of the fallacy of attributing to perfectly solidified lung an increased sound-conducting power, and who are also aware of the extraordinary resonance sometimes observed over partially condensed lung, have jumped to the conclusion that when the lungs are partially deprived of air, they invariably yield a tympanitic percussion-sound, and that hence the source of the phenomenon.[1] But this doctrine is based upon two erroneous assumptions, viz., that "collapsed lungs give a distinctly tympanitic sound"[2] on percussion, and that under ordinary circumstances there exists sufficient distension of the lungs to interfere with and diminish the tympanitic tone of their resonance. No one who has percussed lungs which are thoroughly collapsed, will admit with Skoda that they yield a tympanitic resonance; and although it is true, as already stated, that extreme distension of any air-containing cavity will prevent the occurrence of vibration, and thus may lead almost to dulness on percussion, as may be seen in certain cases of pneumothorax, and may be tried experimentally on tensely inflated lungs, or on a tensely inflated stomach, still it cannot be contended that such a condition exists, or can exist, under ordinary circumstances in the chest. Repeated experiments have convinced me that perfectly collapsed lungs yield a dull sound on percussion; that lungs as ordinarily met with in the dead subject, or inflated to about one quarter of the full extent, yield a clear, but not very full-toned resonance; that inflated to one-half or two-thirds of their full extent, their resonance is as clear, if not more so than before, and is decidedly more full toned or tympanitic; and that it is not until they are distended to their fullest extent, and the whole tissue is to the utmost on the stretch, that any diminution of clearness or ful-

[1] Skoda, "On Auscultation," by Dr. Markham, pp. 13 and 18.
[2] Skoda, loc. cit., p. 19.

ness is perceived. Further, the consideration of certain phenomena connected with respiration establishes the same facts. After deep inspiration, the chest yields a clearer and fuller note on percussion than it does after a forcible expiration when the lungs are in great measure emptied of air; certain it is, moreover, that in most cases of partial tubercular consolidation, when the lung is in great measure deprived of air, the percussion-sound is usually duller than natural; and further, that it is so during recovery from pneumonia, when the lung is still only partially expanded. Indeed, over a lung which has been perfectly solidified by pneumonia, the percussion-note does not regain its normal clearness and fulness until free respiration is again established. As, then, in all these instances in which the lung-tissue is more or less solidified, and contains less than its normal amount of air, the percussion-note is duller than natural; as the same fact is observed as a result of diminishing the inflation of the lung; and as extensive tubercular deposit often occurs, and many cases of pneumonia run their course even to the production of complete hepatization without the occurrence of abnormally increased resonance, it is manifest that mere deprivation of air, whether to a small or to a large extent, will not raise the clearness of the percussion-note.

To what, then, is attributable the increased resonance which sometimes, though rarely, occurs over condensed lung? From a careful consideration of the various circumstances under which this singular phenomenon is met with, I have been led to believe that it arises from the presence of air pent up in a portion of the lung, in the immediate vicinity of consolidated tissue—a condition which prevents the diffusion of the vibrations excited by percussion, and leads to the concentration and intensification of the resonance. Nothing can be more certain than that in many cases of pneumonia, especially when accompanied by some amount of capillary bronchitis, the gorged and distended condition of the capillary vessels, and the effusion existing in the terminal bronchi, block up the passage by which the air would otherwise escape from the air-cells, and thus retain it in them in a state of greater or less elastic tension. The post-mortem examination of persons who have died of pneumonia shows this to be a condition of very frequent occurrence. The lung cannot be compressed by any moderate degree of force; little or no air can be squeezed out of it; but no sooner is its tissue cut by the scalpel, than there issues forth a sanious frothy fluid, or, in other words, a fluid largely mixed with air. The portions of lung in which this condition exists, are those immediately above the point to which hepatization has extended, and below that to which the air has free access. They are those, in short, where crepitation or fine bubbling sounds occur, and are precisely those over which this clear-toned resonance is met with. On several occasions I have traced this resonance shifting its position higher and higher in the chest, as the inflammation has spread upwards, whilst at the same time dulness on percussion has also extended upwards, and has occupied the parts immediately below it, which previously had been the seat of this peculiarly clear resonance. In instances of tubercular deposit, the conditions essential to the retention of air in the lung-tissue are of less frequent occurrence, and abnormal resonance over condensed tubercular lung is met with less frequently; but it does occur in certain instances;[1] and it is easy to conceive how some of the smaller bronchi may become occluded either by tenacious secretion or by the pressure of tuberculous matter; and that the obstruction thus created may, for a time at least, prevent the escape of the air from that portion of the lungs to which these bronchi lead. And, so again, in pleurisy, with an abundant secretion into the pleura. Anything, in short, which serves to occlude or obstruct a bronchus, whilst, as yet, the lung beyond the point of obstruction remains even partially distended with air, will tend to produce this abnormal resonance. It matters not whether the obstructing cause be an effusion into the terminal bronchi, as in certain cases of pneumonia, or whether it be a deposit of tubercle, cancer, or other matter, either in or around the bronchial tubes, or the pressure of any tumor from without, or the pressure against the bronchi leading to the permeable portion of the lung, of lung-tissue already solidified—in all cases, if air is pent up beyond the cause of obstruction, and the adjacent tissue is somewhat condensed, the result is, the production of a peculiar resonance on percussion. If, as in most cases of pneumonia, the mischief commences in the air-vesicles, so that the air is expelled and replaced by solid matter before the bronchi become obstructed, dulness on percussion accompanies the progress of the disease from first to last; whereas, if the mischief implicates the smaller bronchi, and the air-passages become obstructed, whilst, as yet, the pulmonary cells beyond are more or less distended with air, then that condition results which is conducive to abnormally clear though shallow resonance on percussion. Each lobule, in fact, is converted into a multilocular air-distended cavity. The same holds good in regard to tubercular or other deposit as to pneu-

[1] Those gentlemen who were noting my cases in St. George's Hospital, in May, 1857, will remember how remarkably this was illustrated by the case of Mary Wills, who was admitted into Holland Ward on the 6th of that month. See "Hospital Case-book," xx, "Women."

monic exudation; but as the deposition of tubercle usually takes place more slowly than pneumonic exudation, and less frequently leads to obstruction of the bronchi with retention of air in the pulmonary cells beyond the seat of obstruction, it happens that the peculiar resonance in question is seldom so well marked in the former as in the latter disease. In pleurisy, if the effusion into the pleural cavity be copious, and the apex of the lung be adherent to the anterior walls of the chest, it often happens that pressure enough is exerted to obstruct some of the bronchi leading to the pervious portion of the lung, and thus to produce the condition essential to the abnormal resonance.

The accuracy of my conclusions may be tested by the following experiments, which prove that this peculiar abnormal resonance is not due to the sound from the trachea or larger bronchi transmitted through solidified lung, nor, as Skoda asserts, to mere diminution in the quantity of air and to the lung-tissue being less distended in consequence, but is rather referable to the presence of pent-up air having behind it lung-tissue more or less solidified, which prevents the diffusion of the sonorous vibrations.

1st. Take three healthy sheep's lungs. Let one remain in a collapsed condition;[1] inflate the second to about one-half of its full extent, and tie the bronchus so as to prevent the escape of air; inflate the third lung thoroughly, distending it to its fullest extent, and retain it in its distended condition by a ligature on the bronchus. Then on percussing each of them by means of a pleximeter and percussion hammer, it will be found that the first yields a short, dull sound, the second a clear and full sound, the third a sound which varies according to the precise amount of pulmonary distension, but which is usually less clear and far less full than the sound yielded by the second. If the distension be extreme, the sound will be still less clear, and will approach to dulness.

2dly. Let a lung previously inflated with air be injected through the bronchus,[2] and let it be injected to about one-half

[1] I have found that lungs supposed to be collapsed vary immensely in the amount of air they contain (retained, doubtless, by viscid secretion in the bronchi), and consequently in the sound they yield on percussion. It is difficult to obtain, even from the slaughter-house, lungs which are thoroughly collapsed; but it is necessary to do so, if the experiment is to be regarded as of any value.

[2] These experiments are best performed with sheep's lungs, fresh from the slaughter-house. The lungs should be inflated through an injection-tube, fastened tightly into the trachea by means of ligatures, and the injection should not be commenced until the lung is inflated to at least one-half its full extent. Size, mixed with a certain portion of glue, and colored with vermilion, answers well as an injecting fluid.

of its full extent. When the injection has set or become solid, percuss the lung and compare the sound produced with that elicited by percussion of the collapsed lung and of the partially inflated lungs used in the last experiment. It will then be found that if those parts only are percussed in which the superficial portions of the pulmonary tissue remain distended with air, a remarkably clear and shallow sound will be emitted, corresponding to what is termed amphoric resonance, having in some instances more or less of a metallic, ringing character. The resonance is clearer and at the same time shallower than that yielded by any of the uninjected lungs. Percussion over those portions of the lung which lie superficially and contain no air elicits dulness more or less complete, according to the depth at which the air lies beneath.

3dly. Percuss an inflamed lung the lower part of which has undergone pneumonic hepatization. Over the perfectly hepatized portions complete dulness will be elicited by percussion, over those parts at which inflammation is commencing the percussion-note will not be materially altered, whilst over those portions which, though still highly crepitant under the finger, are in a far-advanced stage of inflammation, the percussion-sound will vary according as the air-cells or the terminal bronchi appear to be principally and primarily affected, the sound being more or less dull according as the air-cells are filled to a greater or less extent with solid material, and more or less clear but shallow-toned according as the smaller bronchi are obstructed, and the superficial air-cells still remain distended with air. This may be tested by the spumous, frothy character of the fluid which escapes when the lung-tissue is cut, and by the existence or non-existence of a granular appearance when the lung-tissue is torn. These facts, though not as yet recognized or pointed out by authors, I have verified repeatedly in the dead-house of St. George's Hospital.

CHAPTER VI.

AUSCULTATION—ITS THEORY AND PRACTICE; THEORY OF THE STETHOSCOPE; ADVICE RESPECTING THE FORM OF INSTRUMENT TO BE EMPLOYED; CAUTION TO BE OBSERVED IN THE PERFORMANCE OF "IMMEDIATE" AND "MEDIATE" AUSCULTATION; MODE OF CONDUCTING AN EXAMINATION OF THE CHEST.

THE act of respiration in a healthy chest is accompanied by sounds which are occasioned by the action of the lungs and heart, and which in disease undergo modifications corresponding to the physical alterations produced in the intra-thoracic organs. These sounds, together with those produced by the voice and by the act of coughing, are audible on the external surface of the chest; and the science of auscultation consists in recognizing their character and their various abnormal modifications, and in tracing each to its corresponding physical cause.

Auscultation, like percussion, may be "immediate" or "mediate." It is "immediate" when the ear is itself applied in contact with the chest-walls, either bare or only thinly covered, and when, therefore, the sonorous vibrations are conveyed directly from the chest to the ear; "mediate" when an instrument is interposed between the chest and the ear, and forms a conducting medium between the one and the other. Both these methods of examination have their advantages and disadvantages, and each is deserving of attentive study. The former is attended with the least fuss and parade, and is therefore useful in the case of timid persons or of children who are frightened by the application of a stethoscope;[1] it can be employed at all times, even when no instrument is at hand; it causes no pain or uneasiness to the patient; it is the most convenient method of exploring the posterior parts of the chest in cases not requiring much nicety of examination, and in some instances it affords the means of determining the character of sounds which, heard through the cylinder, are indistinct and indefinite. The latter, on the other hand, enables us to examine the axilla and other regions where the ear cannot be placed in close apposition to the chest-walls; and it provides us with the means of isolating any particular spot, and thus of tracing each sound to its exact

[1] An instrument so named by Laennec. The term is derived from στῆθος, the chest, and, σκοπεῖν, to examine.

position within the thorax. In short, every auscultator should be trained to both methods of examination, and in cases in which both can be employed it matters little which is had recourse to. It often happens that both may be employed advantageously, the one to examine the axillary and other regions which only admit of examination by the stethoscope, the other to explore the inferior lateral regions and the whole of the posterior surface of the chest. But there are so many occasions on which it would be inconvenient, disagreeable, or even indelicate to apply the ear directly to the chest, that every one should take especial care to familiarize himself with the use of the stethoscope.

An endless variety of forms and of materials have been employed in the construction of stethoscopes, some with a view to a fancied superiority of principle, but many more out of a regard to elegance of shape, convenience, and portability. They have been made solid and hollow, rigid, and flexible—of wood, bone, gutta percha, and other substances, and each form and each material has in turn found eager advocates. The truth, however, appears to be that a light-textured vibratile wood, or else gutta percha or vulcanite, is the best material for a stethoscope, and that a rigid stethoscope is the best kind of instrument for ordinary occasions. The only point to be decided, therefore, is the form of instrument to be employed.

A solid cylinder was thought by Laennec to be the form best adapted for the transmission of the chest-sounds; and Dr. Hughes, in furtherance of Laennec's opinion, thinks it "quite evident that it is by the solid walls of the stethoscope that the sound is in all cases principally, and in many cases entirely conducted."[1] On the other hand, Dr. Williams and most other writers attach great importance to the column of air contained in the interior of the hollow stethoscope, and recommend the employment of that form of instrument on acoustic principles quite irrespectively of its convenience and portability. The result of my own experience is in accordance with the latter opinion. Not only in the one instance does a solid piece of wood form the only medium of communication between the chest and the ear, whilst in the other a column of air serves conjointly with the wood as a medium for the transmission of vibrations; but the hollowing of the stethoscope undoubtedly renders the stethoscope more vibratile, and therefore more apt to receive and propagate the sonorous vibrations.

[1] Dr. Hughes, "On Auscultation," p. 80.

It is commonly supposed that the column of air contained in the interior of the hollow stethoscope is set in motion by the vibration of the chest-walls; and Dr. Williams and many other writers have endeavored to point out the precise form of tube and conical excavation which is best fitted to collect and concentrate the sonorous vibrations and convey them unimpaired to the ear. The opinion, however, which forms the basis of these calculations has always appeared to me erroneous; and some ingenious experiments, devised by my friend Dr. Davies,[1] afford demonstrative proof that the vibrations transmitted throughout the substance of the wood which forms the walls of the cylinder produce a greater effect in exciting the air contained within it than do the vibrations of that portion of the chest-walls which is covered by the hollow end of the instrument. Thus the conclusion is forced upon us that although a hollow stethoscope forms the best conductor of the chest-sounds, the precise form of the cavity of the pectoral extremity is immaterial.

The essential points in the construction of a stethoscope are, firstly, that the end applied to the chest be not so thin and sharp as to cause pain to the patient, nor so large as to render it difficult or impossible to apply it evenly and uniformly to the surface of the chest-walls in a moderately thin person, and not so thin as to break easily. Secondly, that the ear-piece be large enough to cover the whole ear, and of a shape to fit and close it—a circumstance of great importance as tending to prevent the dispersion of the vibrations.

For general use I would recommend a stethoscope made of some light vibratile wood such as cedar or cherry, having a shaft from six to eight inches in length, with a clear, polished quarter-inch central canal, an ear-piece two and a half inches in diameter, slightly and but very slightly hollowed,[2] and a thoracic extremity from one inch to an inch and a quarter in diameter, hollowed in a conical form, and having a broad, well-rounded rim. Such an instrument, made of a single piece of wood, or with the ear-piece made to screw, or else slip on to the shaft, is in many respects superior to most of the stethoscopes to be met with in the shops. Practically, however, far more depends on the skill and attention of the examiner than on the form of the instrument, and therefore when a well-made stethoscope has been selected, and an ear-piece chosen which fits the ear comfortably,

[1] Davies, loc. cit., pp. 203, 204.

[2] For most persons the ear-piece should be made only slightly concave; but should the tragus and antitragus be unusually prominent, it must be made concave in a proportionate degree.

the student should keep to that one, and familiarize himself with its use.

Another form of stethoscope, which is sometimes had recourse to for the purpose of exploring the axilla and the lower lateral and dorsal regions, when the patient cannot be readily moved, is that known as a "flexible stethoscope." It is usually made about two feet in length, and is constructed of wire twisted into the form of a hollow cylinder, and covered with silk or worsted thread. There are few cases, however, in which its assistance is really needed, and as it conveys sound less readily than a wooden instrument, it is seldom employed in practice.

A double or binaural stethoscope has been invented by Dr. Leared, and adopted by Dr. Camman, of New York. It consists of two tubes—one for each ear—the thoracic ends of which fit into a hollow cylinder or cup, which is applied to the surface of the chest. It thus enables us to auscultate a particular portion of the chest with both ears at the same moment; and as all sounds are stronger when heard by both ears than when heard by one ear only, it is contended that the delicate sounds which accompany the early stage of phthisis are discoverable by this instrument more readily than by the ordinary wooden one. Unquestionably the sounds are heard louder than through the ordinary wooden instrument; but they are confused by other sounds, which result from the use of the instrument, and which it requires considerable education to distinguish from morbid chest-sounds. On this account, therefore, and also, on the ground that the instrument is more formidable in appearance than the ordinary wooden stethoscope, and is less portable, and otherwise less convenient, it is rarely employed by the practical physician.

Another form of binaural stethoscope has been invented by Dr. Scott Alison, and has been styled by him the "differential stethoscope." Like Dr. Leared's instrument, it consists of two tubes, one for each ear, but, unlike that instrument, each tube has a separate cylinder or cup, which admits of being applied to any part of the chest. Thus, whilst Dr. Leared's stethoscope enables us to hear the sounds emanating from any given portion of the chest with both ears simultaneously, Dr. Alison's instrument enables us to listen to the sounds emanating from two different parts of the chest at the same moment, and with great readiness and accuracy to compare the sounds at any spot on the two sides of the chest by a series of consecutive observations. Both these points are very important. It is no small advantage to be able to examine two portions of the chest without shifting our position, or the position of our patient, or making any movement of the head; for these are precisely the circumstances which are most favorable for the detection of the slightest difference in the sounds on the two sides of the chest. Therefore, if this alone were attainable by the aid of Dr. Alison's instrument, it should be in the hands of every auscultator; but experience proves that it possesses other powers, which render it indispensable to every practitioner. Not only does it enable us to listen at one and the same moment to sounds emanating from two different parts of the chest, but it infallibly determines upon which side of the chest the sounds are loudest and most intense. For, strange as it may appear, it is nevertheless the fact, that when sound is conveyed to both ears simultaneously, but is louder, however slightly, on one side than on the other, the weaker sound is eclipsed or nullified, so that sound is heard on the stronger side, whereas little or no sound is audible on the weaker side. In the early stage of phthisis, when disease is often confined to one side, and occasions a slight difference only in the force and character of the respiratory sounds on the two sides, the value of this differential stethoscope is self-evident.

But whilst its value as an occasional aid to diagnosis is admitted, it must not be regarded as a substitute for the ordinary wooden stethoscope. It is just as formidable in appearance, and in other respects as inconvenient in practice, as Dr. Leared's double stethoscope, and it requires to be handled with extreme nicety. The extremities of the tubes must be fitted closely and equally into each ear, the other extremities applied to the chest in a similar manner, and the elastic portions of the tubes kept equally straight. If these points are not strictly attended to, differences of sound will be necessarily engendered. Further, it has a disadvantage which is fatal to its use by many practitioners, viz., that the sense of hearing in the two ears of the auscultator must be of tolerably equal acuteness.

Another form of instrument, introduced by Dr. Scott Alison, and termed by him a "hydrophone,"[1] deserves a passing notice. It consists of a flat India-rubber bag, about three times the size and of the same thickness as a watch, partially filled with water, and it is used as a medium between the ear and the chest-walls. When the patient is very thin, and the intercostal spaces are deeply sunk, so that the surface of the chest is irregular, and the application of the ear or of the stethoscope is difficult, the hydrophone greatly facilitates auscultation. It fills up the depressions on the surface of the chest, and closes the external ear of the auscultator, and thus enables him to hear sounds

[1] From ὕδωρ, water, and φωνη, voice.

issuing from the chest which would otherwise be inaudible. But it somewhat deadens the sounds as heard through an ordinary wooden stethoscope, and it does not, any more than immediate auscultation, enable the auscultator to isolate any particular part of the chest, nor to examine the axilla or other parts where the ear cannot be placed in close approximation to the chest-walls, so that practically its use is restricted to those cases in which the chest is too thin and the surface too uneven to admit of being examined by the ordinary wooden stethoscope. In such cases, and especially as an aid to the flexible stethoscope, it is of some value.

In the performance of auscultation, just as of percussion, certain precautions are necessary to insure accurate results. The chest should be uncovered, or, if exposure be inadmissible, the covering should be as thin as possible, so as not to offer any material obstruction to the transmission of sound, and of a soft yielding nature, so as not to occasion any friction or rustling which may interfere with or overpower the sounds from within the chest. A calico chemise or nightgown, a soft towel, or an old napkin smoothly spread over the surface, answers the purpose perfectly well. The position of the patient should, if possible, be regulated in the manner pointed out in the chapter on "Percussion,"[1] and, for the reasons already insisted on, great care should be taken to subject each corresponding portion on the two sides of the chest to a precisely similar examination. The posture of the examiner should be easy and unconstrained; and with the view of avoiding the necessity of assuming inconvenient positions, the physician should be able to employ both ears, and should practise both ears accordingly. The examination should be conducted in a quiet room, and all friction between the stethoscope and the clothes should be strictly avoided. Care, too, must be taken not to place the stethoscope over stiff hair on the surface; for the intervention of hair between the chest and the instrument may occasion a fine crepitating sound, greatly resembling the crepitation of pneumonia, and thus may give rise to misconception. So again, if immediate auscultation is had recourse to, the hair should be put aside from around the ear in contact with the chest; for if any portion intervenes between the ear and the chest, a noise is apt to be produced which may be readily confounded with certain morbid sounds emanating from the lungs.

But certain further precautions are specially applicable to the use of the stethoscope. The slightest interval between any por-

[1] See p. 56.

tion of the thoracic end of the instrument and the surface to which it is applied prevents the free transmission of the chest-sounds to the ear, and it is therefore necessary to hold the instrument at right angles to the surface, so that every part of its extremity may be kept closely and evenly in contact with the parietes. If the patient be so thin as to render it impossible for the whole circumference of the extremity of the stethoscope to be placed in close contact with the chest, a pad of soft linen, or a piece of India-rubber, or the hydrophone, may be employed to fill up the depressions between the ribs, and so to produce an even surface for its application; or, if this proceeding be not adopted, the instrument must be discarded, and the hydrophone, or else "immediate" auscultation, had recourse to. Again, the character of the sounds is altered, and their transmission is interfered with, if the ear be not applied to the stethoscope with a sufficient degree of pressure: whilst, on the other hand, it is obvious that forcible pressure on a thin and tender chest cannot fail to occasion pain, and in some measure to interfere with the respiratory movements. Hence the instrument should be applied firmly, yet not heavily or forcibly. In stout persons, however, it is necessary to exert more than usual pressure, in order to compress the loose adipose tissue and thus increase its conducting power; and, if any œdema of the surface exists, an amount of pressure must be employed sufficient to squeeze out all the serum contained in the cellular tissue immediately beneath the stethoscope. For, not only is serum a bad conductor of sound, but, by its escape from the cells under pressure, it may give rise to a fine bubbling, or to a creaking noise of a jerking character, the former of which, except by its continuance when the patient holds his breath, and by its disappearance after continued steady pressure with the stethoscope, is scarcely distinguishable from pneumonic crepitation, and the latter from certain forms of pleuritic friction. The same holds good in a modified degree when the chest-walls are emphysematous, as from a broken rib.

Beyond all this, it is in the highest degree necessary to acquire the power of giving the undivided attention to the sounds emanating from the chest—a power only to be obtained by habit; and it is equally essential to *practise* the mechanical application of the stethoscope; to learn how to place it evenly on the chest; to hold it there firmly; to ascertain by means of the fingers whether every portion of its thoracic end is in close contact with the skin, and, if not, to adjust it properly, and then remove the contact of the fingers; to fit the ear readily to the ear-piece; and to employ sufficient, but not more than sufficient pressure. Nicety in all these matters is only to be obtained by diligent practice, and it is the more necessary because clumsiness or roughness in the use of the stethoscope at once frightens the patient and indicates an inexperienced auscultator.

In conducting a physical examination of the chest, the student will do well to observe a particular order of proceeding. On ordinary occasions the best method is to begin by carefully inspecting the chest, its form, size, and movements; next, to employ the sense of touch, ascertaining in this way the amount of movement and the character of the vibrations in different parts, and then to have recourse to percussion, first on the clavicles, then on the anterior surface of the chest, proceeding from above downwards, next on the lateral regions, and lastly, on the posterior part of the chest. The stethoscope should then be made use of, and the parts examined by its means in the same order. The various methods of examination will thus severally furnish evidence as to the condition of the parts within the chest, which, taken together, will often justify our forming conclusions not warranted by the signs obtainable by any single method of examination.

It is sometimes objected that such an examination is tedious to the physician and wearisome to the patient, and therefore practically ought not to be employed. But an attentive student will soon perceive that if the examiner is skilful and well practised, and is systematic in his method of proceeding, a few minutes will usually suffice to elicit information not otherwise attainable. Even when, in obscure and difficult cases, it is necessary to devote more time and attention to the investigation of the disease, it certainly is our duty to sacrifice such time, and for once to run the risk of fatiguing our patient, rather than to undertake the treatment of the case in doubt or ignorance as to our patient's disorder. In the simplest case, mischief often occurs in situations where the general symptoms least betoken its existence; and therefore, if the whole chest be not examined before an opinion is expressed, or a line of treatment decided on, a grievous error may be made.

The examination should be commenced whilst the patient is breathing naturally; he should then be desired to take a deep inspiration, then to speak, then to cough, and then again to breathe naturally as is his wont. By these means, as will be explained hereafter, it will often happen that much valuable information not otherwise attainable may be acquired, and the real condition of the lungs ascertained.

One difficulty occasionally presents itself, which is not to be overcome without much ingenuity and patience. Some persons, when under examination, seem incapable of breathing naturally, and equally so of taking a deep inspiration. When told to do either the one or the other, they commence a series of unnatural, awkward movements, which materially impede the entry of air into the lungs, or modify the sounds to which it gives rise: they open the mouth, raise the shoulders, and fix the chest, without the least attempt at full inspiration; or, if they do breathe efficiently, they draw in the air through the compressed lips, or else open the mouth and relax the fauces in such a manner as to occasion noises which overpower the sounds produced within the chest. Thus, not unfrequently considerable difficulty is experienced in making a patient breathe in such a manner as to enable us to judge of the condition of the lung. In some such instances our object may be attained by telling him to sigh, or to fill the chest, or to draw a deep breath, exemplifying the action by breathing deeply and noiselessly several times in succession, and desiring him to imitate our action. But not unfrequently even these expedients fail; and then the only means of effecting our object is by directing him to speak or cough for some moments consecutively, when, after the repeated short expirations which accompany those acts, a full, noiseless inspiration follows, and he does involuntarily what his previous efforts had failed to accomplish.

CHAPTER VII.

RESPIRATORY SOUNDS.

THE next points for consideration are, the cause and nature of the respiratory sounds, and the modifications they undergo in disease.

It may be well to premise that two sounds are audible by auscultation accompanying each act of respiration—the one corresponding to the act of inspiration, the other to that of expiration, and named accordingly the "inspiratory" and "expiratory" sounds. These sounds are heard of varying duration, rhythm, and character in different portions of the respiratory tract. Along the pharynx, larynx, and trachea, they are of nearly equal duration, the latter being somewhat the most prolonged;[1] a dis-

[1] The difference in point of duration is very slight, and, frequently, is not perceptible. When it does exist, it is attributable, I believe, to the fact, that after the completion of the respiratory act a pause ensues, which admits

tinct interval occurs between the cessation of the former and the commencement of the latter; and they are loud, dry, hoarse, and hollow, as if caused by a rush of air through a tube of considerable diameter, rough and irregular on its internal surface. This is the character of what is termed "tracheal respiration."

Beneath the upper bone of the sternum, and in thin persons between the scapulæ at a point corresponding to the bifurcation of the trachea and the origin of the larger bronchi, the rhythm and character of the sounds are altogether changed. There is no longer the same interval between them,—they follow each other more closely; the inspiratory sound is considerably longer and louder than the expiratory, and both are softer, less loud and dry, less hollow and less blowing than the sounds heard over the trachea. These points characterize what has been termed "bronchial respiration," or "tubular breathing."

Passing still further along the respiratory tract to those parts of the chest which contain only the ultimate ramifications of the bronchi and the vesicular structure of the lung, the character of the sounds changes again. There no longer exists any appreciable interval between the inspiratory and expiratory sounds; the former is above twice as long as the latter, and is heard of a higher pitch—so that whilst the inspiratory sound accompanies the whole inspiratory act, and is soft, breezy, gradually developed, and continuous, the expiratory sound is short, weak, and in some persons almost, if not quite inaudible, especially on the upper part of the left side of the chest. These are the peculiarities of what has been termed "pulmonary respiration," and the sound is known as the "respiratory" or "vesicular" murmur.

The source of these respiratory and expiratory sounds, and the cause of their modification at different parts of the respiratory tract, admit of satisfactory explanation. The sounds themselves are traceable to the vibrations excited in the air by the irregularities of the surface over which it passes during each act of inspiration and expiration, as also, probably, in some slight degree to the alternate expansion and collapse of the lung-tissue; the difference in their duration, rhythm, character, and loudness in the various portions of the respiratory tract, to differences in the velocity and volume of the current of air in the different parts, to the nature of the channels through which it travels, to their position as regards the chest-walls, to the direction of the respiratory blast, and the consequent readiness with which the

of the uninterrupted continuance, for a few seconds, of the sonorous vibration excited during *expiration*; whereas the expiratory blast commences immediately on the cessation of inspiration, and thus destroys or prevents the continuance of the sonorous vibrations excited during that act.

sounds are conveyed to the ear. If the several conditions attending the production of sound and its transmission to the ear in different portions of the respiratory tract are carefully considered, the cause of the observed varieties will be at once apparent.

To begin with the inspiratory sounds. What are the precise conditions observed in the first portion of the respiratory tract? The cartilaginous rings of the trachea are strongly marked, and its walls, therefore, are rough and irregular; the calibre of the tube is considerable, the volume of air passing through it is large, and as the current is drawn in with immense force, it traverses this portion of its course with great velocity. Hence the resulting sound is loud, dry, hoarse, and hollow, conveying the impression of air rushing through a tube of considerable diameter, rough and irregular on its internal surface. The trachea being superficial in position, the sound reaches the ear in all its loudness, roughness, and hollowness.

Below the bifurcation of the trachea the bronchi subdivide into smaller and smaller tubes, and the cartilaginous rings which enter into their composition become less and less pronounced, until in the terminal ramifications of the bronchi they cease to exist, and the tubes are smooth on their internal surface. Thus differences are at once perceptible in the causes of sound in the different portions of the respiratory tract. The rush of air is the same in all, but the calibre of the tubes through which it has to pass decreases gradually, and the nature of the surface over which it travels varies greatly in the different parts of its course. In the trachea and upper air-passages the air meets with little or no impediment beyond that presented by the roughness and irregularity of their walls: in the larger bronchi it has to encounter the obstacle presented by the subdivision of the bronchial tubes; whilst in the third part of its course, where it passes along the terminal bronchial ramifications into the vesicular structure of the lungs, it no longer meets with the roughness presented by the existence of the cartilaginous rings, but it has to overcome the contractile power of the air-passages and is thrown into vibration by impinging upon the uneven, irregular surface of the pulmonary vesicles. This further difference is also to be remarked, that whereas the larger bronchi are well calculated, from the firmness and elasticity of their walls, to reflect and propagate any sound which may be generated within them, and in some situations are close to the parietes of the chest, the smaller bronchial ramifications are devoid of cartilaginous rings to keep their sides asunder, and are mostly surrounded by the spongy tissue of the lung—a bad conductor of sound. The result is just what might have been anticipated, viz., that the inspiratory sound emitted from the larger bronchial tubes, though less loud and rough, less hollow and less blowing than the tracheal inspiratory sound, is far louder, more rough than, and of a different character from the soft, breezy, vesicular murmur which arises from the terminal portions of the respiratory tract; and further, that this vesicular murmur, which is superficial in its situation and reaches the ear without difficulty, is alone heard in every portion of the chest where the bronchi do not come in close proximity to the thoracic walls. Nor can the latter circumstance excite surprise after what has been stated in a previous chapter[1] relative to the conducting power of different media; for, independently of the fact that the superficial position of the vesicular murmur necessarily leads to its obscuring or overpowering any more deeply seated sound, it is obvious that, except when the bronchial tubes are placed in close proximity to the thoracic walls, all vibrations originating in the bronchi must pass from air to membrane, and from membrane to air, many hundred times before they can reach the surface of the chest. Thus the medium of their transmission is constantly changed, and at each change their intensity is diminished.

The duration of the inspiratory sound in different portions of the respiratory tract is also regulated in accordance with fixed laws. The smaller the size of the air-passages through which the current of air has to pass, the weaker, *cæteris paribus*, the current which will produce sonorous vibrations therein. Thus the *in*spiratory blast will occasion sound in the vesicular structure of the lung, and in the smaller bronchial passages, before it has acquired sufficient intensity to cause sonorous vibrations in the trachea and larger bronchi, and it will serve to keep up such vibrations in the pulmonary vesicles and the terminal branches of the bronchial tubes for some time after it has ceased to occasion sonorous vibrations in the larger bronchi and the windpipe. Hence it occurs that the vesicular inspiratory sound is more prolonged than the tracheal and the larger bronchial inspiratory sounds.

So again, in regard to the rhythm of respiration, and to the duration, character, and loudness of the expiratory sound in different portions of the respiratory tract. Many of the causes which give rise to sound during inspiration produce no such effect during expiration, and this holds good especially in regard to those causes of sound which exist in the pulmonary portion of the respiratory tract. Thus during expiration the current of

[1] Chap. V, p. 46.

air does not meet with resistance from the contractility of the air-passages as it does in ordinary inspiration, and being no longer inwards and against the irregular surface of the air-vesicles and the edges of the bronchial subdivisions, but outwards, in the same direction with those causes of vibration, it is in no way acted on by them.[1] Hence, as the expulsion of air from the air-vesicles takes place gradually, as the current is necessarily small and weak, and as in their natural condition the walls of this portion of the air-passages are smooth, so that the air meets with no obstacle in traversing them, the expiratory sound over healthy lung-substance is soft and weak. Such as it is, however, it takes its origin in the collapse of the lung and the rush of air from the air-vesicles, and commencing, therefore, with the commencement of expiration, it immediately succeeds the cessation of inspiration. But it is necessarily much shorter than the inspiratory sound; for the resistance to the exit of air from the chest is less than that to its entrance into the lungs, whilst the expiratory muscles, aided by the natural resiliency of the lungs,[2] and by the elasticity of the costal cartilages, form an expulsive power about one-third greater than that concerned in the act of inspiration.[3] And to such an extent do these causes operate in influencing the duration of the act, that pulmonary expiration is found to occupy not much above one-third of the time occupied by the act of inspiration.

In the other portions of the respiratory tract, a material difference is observed in the rhythm of respiration and in the character and duration of the expiratory sound. The more remote any portion of the respiratory tract from the air-vesicles whence the air is expelled, the larger, of course, the calibre of the tubes through which the air has to pass; the larger and stronger must be its current, in order to excite sonorous vibrations therein, and the longer, consequently, the time which must elapse before such vibrations can be produced; in other words, the longer must be the interval between the cessation of the inspiratory and the commencement of the expiratory sound.

[1] The difference resulting from this arrangement may be illustrated by the effect of blowing first *against the edges* of a sheet of paper, and then *along* its surface.

[2] The power of this resiliency of the lungs has been shown experimentally by Dr. Carson, in a paper published in the "Philosophical Transactions" for 1820, and its importance on the mechanism of respiration is manifest in the symptoms observed in emphysema, in which this elasticity of the lungs is lost.

[3] See a paper by Dr. Hutchinson, "Med.-Chir. Trans.," vol. xxix; and also Valentin, "Lehrbuch der Physiologie des Menschen," p. 529, quoted by Dr. Davies.

Hence, whilst in the pulmonary portion of the respiratory tract the expiratory succeeds immediately to the inspiratory sound, a distinct interval occurs between the two sounds over the bronchi, and a still longer interval over the trachea. Moreover, in the bronchi, as in the vesicular portion of the respiratory tract, the causes of sound during expiration are far less numerous than those which operate during inspiration, for the vibrating tongues formed by the subdivision of the air-passages are not, as during inspiration, opposed to the current of air, but are in the same direction with it. Thus it happens that the bronchial expiratory sound, though louder and rougher than the vesicular expiratory murmur, is neither so loud nor so rough as its corresponding inspiratory sound, nor, for the reasons already stated, is it nearly so prolonged.

In the third portion of the respiratory tract the calibre of the air-tubes (the trachea and larynx) is still larger, and consequently a longer time elapses after the cessation of inspiration before the expiratory blast acquires sufficient volume and strength to throw the air into sonorous vibration. But when once this point has been reached, the force of the blast, the roughness of the internal surface of the passage, and the tension of its walls, are such as to insure the production and maintenance of sonorous vibration as long as the expiratory effort is continued. Indeed, the causes of sound in this portion of the respiratory tract are almost the same during expiration as during inspiration. Thus it is found that the tracheal expiratory sound, though commencing later, or, in other words, after a longer interval than the bronchial expiratory sound, is of longer duration than that sound, and indeed is equal in that respect to the tracheal inspiratory sound.

The importance of these facts, as bearing on the diagnosis of disease, will be manifest when it is stated that the respiratory sounds undergo modifications corresponding to any physical alteration produced in the condition of the lung-substance or the air-passages, or in the freedom of ingress or egress of the air. At present it is sufficient to mention the circumstance, which will be illustrated and explained as each separate modification of breathing is reviewed.

And first in regard to the varieties of respiration which are compatible with a healthy condition of the lungs. What is their nature, and what are their causes?

The respiration already described is that of a healthy adult of middle age. In infancy, in childhood, and in old age, the conditions are somewhat different, and the respiratory sounds vary accordingly. In infancy, the inspiratory muscles are weak, and have a difficulty in overcoming the resistance offered to a full inspiratory effort by the elasticity of the chest-walls and the resiliency of the lungs. Hence, as the diaphragm is the most powerful of the inspiratory muscles, the respiration is almost wholly abdominal; and as the resistance offered by the elasticity of the chest increases rapidly with the increase of thoracic expansion,[1] the breathing is necessarily short and hurried, and the lungs being imperfectly expanded, the respiratory sound is faint. In childhood, or, in other words, when the strength has increased, and the muscles have become more fully developed, the chest expands freely at each inspiration; indeed, the thoracic walls, being extremely elastic, admit of great freedom of action in the lungs: the resiliency of the lung-tissue is remarkable; the pulmonary air-vesicles are small, so that a larger number of them exist within a given space than are to be found within the same space in adult life, whilst, as the walls of the chest are thin, the respiratory sounds are transmitted with great readiness to the ear. The result of these various circumstances is, that the faint respiratory sounds of infancy are replaced by breathing sounds which are louder and more intense than those heard at a more advanced age, though their character remains unchanged. This modification of breathing is termed *puerile* respiration.

In old age, on the contrary, the costal cartilages are stiffened, the elasticity of the lung-tissue is diminished, and consequently the play of the lungs is lessened. The inspiratory sounds, therefore, are usually less intense than in early adult life, and the expiratory sound is more prolonged. This modification is termed *senile* respiration.

In youth the respiration is more frequent or rapid than in adult life, and in all persons, particularly in women, and in those especially who are accustomed to tight lacing, the pulmonary respiratory sounds are fuller in the upper than in the lower part of the chest, in consequence of the greater play and more thorough inflation of that portion of the lungs.[2] They

1 "Med.-Chir. Trans.," vol. xxix.

2 This is strictly true, for tight lacing, or, indeed, anything which restrains the action of the lower ribs, must necessarily occasion increased action of the upper part of the chest. But something more than this is needed to explain the remarkable action of the upper part of the chest in women, for the contrast between the respiratory movements in the male and female is almost as marked in naked savages as in tightly clad and civilized Europeans. Probably the cartilages of the ribs are endowed with extraordinary elasticity, and the intercostal muscles with unusual power and freedom of action in the female, with the view of obviating the great interference with respiration, which would otherwise result from the enlargement of the gravid uterus, and the consequent obstacle to the descent of the diaphragm.

are often particularly loud and intense in nervous persons suffering from hysteria and other similar affections, for the reason, I believe, that, when under the excitement of examination, such persons breathe more deeply and forcibly than under ordinary circumstances; whilst in certain individuals they are unusually weak, in consequence, I believe, of the large relative size of the lungs, and the less necessity which therefore exists for the full inflation of the air-vesicles.[1] In neither instance, however, is the character of the sound altered, nor is the peculiarity alluded to attributable to disease. Sometimes, even when the inspiratory sound is as loud as usual, the expiratory sound is very weak, and in certain instances is unattended by any audible sound, especially in the left infra-clavicular region. Again, when corresponding portions of the chest are examined, the *inspiratory* sound is usually found to have the same character on both sides, though it is sometimes louder on the left supra- and infra-clavicular regions than it is in the corresponding regions on the right side. But the *expiratory* sound is commonly louder and more prolonged in the supra- and infra-clavicular regions on the right side, and in the superior scapular and inter-scapular regions on the same side, than it is on the left, the difference being more perceptible at the inner than at the outer part of these regions, and more strongly marked in females than in males. Indeed, if the observation be confined to the right infra-clavicular region, it may be stated almost as a rule that the intensity and duration of the expiratory sound are greater on the right than on the left side, the difference being most marked in thin, nervous females. Again, in certain instances, the sounds of respiration on both sides of the chest are either weak and indeterminate in character, or else are replaced by an indistinct humming.

Now, it must be distinctly understood that the existence or non-existence of the varieties of breathing already described affords no criterion as to the presence of disease. They may not exist, and the patient may be perfectly free from disease; or they may be strongly marked, and he may be healthy nevertheless. But it is otherwise in regard to certain modifications of respiration, to which we have now to direct our attention. Those already mentioned derive their chief importance from the fact

[1] Direct experiment with the spirometer has convinced me that in the majority of such cases, the vital capacity is above the average, or, in other words, that the size of the lungs relatively to the body and the circulatory apparatus is unusually large; and nothing can be conceived more directly conducive to weak respiratory sounds during tranquil respiration than a breathing apparatus which, being unusually capacious, is imperfectly or partially expanded at each inspiration.

that, if not rightly interpreted, they might often lead to a presumption of disease: those now to be described are always indicative of an unhealthy condition of some portion of the respiratory apparatus. They consist of changes in the intensity, rhythm, and quality of the breathing sound, and it will be seen that in the first two varieties the duration of the sounds occupy an important place.

1st. In *Intensity*, the respiratory murmur may be—

(*a*) Exaggerated or increased.
(*b*) Weak or diminished.
(*c*) Weak and indeterminate.
(*d*) Absent or suppressed.

2d. In *Rhythm* (*a*), the respiration may be jerking or interrupted (the *respiration saccadée* of Laennec).
(*b*) An interval of greater or less duration may intervene between the inspiratory and expiratory acts.
(*c*) The relative duration and intensity of the inspiratory and expiratory sounds may be altered.

3d. In *Quality*, the sounds may be—

(*a*) Coarse, harsh, or rough.
(*b*) Blowing or whiffing.
(*c*) Of a hollow character.
(*d*) Accompanied by a metallic resonance.

1st. *Intensity.*—(*a*) Exaggerated breathing, as contrasted with the natural pulmonary respiration, is characterized by the greater intensity and duration of its sounds. It is sometimes termed "*puerile*" respiration, from its resemblance to the loud respiration of children; and "*supplementary*," from the fact that it is not abnormal in character, but is simply caused by an inordinate action of certain healthy portions of the lungs, set up to supply the deficiency of respiration consequent on the inactivity of other portions which are destroyed or affected by disease.[1] Thus, in fact, it is due to locally increased pulmonary

[1] The term "puerile," as applied to this variety of respiration, is singularly inappropriate, inasmuch as "puerile" breathing is a *healthy* type of respiration occurring in children, and "exaggerated" breathing is commonly met with in adults, and is essentially dependent on the presence of disease in some other portion of the lungs. The term "supplementary" is at once appropriate and significant of the true nature of this variety of breathing.

action; and its peculiarity is derived partly from the unusually large number of air-vesicles in any given space inflated at each act of inspiration, partly from the completeness of their distension, and partly from the increased volume of air and the unwonted force and rapidity of its current in those portions of the lungs where the exaggerated breathing sound exists. Unlike the puerile respiration of children, and the naturally loud breathing which is sometimes met with in the healthy adult, exaggerated breathing is never heard diffused entirely throughout the chest.[1] It exists at one spot and not at another, according to the position and extent of the mischief in which it takes its origin; its intensity is greatest close to the actual seat of disease; and, after a time, it may become inaudible where it had previously existed in well-marked character. Owing to the close proximity of disease, which may obstruct the bronchi, and thus offer some impediment to the egress of air, the expiratory sound is often prolonged relatively as well as absolutely; but this is a modification which, though frequently coexistent with exaggerated breathing, is essentially distinct from it in its mechanism, and is attributable to causes which are not necessary to its production.

Thus, then, as anything which prevents the free play and perfect action of any portion of the pulmonary tissue may give rise to exaggerated breathing in other portions of the same or opposite lung, its presence should lead us to search out and discover the cause and seat of the obstruction of which it is the certain index. The mere fact of the existence of exaggerated breathing is positive proof of pulmonary obstruction: the cause and seat of the obstruction may vary according to circumstances. Fluid effused into the pleural cavity by compressing the lung and thus interfering with its action; aneurismal or other tumors, by pressing on and thus obstructing the bronchi; mucus or other matters blocking up the air-passages; pneumonic exudation; sanguineous effusion; tuberculous or other deposits, by condensing the lung-tissue; or, on the other hand, vesicular emphysema, by diminishing its elasticity, and thus impairing the action of circumscribed portions, may, one and all, produce exaggerated breathing of greater or less extent and intensity in the adjoining healthy pulmonary tissue. The only other condition in which it occurs is that of spasmodic asthma, in which the healthy tissue of the lung, suddenly released from bronchial spasm, is

[1] The only exception to this rule is in "spasmodic asthma," when the nature of the complaint is well marked and obvious, and the exaggerated breathing is of only temporary duration.

immediately inflated by a deep and forcible inspiration. In such a case as this the exaggerated breathing is only of temporary duration, and the nature of the malady is sufficiently obvious; but in every other instance its nature, seat, and extent require careful investigation.

(*b*) *Weak respiration* contrasts strongly with exaggerated breathing. Its sounds, though not altered in character, are diminished in intensity and duration. Referable simply to deficient expansion of the pulmonary tissue, it may arise from any cause which directly or indirectly interferes with or controls the inflation of the lungs. Hence its causes are numerous, and require care and judgment in their discrimination. Malformations of the chest, from whatever cause arising, may give rise to this modification of breathing, and so may deficient muscular action, whether from debility, exhaustion, or paralysis, on the one hand, or as the result of pain, as in pleurisy, intercostal rheumatism, or peritonitis, on the other. Defective pulmonary elasticity, as in vesicular emphysema, is another fertile source of weak respiration, and any obstruction of the air-passages, whether from internal causes or from causes external to the lungs, will also give rise to it;—the chief internal causes being certain diseases of the larynx; thickening of the mucous membranes lining the bronchi, with consequent diminution in the diameter of the air-tubes; an accumulation of mucus or other secretion in the trachea or bronchi; and the presence of foreign bodies or the deposit of tuberculous or other matter in the interior of the air-passages;—the external causes being the pressure of enlarged bronchial glands, or of aneurismal or other tumors; moderate effusions into the cavity of the pleura or pericardium; pneumothorax; or, in short, anything which removes the lungs from the chest-walls, or so far presses upon the pulmonary tissue as to interfere with the due performance of pulmonary respiration, without altogether obstructing or annihilating it. Sometimes weak respiration is of temporary duration only, or recurs in an intermittent form, indicating an equally temporary or occasional interference with the entry of air into the lungs. Thus, it results from the partial closure of the glottis which occurs in whooping-cough, and is met with in spasmodic asthma, pleurodynia, and other affections, and its intermittence constitutes an important element of diagnosis in cases in which aneurismal or other movable tumors press upon the larger bronchi, and in those also in which obstruction is caused by the presence of foreign bodies in the main bronchi on either side of the chest.

(*c*) Under the head of weak respiration may also be ranged certain cases in which the breathing sounds are either weak and indeterminate in character, or else are replaced by an indistinct humming. This condition of respiration is not necessarily limited to unhealthy persons, as I have met with several well-marked instances of it in persons enjoying robust health.

(*d*) *Absence or suppression of respiration.*—The total absence of respiratory sounds denotes the existence of causes which either obstruct the bronchi completely, or else prevent the action of the lungs. The one essential condition is that air shall not permeate the lung-tissue. Thus, the causes already mentioned as productive of weak or diminished respiratory sounds may lead, when still further developed, to their entire annihilation. Excessive tuberculous, cancerous, or other infiltration of the lung, or of any portion of it; an extreme degree of vesicular emphysema; complete occlusion of a bronchus, whether from internal obstruction, or from external pressure, as from aneurismal or other tumors; large collections of fluid or of air in the chest, compressing the entire lung and forcing it back, a carnified, impervious mass, against the vertebral column,—these are some of the causes of absent respiration, the two latter being of most frequent occurrence.

2d. *Rhythm.*—During healthy respiration, *in*spiration and *ex*piration take place evenly; the expiratory follows closely upon the inspiratory movement, and a brief interval of repose follows the completion of each respiratory act. These general characters hold good, whatever may be the increase or decrease in the frequency of the respiration, so that the expiratory sound maintains its due relative bearing to the inspiratory, the ratio of the one to the other being about 1 to 4. But in disease various modifications are observed. Thus—(*a*) the mode of evolution of the sounds may be altered, and instead of being even and continuous from their commencement to their close, both sounds, but especially the inspiratory, may become jerking or interrupted—evolved, as it were, by instalments. This is the *respiration saccadée* of Laennec,—the cogged-wheel respiration of some English authors. (*b*) Expiration, instead of following closely upon inspiration, may be separated from it by a distinct interval or pause. And (*c*), instead of bearing their due relative proportion to each other, the sound of expiration may be increased in intensity and duration until it becomes louder than the inspiratory murmur, and extends over the normal interval of repose, so that inspiration succeeds expiration with scarcely an appreciable interval of silence.

(*a*) The first-named variety, when existing generally over one or both lungs, is indicative of something giving rise to irregularity in the act of inspiration. Thus, timidity or nervousness whilst the patient is undergoing examination; the pain of pleurisy or intercostal rheumatism; spasmodic affections of the air-passages; or, indeed, any cause which will produce occasional checks during the act of respiration, may give rise to jerking interruption in the sounds. And as the cause of such irregularity in any particular instance must be sufficiently obvious, the existence of generally diffused jerking respiration is comparatively unimportant in a diagnostic point of view. But when it exists in small portions only of the lungs, and when, more especially, it is met with in the upper part of the chest, its practical significance can hardly be overrated, inasmuch as the finely divided jerking respiration, or, as it has been termed, the "cogged-wheel" rhythm of respiration, heard in that situation, is a frequent accompaniment of incipient tuberculization of the lungs. Its immediate cause is not quite determined. It is commonly supposed to be referable to the presence of thick, tenacious mucus, which presents some obstruction to the free ingress and egress of the air. But although this is doubtless a frequent cause of it, I am strongly inclined to the belief that it may arise independently of mucus in the air-passages, as a result of obstruction, consequent on pressure caused by the presence of tubercles or some other form of consolidation. Not only is it easy to conceive how the presence of tubercles or other solid matters may press upon the smaller bronchi and produce such obstruction, but I have so often met with jerking respiration, during the early stage of tubercle, in persons who at the time were quite free from cough, expectoration, râles, and every other evidence of mucus in the air-tubes, that I cannot regard it as invariably, or even generally, referable to the presence of mucus. It is always due to temporary local obstruction of the air-tubes, of such a nature as will yield to the force of the inspiratory blast as soon as the expansion of the lungs has reached a certain point; but I believe that the obstruction is usually occasioned by thickening of, or pressure on, the walls of the air-passages, and not by the presence of thickened mucus.

(*b*) The second variety in which a distinct interval or pause occurs between the close of inspiration and the commencement of expiration is of less importance in a diagnostic point of view. It may depend upon any cause which puts a stop to the production of sound before the close of the respiratory act; or which, on the other hand, prevents the production of sound during the commencement of expiration. Thus, the inspiratory sound, in cases of pulmonary consolidation, ceases before the accompanying expansion of the chest; whilst, on the other hand, in emphysema, when the elasticity of the pulmonary tissue is impaired, the contained air is not expelled from the air-vesicles during the first portion of the expiratory act. In either case an interval occurs between the cessation of the inspiratory and the commencement of the expiratory sound; but, in the one case, this is caused by the inspiratory sound being prematurely checked, in the other by an abnormal delay in the commencement of the expiratory sound.

(*c*) The third variety is characterized by an alteration in the relative duration and intensity of the inspiratory and expiratory sounds. The proportion which the former bears to the latter is about four to one in a state of health, and may become as one to five under the influence of disease. This excessive relative prolongation of the expiratory sound is met with only in emphysema, and is referable in part to the absolute prolongation of expiration consequent on the diminished resiliency of the lung and the swollen and thickened condition of its mucous lining membrane, and in part also to the abnormal shortening of the sound of inspiration, the sound in many cases of emphysema being heard only towards the close of the inspiratory act. But simple prolongation of the expiratory sound is of common occurrence, and its essential cause is the same in all instances, viz., a want of freedom in the egress of the air from the lungs. It may exist to a greater or less degree, and many causes may conduce to its production. One of the most common is the thickened and swollen condition of the mucous membrane of the air-passages which exists in chronic bronchitis; another, the pressure of aneurismal or other tumors on the bronchi; and another, consolidation of the lungs, from whatever cause arising—whether from internal deposit or from external pressure, as in pleurisy or pneumothorax. But the most common, and therefore practically the most important, is the presence of tubercles in the lungs. Tubercular matter, when deposited in the lungs, is found either on the free surface of the bronchial mucous membrane, or in the tissue external to the air-vesicles. In either case it gives rise to more or less swelling of the bronchial lining membrane, and causes projections into the air-passages, with a corresponding diminution in their calibre; impairs or interferes with the resiliency of the lung, and creates more or less impediment to the free egress of the air. Thus, the conditions essential to the existence of prolonged expiratory murmur are often present at an early stage of consumption, and as many of the more characteristic signs of tubercle are at that time wanting, prolonged expiration becomes a valuable diagnostic sign. It is insufficient of itself to afford evidence of the disease, but it serves to confirm in the strongest manner any evidence derived from other sources.

3d. *Quality.*—Anything which interferes with the normal condition of the lung, whether by solidifying or otherwise altering the density of its tissue, by producing alterations in the calibre of the air-passages, or by affecting the dryness of the bronchial mucous membrane, and thus increasing the friction encountered by the inspired air, will interfere with the soft breeziness of the sound which has been described as characteristic of healthy respiration. The earliest deviation from the natural standard *quality* of respiration is roughness, coarseness, or harshness, occasioned sometimes by rarefaction, but more commonly by solidification of the pulmonary tissue, with some degree of swelling of the bronchial mucous membrane. This variety of abnormal breathing, especially when strongly pronounced, is marked by the peculiarly blowing or whiffing character of the sounds, and by their increased duration and intensity,—characters which, though common to both sounds, especially in their highly developed types, are first perceived and are most strongly marked in the sound of inspiration. But the greater the amount of local change the greater the alteration in the sounds of respiration; and thus it happens that when the pulmonary tissue is condensed, and the sounds emanate from the large bronchi or from hollow spaces in the lungs, the character of *hollowness* is superadded to the other characters already described, and the breathing is not only rough, harsh, and blowing, but is of a distinctly hollow character.[1] Changes in the density of the pulmonary tissue may thus be traced by the alterations which occur in the respiratory sounds. Condensation, for instance, when only slight, and confined to the air-vesicles and smaller bronchial tubes, is simply accompanied by a roughness, coarseness, or harshness of respiration, marked by a blowing or whiffing character, such as is not observed in a state of health; when the structure of the lungs is more deeply implicated, and the smaller tubes are rendered impervious, so that the vibrations

[1] Some theoretical considerations relative to the reinforcement of the respiratory sounds in certain instances of hollow breathing have been lately forced into more or less prominence by Skoda, of Vienna. He has attempted to ignore the effect of pulmonary consolidation in rendering more audible on the chest-walls any sounds emanating from the bronchi, and has endeavored to explain the admitted increase in the loudness of such sounds under certain circumstances by reference to the doctrine of consonance. As the matter will be discussed in connection with increased vocal resonance, to which the same theory applies, it is needless to enter on the question at present, but I would refer the reader to pp. 115–18 of this treatise. Whatever is there stated relative to increased vocal resonance applies equally to the reinforcement of the sounds of respiration.

conveyed to the ear are derived from bronchial tubes of a larger size, the sound, though still rough, coarse, or harsh, is more dry, more blowing, and more prolonged than in the earlier stage of the disease; whilst at a still later stage, when even greater condensation of the lung-tissue has occurred, and the sounds emanate chiefly from the larger bronchi, the breathing assumes a decidedly hollow character, and somewhat resembles that which is heard in health over the larger bronchial tubes. Hence this variety of breathing has been termed "bronchial" breathing. When solidification is complete,—when in fact the lung is hepatized, as in pneumonia, and the walls of the bronchi are dry and tense, forming good reflectors of sound, a metallic resonance is superadded to the mere bronchial character, the blowing is less diffused, and a sound results which greatly resembles that produced by blowing sharply through a brass tube. This form of breathing has been called "tubular" breathing. When again, as is sometimes the case, a large bronchus, or a bronchial tube which has undergone dilatation, or a small empty vomica with tense walls, is surrounded by condensed and homogeneous lung-tissue, the air passing through or into it produces a sound resembling that produced by air passing rapidly into a small cavity. This form of respiration has been styled "cavernous" respiration. When, again, the cavity is still larger, and is bounded by tense walls, as, for instance, the pleural cavity, or, as in some rare cases, a large excavation in the lung, the air passing into it occasions a sound like that of air passing rapidly into an empty jar or small cask, and hence this variety of respiration has been termed "amphoric" respiration.

It will be perceived that all these varieties of hollow breathing merge gradually into each other, and differ only in the size and condition of the air-passages or spaces in which the sounds take their source, and in the density, homogeneity, and conducting power of the surrounding parts. They are not peculiar to any particular form of disease, nor to any one form of structural mischief, but are each indicative of a certain physical condition of the pulmonary apparatus, the nature and true significance of which must be determined by other means. Thus, for instance, a large or a dilated bronchus, surrounded by condensed pulmonary tissue, may produce a sound which cannot be distinguished from the so-called cavernous breathing, whilst, on the other hand, a small cavity, or a number of small, empty vomicæ in the lungs, may occasion respiratory sounds closely resembling the so-called bronchial and tubular breathing. So gradually do the different varieties of hollow breathing merge into each other, that it is sometimes difficult to decide whether a given sound arises from a bronchus or from an abnormal cavity in the lung-tissue. All that is essential to the production of any form of hollow breathing-sound is an empty space of a certain size, in free communication with the upper air-passages, and surrounded by consolidated or homogeneous, yet vibratile lung. The precise nature of the hollow space and the exact cause of the consolidation or increased homogeneity of the surrounding tissue cannot be determined solely by the auscultatory signs, but must be decided, if at all, by the history of the case, the condition of the patient, the situation of the spot at which the sound is most audible, and by the signs elicited by other means of examination. The terms, "bronchial," "tubular," and "cavernous," therefore, are objectionable, as implying a theory as to the source of the sounds, which, in many instances, may be erroneous, and in its practical application has often misled inexperienced auscultators. If, for convenience sake, the terms are retained, their true significance must be borne in mind, in order that undue importance may not be attached to them. Neither of them conveys the slightest information which is not afforded by the term "hollow," and I am strongly of opinion that we should do well to expunge such fallacious words from our phraseology, and employ the term "hollow" as applicable to all abnormal breathing-sounds of a hollow character.

It follows, from what has been already stated, that deviations from the healthy quality of respiration are to be found in the early or dry stage of bronchitis, in which the air-tubes are congested and devoid of secretion; in congestion of the lungs, with distension of the vessels beneath the pulmonary vesicular membrane, causing narrowing of the air-passages or irregularity of their surface; in dilatation of the bronchi, and in excavations of the lung-substance, from whatever cause arising; in vesicular emphysema, and in pulmonary consolidation, whether caused by the infiltration of tubercle or other matter, by the contraction of inflammatory exudation, or by the pressure of fluid or of solid or gaseous materials in the pleural cavity—the only exceptions being when sufficient healthy lung-tissue intervenes between the ear and the part affected to mask the morbid sounds by its own natural ones, and when the amount of foreign matter in the pleura is such as to put a stop to all pulmonary sounds, or prevent their conduction to the ear. In the latter case, the lung is pushed upwards and backwards against the spinal column, and a hollow breathing-sound will still be audible about the middle of the back, close to the spine, or, in other words, in the position of the main bronchi.

Further, it is obvious that, according as the changes capable of altering the quality of the respiratory sounds are limited to one spot, or are generally diffused throughout one or both lungs, so will be the extent of the area over which the altered breathing-sound is heard; and according as the changes are of a transient or enduring nature, so will the alteration in the sound prove more or less persistent. These are points which, viewed in connection with the history of the case, and the position and character of the sound itself, not infrequently lead to a correct diagnosis as to the nature of the existing mischief.

The following tables will show at a glance the varieties of respiration in health and disease, their special characters, their mode of production and usual seat, as also the forms of disease with which the morbid varieties are usually associated.

TABLE I.

RESPIRATION IN HEALTH.

Varieties.	*Synonym.*	*Character of the sound.*	*How produced.*	*Its usual seat.*
Pulmonary respiration.	Respiratory or vesicular murmur.	*Inspiration.*—A soft, diffused sound, of a breezy character, gradually developed and continuous, accompanying the whole respiratory act: followed without any appreciable interval by the *expiratory* sound, which is short, weak, and, in some persons, almost inaudible.	*Inspiration.*—By vibrations excited in the inward current of air by its friction against the walls of the air-passages; by the obstacles presented by the subdivision of the bronchi; and by the irregularity of the surface of the pulmonary vesicles on which the air impinges. *Expiration.*—By the vibrations excited in the expired air by its friction against the walls of the air-passages.	The air-vesicles and terminal bronchi, and therefore heard in all parts of the chest except the upper part of the sternum and the space between the scapulæ, corresponding to the roots of the larger bronchi.
Puerile respiration.	Exaggerated vesicular murmur, or puerile breathing.	The same as the pulmonary vesicular murmur, but exaggerated or intensified in degree.	By the same causes as produce the ordinary vesicular murmur, but the sound acquires intensity in consequence of the great freedom in the action of the lungs in early childhood, the remarkable resiliency of the lung tissue, and the small size of the air-vesicles which leads to there being a larger number of them in a given space than in adult life.	In children, in all parts of the chest where the ordinary vesicular respiration is audible.
Bronchial respiration.	Bronchial breathing.	Both the respiratory sounds are louder, rougher, and of a higher pitch than their corresponding vesicular murmurs; they are of a somewhat hollow, blowing character, more rapidly evolved, and follow each other less closely, so that there is an appreciable interval between the close of inspiration and the commencement of expiration. This variety of breathing merges gradually into the next variety, tracheal or tubular breathing.	Its peculiar character is referable to the size and roughness of the air-passages in which it originates, and to their sound-reflecting power.	The upper part of the sternum and the space between the scapulæ corresponding to the roots of the bronchi.
Tracheal respiration.	Tracheal or tubular breathing.	A loud, dry, hoarse, and hollow sound, as if caused by the rush of air through a tube of considerable diameter, rough and irregular on its internal surface. The inspiratory and expiratory sounds are of nearly equal duration, and a distinct interval occurs between the cessation of the former and the commencement of the latter.	Its peculiar character is due to the size, roughness, and sound-reflecting property of the tubes in which it originates.	Over the larynx and trachea.

CHANGES WHICH OCCUR IN THE

Connected with changes.	*Variety.*	*Synonym.*	*Character of the sound.*
In intensity.	Exaggerated or increased.	Supplementary or puerile* breathing.	The same as the ordinary vesicular murmur of health, but exaggerated or intensified in degree: identical in character with the puerile respiration of healthy children.
	Weak or diminished.	Feeble respiration.	The ordinary vesicular murmur, not altered in character, but simply diminished in intensity and duration.
	Weak and indeterminate.	Indeterminate breathing.	The precise character of the sound cannot be defined: it is sometimes weak and indeterminate, and at others is replaced by an indistinct humming.
	Absent or suppressed.		No sound is heard.
In rhythm.	Jerking or interrupted.	Cogged-wheel respiration, or respiration saccadée of Laennec.	The character of the sounds need not be materially altered, but both sounds, especially the inspiratory, instead of being even and continuous from their commencement to their close, are jerking or interrupted—evolved, as it were, by instalments.
	A distinct interval of varying duration, may occur between the inspiratory and expiratory acts.	. . . .	Expiration, instead of following closely upon inspiration, may be separated from it by a distinct interval of varying duration.
	The relative duration and intensity of the inspiratory and expiratory sounds may be altered.	. . . .	The inspiratory sound may be relatively and absolutely shorter than in health, and the expiratory sound relatively and absolutely louder and more prolonged. Sometimes, though the duration and intensity of the *inspiratory* sound are not perceptibly affected, the *expiratory* sound may be prolonged and intensified.
In quality.	Coarse, harsh, or rough. Blowing or whiffing.	. . . .	Coarse, harsh, and rough, in the first instance, but when strongly developed of a blowing or whiffing character. The change is perceptible in both sounds, which are increased in duration and intensity, but it is most marked in the sound of expiration. It gradually passes into the next variety.
	Of a hollow character. Bronchial or tubular.	Bronchial or tubular breathing.	To the blowing or whiffing character of the sound last described is superadded the character of hollowness, so that the breathing is not only rough and blowing, but of a distinctly hollow character.
	Of a hollow character. Cavernous.	Cavernous breathing.	
	Of a hollow character. Amphoric.	Amphoric breathing.	
	Accompanied by a metallic resonance.	. . . .	A metallic character is superadded to the hollowness of the sound whenever, under the conditions productive of hollow breathing, the walls of the air-passages are tense and possess a great sound-reflecting power.

* Puerile is a wrong term as applied to any form of *morbid* breathing.

SOUNDS OF RESPIRATION IN DISEASE.

How produced.	*Its usual seat.*	*Disease with which it is usually associated.*
By the inordinate action of certain healthy portions of the lungs, set up to supply the deficiency of respiration consequent on the inactivity of other portions which are destroyed or affected by disease.	Not peculiar to any portion of the lungs, and not diffused generally throughout the chest, like puerile breathing (except in spasmodic asthma, when it is of temporary duration), but limited to certain spots in the vicinity of diseased portions of the lungs.	Pleurisy; aneurismal, or other tumors pressing on certain bronchi; pneumonia; tubercles on the lungs; vesicular emphysema; spasmodic asthma.
By any cause which interferes with and prevents the full inflation of the lungs.	Not peculiar to any portion of the lungs; often diffused throughout the chest, but sometimes limited to a part of the lung supplied by one or more particular bronchi.	Deficient muscular action, from whatever cause arising; defective pulmonary elasticity, as in vesicular emphysema; and any obstruction of the air-passages, whether from internal or external causes.
Not clearly ascertained.		
By any cause which prevents the air from permeating the lung-tissue.	May occur in any portion of the chest, but always limited to one or more parts, and usually to the whole or some portion of one lung only.	Its most common causes are extreme emphysema, and excessive effusions of air or fluid in the pleura, as the result of pleurisy or pneumothorax.
When generally diffused over the surface of the chest, it is produced by something which occasions a jerking interruption in the act of breathing, as nervousness, the pain of pleurisy, or spasmodic affection of the air-passages. When confined to particular portions of the lungs, it is due to some local obstacle to the free ingress or egress of the air, usually to the pressure of tubercular deposit, or to the presence of thick mucus in the air-passages.	Not necessarily confined to any particular portion of the chest, but of greatest clinical significance when confined to the apices of the lungs.	The early stage of phthisis, when the sound is confined to certain portions of the lungs; spasmodic affections of the air-passages or nervousness, when it is generally diffused over the chest.
Either by the cessation of the inspiratory sound before the close of the inspiratory act, as occurs over any portion of the lungs in which the air-vesicles are obstructed; or	Not peculiar to any portion of the chest.	Extreme pulmonary consolidation, from whatever cause arising.
By the absence of sound during the first portion of the expiratory act, as occurs when there is any obstacle to the collapse of any portion of the lung-tissue.	Not peculiar to any portion of the chest.	Emphysema, or extreme pulmonary consolidation.
By any cause which affects the resiliency of the lung, and produces a swollen and thickened condition of the mucous membrane of the bronchi, or which otherwise causes an obstruction in the air-passages.	Not peculiar to any portion of the chest, but when the sound is confined to the apices of one or both lungs it is suggestive of tubercular deposit.	Emphysema; chronic bronchitis; tubercles in the lungs; and the pressure of aneurismal or other tumors.
By anything which alters the density of the lung-tissue, produces alterations in the calibre of the air-passages, or affects the dryness of the bronchial mucous membrane.	Not confined to any portion of the chest; most significant at the apices of the lungs in the early stage of phthisis.	The first stage of bronchitis; dilatation of the bronchi; and pulmonary consolidation, whether from the deposit of tubercle, inflammation, or any other cause.
By whatever leads to condensation of the vesicular structure of the lungs and the transmission of sound from the larger bronchi, or from any other space or cavity in the chest. The lower types of it are connected with the existence of small spaces—whether bronchi or vomicæ, in the pulmonary tissue—surrounded by consolidated tissue; the more highly developed forms, with large spaces, having good sound-reflecting walls.	May occur abnormally in any portion of the chest.	In pulmonary consolidation, from whatever cause arising; the presence of enlarged bronchi or tuberculous cavities; and the existence of pneumothorax, with a fistulous opening between the lung and the pleural cavity.
By whatever, under the conditions productive of hollow breathing, imparts excessive tension and increased sound-reflecting power to the walls of the air-containing space.	The posterior and inferior portions of the chest.	Pneumonic hepatization; tuberculous cavities, with tense walls; pneumothorax, with a fistulous communication between the lung and the pleural cavity.

CHAPTER VIII.

ON THE RESONANCE OF THE VOICE IN HEALTH AND DISEASE—BRONCHOPHONY, PECTORILOQUY, AND ÆGOPHONY; AND ON THE DOCTRINE OF CONSONANCE AS APPLIED TO SOUNDS EMANATING FROM THE CHEST; AUTOPHONY; AND TUSSIVE RESONANCE.

Auscultation of the voice, as transmitted through the chest-walls, is another method of obtaining information as to the condition of the lungs and their investing membrane. Varying in pitch according to the varying action of the muscles which regulate the larynx, the sounds formed by the vibrations of the vocal cords are propagated, not only upwards through the mouth, but downwards through the bronchi into the air-cells of the lungs. Accordingly, if the ear be applied directly to the chest-walls, or if the stethoscope be used as a medium of conduction, the voice-sounds may be heard reverberating in the chest. And just as might have been expected *à priori*, it is found that their character is not necessarily proportioned to the loudness of the voice, but, *cæteris paribus*, depends on the condition of the air-passages and of the surrounding pulmonary structure. Thus it becomes evident that if we make ourselves familiar with the sound of the voice as heard in the different portions of a healthy chest, and can discover the conditions which lead to an alteration in its character, auscultation of the voice will furnish a valuable index to the existence of various forms of disease.

The first points to be determined are the peculiarities of character belonging to the vocal resonance during health in different portions of the respiratory tract, and the mechanism or precise local conditions to which these peculiarities are attributable.

If the stethoscope be applied over the larynx of a healthy person while speaking, the voice will be transmitted through the instrument to the ear with greater force and loudness than it reaches the other ear, inasmuch as in the one case the sound is concentrated within the narrow tube—the larynx—to which the stethoscope is applied, and in the other is diffused in a large space, and thereby necessarily loses in intensity. This resonance of the voice is termed natural laryngophony, while that which is heard along the trachea, whether in the mesian line in front or at the lateral parts of the neck, and, to some extent, even over the spinous processes of the vertebræ behind, is called natural tracheophony. In laryngophony and tracheophony the voice seems not only to be produced but to be concentrated immediately beneath the stethoscope, and conveys the impression of passing through the instrument so as to strike directly and forcibly on the ear.

At the upper part of the sternum and on the posterior surface of the chest, between the spines of the scapulæ, where the larger bronchi are given off, the voice is heard less loud, less concentrated, and less distinctly articulated; it appears to be produced at a greater distance, to be more diffused, and to strike less directly against the ear, so that the words seem to be at the end of the instrument, instead of passing through it into the ear. This being the sort of vocal resonance heard over the larger bronchial tubes, is termed natural bronchophony.

Over smaller bronchi the articulation is still more imperfect, and the resonance is considerably weaker and less concentrated, whilst over the whole of the parietes of the chest under which lies the vesicular structure of the lung the vocal resonance is even more diffused, is no longer articulated, and, in fact, does not amount to more than an obscure buzzing or humming. Even this buzzing is sometimes absent.

But there are other modifications which require notice. As a general rule, vocal resonance is most distinct in thin persons, and is obscure or indistinct in persons whose chests are well covered with fat and muscle, and in whom consequently an imperfect conducting medium is interposed between the sound and the ear of the observer; it is strongest when the voice is grave or low-toned; clearest and most distinctly articulated when it is sharp or of a high pitch; it is louder in front than behind, except just in the interscapular regions over the root of the main bronchi, and is more developed in the upper than in the lower parts of the chest; it is more intense on the right side than on the left, especially under the clavicles and in the interscapular regions, and it varies greatly in intensity in healthy persons apparently of the same physical development,—a fact analogous to the greater intensity of the respiratory murmur in some persons than in others, and, like that, attributable, I believe, to variations in the relative size of the pulmonary and circulatory apparatus, and to the consequent greater or less degree of pulmonary expansion. To such an extent, however, does this hold good, that in some perfectly healthy persons the vocal resonance over the vesicular structure of the lungs is louder than it is in others equally healthy and of the same physical development, over moderate-sized bronchi; whilst in others, again, apparently under similar physical conditions, the voice-sounds will be either altogether inaudible or will be heard only as an indistinct humming.

It is obvious, then, that no correct inference as to the condition of the lungs can be drawn merely from the character of the vocal resonance in any given case: in other words, there is no standard of health in this respect. The vocal resonance, which may be either weak or exaggerated, if judged of by an average standard, may be perfectly normal in the case in question, and cannot be regarded as indicative of disease, unless it be found to vary at different periods. So, too, its character in any given portion of the chest cannot of itself afford trustworthy evidence of mischief; it is only by comparison of corresponding parts on both sides of the chest, due allowance being made for the difference which naturally exists between them, that any satisfactory conclusion can be arrived at.

Further, as the resonance is found to vary with the pitch and tone of the voice, it is essential, with a view to trustworthy conclusions, that the same word, or series of words, should be repeated in a monotonous manner; and as the use of different words gives rise to marked modifications in the character of the vocal resonance and is apt to confuse the ear, it is advisable to direct the patient to repeat a single word, such as "twenty," rather than to count one, two, three, four, as is sometimes recommended. The louder the patient's utterance the greater, *cæteris paribus*, the intensity of the vocal resonance, and, therefore, as a general rule, he should be directed to speak in a full, clear voice, averting his head from the auscultator. The stethoscope should be applied to the chest-walls firmly but lightly: if too much pressure be made, vibration will be checked, and the vocal resonance will be diminished, whereas, if the pressure be not sufficient to keep the instrument in firm apposition with the integuments, a tremulous, bleating character may be imparted to the resonance, and may lead the inexperienced to an incorrect diagnosis.

These points, then, being borne in mind, I will endeavor to describe the various modifications of vocal resonance which are met with in disease, and to point out their causes and precise clinical significance.

It may be stated generally that the natural vocal resonance, without material alteration in its quality, may be either increased or diminished in any and every portion of the respiratory tract, and that this increase or diminution of intensity is attributable to the physical condition of the parts whence the sound passes to the ear. But further, the *quality* of the resonance may be altered, as well as its intensity, and, instead of the natural voice-sound, there may be a ringing, squeaking, quavering, or bleating resonance of a metallic or an amphoric character. These varieties are attributable to differences in the physical condition of the thoracic viscera. Placed in a tabular form, the varieties of abnormal vocal resonance may be arranged as follows:

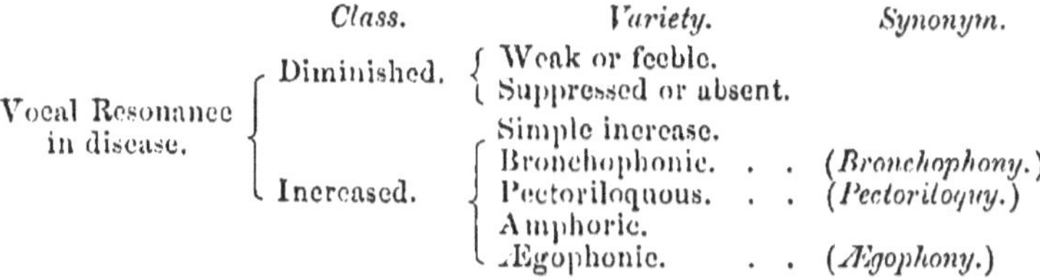

	Class.	*Variety.*	*Synonym.*
Vocal Resonance in disease.	Diminished.	Weak or feeble.	
		Suppressed or absent.	
	Increased.	Simple increase.	
		Bronchophonic. . .	(*Bronchophony.*)
		Pectoriloquous. . .	(*Pectoriloquy.*)
		Amphoric.	
		Ægophonic. . .	(*Ægophony.*)

In either of the four last-named varieties, the resonance may be of a metallic, ringing character.

The varieties included under the class of "diminished resonance" require little explanation. They are characterized by nothing more than weakness or diminution in the force or intensity of the resonance, and are met with of every degree ot feebleness. Sometimes, though weak, the resonance remains clear; sometimes an indistinct humming only is heard; and at others, again, the sound of the voice fails altogether to reach the surface of the chest. But, whether the resonance be altogether absent, or be only diminished in intensity, the cause of the alteration is in either case the same, viz., either weakness of the voice, or the occurrence of some physical change in the parts beneath the stethoscope calculated to interfere with the transmission of the vocal vibrations. This, of course, may vary in different cases. Sometimes bronchitis, with copious secretion, will lead to diminished resonance by obstructing the bronchi; tumors pressing upon the main bronchus will give rise to a similar result; sometimes the same effect will be produced by extensive pulmonary solidification, whether pneumonic, tuberculous, or cancerous,—consolidation so extensive, that the small stream of air which the lungs still admit is unable to throw the solid matter into vibration; sometimes by the presence of cavities which have not a free communication with the bronchi; sometimes by emphysema; and at others by tumors external to the lungs, or by a thin layer of fluid in the pleura. Each of these causes, if proceeding beyond a certain extent, may not only impair, but altogether prevent the transmission of the voice-sounds to the surface; but entire absence of the vocal resonance is most commonly produced by extensive pleuritic effusion, pneumothorax, or the presence of very large tumors lying between the lung and the surface of the chest.

Increased resonance is a subject which requires more detailed consideration. It is always referable to one of two causes,—either to consolidation of the lung-tissue around the air-tubes, whereby the diffusion of the local vibrations through the ultimate ramifications of the bronchi and the air-vesicles is prevented, the sound-reflecting power of the tubes is increased, and the pulmonary parenchyma is rendered more homogeneous, and a better conductor of the vocal vibrations to the surface of the chest;[1] or else to dilatation of the bronchi, or the formation of abnormal spaces in the lungs capable of concentrating and reflecting sound with unusual intensity. In either case the condition specified may exist to a greater or less degree, and *cæteris paribus*, the intensity and precise character of the vocal resonance will vary to a corresponding extent; so that in different cases or at different periods of the disease, or in different parts of the chest in the same case and at the same period, every possible variety of tone may be met with, from the slightest augmentation of the natural voice-sounds up to the loudest and most intensely developed form of increased vocal resonance. Five varieties, however, may be distinguished, as being sufficiently distinctive, and having each its own practical significance. The vocal resonance may be simply increased; the voice-sounds, unaltered in quality, may be heard louder than natural, but diffused over the surface of the chest, just as they are in health; or 2dly, their quality as well as their intensity may be altered, and they may be heard louder and clearer than natural, not distinctly articulated, and not appearing to pass through the stethoscope into the ear, but concentrated as it were beneath the stethoscope (bronchophonic resonance); or, 3dly, they may be clearer and more distinctly articulated than natural, and may appear to be concentrated in

[1] In speaking of consolidation and consequent increased homogeneity as a cause of increased vocal resonance, I would confine the observation to those cases in which the effused matter is of an elastic vibratile nature; for the presence of non-elastic material in the lung may actually diminish or altogether abrogate its voice-conducting power. This may be tried experimentally by injecting one lung with size or some other elastic material, and another with tallow or other non-elastic matter, and then listening through a stethoscope applied to one part of the lung, whilst a second person speaks through a stethoscope applied to another part of the same lung. The differences observable between the voice-conducting power of the lung, according as it is solidified by one material or another, will then be strikingly apparent. I have repeated this experiment with a variety of materials; and I entertain no doubt that the varying sound-conducting power of different lungs, apparently equally condensed, is attributable to the varying elasticity of the matter to which their consolidation is due.

the greatest degree, and to pass through the stethoscope directly into the ear, producing, when loud, a painful concussion on the ear of the observer (pectoriloquous resonance); or, 4thly, they may be loud, and of a metallic, ringing character, imperfectly or not at all articulated, not transmitted forcibly through the stethoscope, and conveying the impression of being produced in a large hollow space bounded by tense walls (amphoric resonance); or, again, 5thly, they may be of a tremulous, bleating, or quavering character, forming a sort of Punchinello voice, synchronous with, but of a higher pitch than the voice of the patient, or else following it like a feebly whispered echo, and rarely traversing the stethoscope (ægophonic[1] resonance).

Now the first of these five varieties of increased vocal resonance, or that in which the voice-sounds, unaltered in quality, and diffused, as usual, over the surface of the chest, are heard louder and of greater intensity than natural, is met with whenever anything occurs to prevent the diffusion of the vocal vibrations through the smaller ramifications of the bronchi and the pulmonary vesicles, and to increase the homogeneity and vibratile power of a superficial portion of lung, the larger bronchi of which remains pervious. The obstruction to the diffusion of the vocal vibrations increases the number and strength of the vibrations in the bronchi which are still pervious, and the density of the surrounding tissue renders the bronchial tubes capable of reflecting sound more strongly than usual; whilst the increased homogeneity of the pulmonary tissue favors the transmission of the sound to the chest-walls, and so to the ear of the observer. The conditions, therefore, under which simple increase of vocal resonance occurs, are slight infiltration, whether solid or semifluid, of the lung-tissue, as from tubercle, cancer, pneumonia, pulmonary congestion, and slight œdema of the lungs. Indeed, it would appear, that according as consolidation takes place to a greater or less extent, so the resonance of the voice increases, until it ultimately loses the character of simple increased resonance, and acquires that of the second variety.

The second variety (the bronchophonic), in which the sounds are heard louder, clearer, and more vibratory than natural, and concentrated, as it were, at the end of or beneath the stethoscope, may be produced by anything which favors the transmission to the chest-walls of sound emanating from spaces of larger calibre than the small bronchial ramifications usually existing on the surface of the lung. Hence it may arise from a more advanced stage of pulmonary consolidation than that which gives rise to

[1] From *αιγος*, the genitive of *αιξ*, a goat, and *φωνη*, voice.

simple increased resonance; from partial condensation of the pulmonary tissue, as the result of external pressure; from enlargement or dilatation of the bronchial passages—a condition usually attended by thickening and induration of their walls, and by condensation of the surrounding tissue; from the presence of small empty vomicæ in the lungs; from large cavities, the walls of which are flaccid or irregular, and do not reflect sound strongly; and from dilatation of the bronchi and distension of the air-cells, in certain cases of vesicular emphysema. In all these cases, the air-containing spaces in the lungs are abnormally large, and the resonance emanating therefrom is in consequence unusually loud; or else the occurrence of condensation of the lung-tissue, by causing obliteration of the air-vesicles and terminal bronchi, not only checks the diffusion of the vocal vibration, but increases the sound-reflecting power of the bronchi, intensifies the resonance, and facilitates the transmission of sound from the larger bronchi to the surface of the chest. The only exception is when the extent and degree of consolidation are such that the vocal vibrations conveyed along the air-passages which are still pervious, prove insufficient to throw the condensed mass into vibration.

The third variety (the pectoriloquous—the chest-voice—from *pectus*, a chest, and *loquor*, to speak)—in which the sounds of the voice, clearer, more ringing, and more distinctly articulated than natural, and concentrated in the highest degree, seem to pass directly through the stethoscope into the ear of the observer —is met with whenever, coincidently with great homogeneity and elasticity of the intervening lung-tissue, there exist near the surface of the chest a space or spaces capable of concentrating the voice-sounds, and reflecting them with great intensity. It matters not what the nature of the space—whether a bronchus, or a pneumonic, tuberculous, or other excavation, the only conditions essential to the production of pectoriloquous resonance are those above stated. Its peculiar and distinctive character is, that the merest whisper will reach the ear clearly and distinctly articulated. Its most common source is an empty tuberculous excavation in the lungs; but a large bronchus, surrounded by condensed lung, is not an infrequent cause of it. Pneumonic abscess and cancerous excavations being of rare occurrence, are quite exceptional causes of the phenomenon.

The fourth variety (the "amphoric," from *amphora*, a jar)— in which the sounds, though of a metallic ringing character, do not forcibly traverse the stethoscope, are very imperfectly, if at all articulated, and convey the impression of being produced in a large hollow space bounded by tense walls—derives its name from the resemblance of the sound to that produced by speaking into an empty jar. It is met with most commonly in cases of pneumothorax; but it may also arise in instances of large tuberculous excavations in the lungs, especially when the cavity communicates freely with the bronchi, and its walls are thin, smooth, and tense, and adherent to the anterior parietes of the chest.

Now, it is obvious that bronchophonic, pectoriloquous, and amphoric resonance are merely degrees or progressive developments of one and the same phenomenon, and that just as simple increased resonance is a gradual transition from natural resonance, so simple increased resonance may pass quite gradually into bronchophonic resonance, which in its turn may merge just as gradually into pectoriloquous or even into amphoric resonance. No precise line of demarcation can be drawn between the different varieties beyond those already described; and although amphoric resonance is not likely to be mistaken for the other varieties, it is often difficult to decide whether the resonance in any particular case is sufficiently concentrated to deserve the title of bronchophonic, or whether, again, in another instance, it should be termed bronchophonic or pectoriloquous. Fortunately, therefore, the decision of the question is not of material consequence; for no positive inference can be drawn as to the condition of subjacent parts simply from the existence of one or other variety. They all indicate the existence of large air-containing spaces in the lungs, an extraordinary reflecting power in the walls of the spaces in which the air is contained, and, usually, increased homogeneity of the surrounding lung-tissue; but, except in extremely well-marked cases, it is impossible, from the vocal resonance alone, to decide even approximately whether pectoriloquous resonance emanates from a large bronchus surrounded by condensed lung-tissue, from a dilated bronchus, or from a vomica in the lungs.

Laennec stated that pectoriloquous resonance is a certain sign of an excavation in the lung-substance, and even in the present day there are too many practitioners who so regard it, and thus are led into errors of diagnosis. Even at the apex of the lung, where no large bronchus exists, it is not of itself sufficient to warrant a conclusion as to the existence of a vomica, for loud pectoriloquous resonance may be transmitted through consolidated lung from a large neighboring bronchus. When its character is hollow, ringing, and metallic, the presumption is strongly in favor of such a cause, though the evidence derivable merely from the sound is still insufficient to decide the question. Some of the most strongly marked instances of loud, ringing, pectoriloquous resonance I ever met with have occurred at the apex of the lung, in connection with pneumonic consolidation of the upper lobe; and in some few instances I have observed this resonance in a strongly developed form over small extra-pulmonary tumors. Over large or dilated bronchi, surrounded by consolidated lung, it is by no means of unfrequent occurrence.

In regard to the less developed forms of increased vocal resonance, still greater uncertainty exists. Not only may considerable increase of vocal resonance, whether simple or bronchophonic, result from comparatively slight consolidation of the lung, provided such consolidation be superficial in its position; but, on the other hand, the faintest bronchophonic resonance—not exceeding that which would be produced under certain circumstances by slight pulmonary consolidation, may result from a cavity even of considerable size, provided that its walls are thick, flaccid, and irregular, and therefore incapable of reflecting sound strongly; that it still contains fluid; or, that it is seated deeply in the chest, with healthy, pervious, imperfectly conducting lung-tissue intervening between it and the surface. Nay, more, if its communication with the bronchi is not free, vocal resonance may be inaudible over it, or, if occurring at all, may be transitory and intermittent; so that, on the one hand, the presence of pectoriloquous or bronchophonic resonance does not afford proof of a cavity in the lung, neither does its absence furnish evidence of the non-existence of a cavity.

In some instances, however, the character of the sound and the conditions under which it is heard suffice to indicate with tolerable certainty the source of the abnormal resonance. Thus, if at the apex of the lung the vocal resonance be hollow, ringing, and metallic, and a *mere whisper* proves markedly pectoriloquous, and passes through the stethoscope into the ear, clearly and distinctly articulated, there can be little doubt that the sound emanates from an empty space occupying a very superficial position in the chest, having a free communication with the bronchi, and bounded by thin, tense, elastic walls, capable of concentrating and reflecting sound. From no other sort of space, and from no space in any other position, could a hollow, ringing, pectoriloquous resonance be conveyed to the ear under such circumstances; and as no such space exists naturally in such a position in the lungs, it is fair to conclude that the sound in question emanates from a vomica.

Some assistance in determining the cause of increased resonance may be derived in many instances from an investigation of the extent of area over which the abnormal sound can be heard. When it is of very limited extent and sharply defined, it probably results from a dilated bronchus, from a small vomica in the lungs, or from the pressure of a small extra pulmonary tumor; whereas when it is more extensive, and may be traced over a considerable portion of the chest, it is more probably due to pulmonary infiltration, whether pneumonic, tuberculous, or cancerous, and a gradual diminution in its intensity will be perceptible as the stethoscope is carried along the surface of the chest towards the healthy portions of the lung. A loud, hollow, metallic resonance, whether bronchophonic or pectoriloquous, if heard over an extended area, is almost necessarily derived from large bronchi surrounded by consolidated lung. The only other source of it is an enormous cavity.

There are some peculiarities attendant on the more highly developed forms of increased vocal resonance which deserve a passing notice, not on account of any practical value which attaches to them, but in consequence of a theory which has been founded upon them by Skoda and his followers.[1] I allude to the fact that, in certain instances, the voice heard through the stethoscope applied to the chest-walls is as loud if not louder than the voice heard in the same manner over the larynx or trachea. Thus, when the cylinder is applied over a cavity in the lungs, the voice occasionally seems louder, and strikes with greater intensity on the ear than it does when auscultated over the larynx, which is its seat of production. What, then, is the cause of this reinforcement of the voice? It is not my intention to discuss the question fully, inasmuch as it is one which requires lengthened consideration, and does not lead to any practical result; but a few remarks on bronchophonic resonance may not be out of place. First, then, it may be stated that Laennec attributed the variations perceived in the strength and clearness of the thoracic voice to variations in the sound-conducting power of the lung-tissue. He considered that lung-tissue, in its normal condition, being non-homogeneous, is a bad conductor of sound, and that, infiltrated or condensed, and thereby rendered more homogeneous, its sound-conducting power is increased; and that to such increased homogeneity and consequent increased sound-conducting power bronchophonic resonance is really attributable. But I have shown by direct experiments that the increase of homogeneity of the lung-tissue, resulting from consolidation, will not lead to increased sound-

[1] Whatever is here stated relative to the cause of increased vocal resonance applies equally to the intensification of tussive resonance and the respiratory sounds.

conducting power in the lung, unless the consolidating matter be elastic or vibratile; and even if it were otherwise, it is obvious that mere consolidation or increase of homogeneity will not serve to explain the positive reinforcement or intensification of the voice, which is met with in certain instances of disease; neither will it suffice to explain the difference in pitch which is sometimes observed between the bronchophonic and laryngeal voice. Some further explanation must therefore be sought. Skoda, blinded by the result of imperfect experiments on the dead body, and in defiance of many facts observed in the living, asserts[1] "that sound is heard somewhat further through healthy than through hepatized lung;" and maintaining that "variations in the strength and clearness of the thoracic voice cannot be explained by the differences in the sound-conducting power of normal and abnormal lung parenchyma," rejects Laennec's theory *in toto*, and substitutes for it the doctrine of consonance —a doctrine at once fanciful and inconsistent with acknowledged facts. Not content with drawing attention to the possible increase of the voice-sounds by consonance, under certain rare and exceptional conditions, the Viennese professor confidently proclaims that consonance is a principle of general application in the intensification of sounds audible on the chest-walls, and that bronchophonic resonance and other phenomena, which Laennec attributed to the varying conducting power of healthy and diseased lung-substance, are explicable only under this supposition. But those persons who are familiar with the laws of sound, must be aware that the conditions essential to the production of consonance must be necessarily of rare occurrence in the air within the bronchi; and those who will observe carefully the phenomena attendant upon the sounds which Skoda has designated "consonating," will soon perceive that they commonly present characters which are quite inconsistent with the recognized acoustic phenomena of consonance.[2] Therefore, without denying that consonance is sometimes instrumental in intensifying or reinforcing the vocal resonance, I am satisfied that it cannot often prove so, and that it is quite inadequate to explain the occurrence of bronchophonic resonance under ordinary circumstances.

What, then, is the cause of the phenomenon? When it occurs in connection with consolidated lung, I believe it to be referable mainly to three causes—1st, the non-diffusion of the vocal vi-

[1] Skoda, loc. cit., p. 39.

[2] For an excellent disquisition on the facts bearing upon Skoda's doctrine, see Walshe, loc. cit., pp. 142 to 152.

brations, which results from the consolidation of the pulmonary vesicles and terminal bronchi; 2d, the increased power possessed by the bronchi which are surrounded by condensed lung-tissue of reflecting, and thereby intensifying the sound as heard by the ear; 3d, the increase of conducting power acquired by the pulmonary parenchyma under certain conditions of consolidation and homogeneity. The effect of consolidation in preventing the diffusion and consequent weakening of the vocal vibrations, and in augmenting the reflecting power of the air-passages, and thereby of inducing concentration and intensification of the voice sounds, must be obvious to all who consider the question; and although it is possible to conceive consolidation to result from infiltrated material of such an inelastic nature as to deaden all vibration, and to hinder rather than facilitate the transmission of the voice-sound to the surface of the chest; nay more, though observation and experiment[1] clearly prove that infiltrated matter does vary greatly in its vibratile power, and therefore in its powers of transmitting the voice—there is, nevertheless, abundant evidence to establish a vast increase in the conducting power of lung in certain stages of condensation and homogeneity.[2]

But other circumstances are necessary to explain the positive intensification or reinforcement of the voice, which is often heard over vomicæ, dilated bronchi, &c., and also the difference of pitch observable, in some instances, between the laryngeal and thoracic voice. Here other causes come into operation which, as regards the lungs, must be regarded as exceptional. I refer to those sources of increased or modified sound which are known as consonance, unison resonance, and echo. I quite agree with Dr. Walshe in believing that each of these agencies may come into play in the intensification of the voice-sounds, under different conditions of the thoracic organs; but I am convinced that the two latter are those to which the phenomenon is most commonly attributable. For although it is conceivable that consonance should occur under certain rare and exceptional conditions, the laws which regulate its production are such as to render its frequent occurrence in the lungs well-nigh impossible; whereas, unison resonance and echo may readily occur in many states of the pulmonary apparatus. Those who are curious in this matter, may refer to Dr. Walshe's pages, where the subject is fully dis-

[1] See note, p. 110.

[2] Dr. Scott Alison has pointed out that the increase of vocal resonance, as heard through a stethoscope over solidified lungs, is partly due to the fact that sonorous vibrations are transmitted more readily from solids to solids than from air to solids.

cussed; and therefore, without going further into the matter, I will merely reiterate my opinion, that bronchophonic resonance is ordinarily referable to the three cuases already specified, viz., non-diffusion of the vocal vibrations consequent on the obstructed condition of the air-cells and the terminal bronchi; increased reflecting power of the bronchi; and augmented conducting power of the lung: but that where its intensity exceeds that of the laryngeal voice, or its pitch differs from that of the laryngeal voice, echo, or unison resonance, or both combined, are usually concerned in modifying the phenomenon. Probably echo is the cause of its unusual intensification in certain examples of pulmonary consolidation, and unison resonance in cases of emphysema, whilst unison resonance and echo combine to produce the effects, in certain instances of vomicæ in the lungs. True consonance can only come into play when "the tones of the laryngeal voice chance to bear a certain mathematical relationship to the fundamental note of a resounding space in the chest."

The fifth variety of increased vocal resonance (the ægophonic), or that in which the voice as heard on the chest-walls is of a tremulous, bleating, or quavering character, forming a sort of nasal Punchinello voice, is closely allied to bronchophonic resonance, being in fact nothing else than ordinary bronchophonic resonance modified, and rendered tremulous in its character. It is clear in tone, superficial in situation, and usually synchronous with the patient's voice, though sometimes following immediately after it, resembling a weak, bleating, silvery echo, vibrating as it were on the surface of the lungs; it is usually of a higher pitch than the natural voice, and is not produced by every word, even though pronounced with equal force and with the same pitch; it rarely traverses the stethoscope; is usually persistent; and though seldom continuing for longer than a few days, may endure for a period of several weeks. It is most commonly heard at the lower and posterior parts of the chest, in the vicinity of the larger bronchi, is audible over a very limited surface, and is often confined to a small space near the inferior angle of either scapula. Sometimes, however, it extends to the nipple in front, and occasionally may be traced rising gradually in the chest until it is heard at the very apex of the lung. Further, it is capable in some instances of being altered in position, or even temporarily got rid of by a change in the posture of the patient.

The cause of ægophonic resonance has proved a fertile source of theory and conjecture. Practically, this resonance is usually found as an accompaniment of effusion into the pleural cavity; is most marked when the fluid is small in quantity; rises higher in the chest with the progress of effusion, and is only heard about on a level with its surface; disappears when a large quantity of fluid has accumulated, and returns when absorption of the fluid has proceeded to a certain extent. Hence Laennec concluded that its presence is characteristic of a thin layer of fluid in the cavity of the pleura, and suggested that its peculiar character is sufficiently accounted for by flattening of the bronchi as a result of compression, and by the vibration of the thin layer of effused fluid. This opinion he fortified by reference to the peculiar character of the sound produced by the flattened mouthpieces of the bassoon and oböe, which he considered to be represented by the flattened bronchi, and, further, by the experiment of listening to the voice through a bladder half filled with water placed on the interscapular region of healthy persons where natural bronchophony is very intense—a contrivance which he said imparted to the vocal resonance a tremulous ægophonic character. But even admitting the flattening of the bronchi suggested by Laennec (and its existence is more than problematical), such flattening cannot be regarded as analogous to the mouthpiece of a clarionet, which contains a thin vibrating plate; and the experiment of listening to the voice through a half-filled bladder has not succeeded in the hands of others. I have listened through a bladder half filled with water, applied not only over the interscapular regions, but over the larynx and trachea where vocal resonance is most intense, and have uniformly failed in obtaining a modification of the voice-sounds in any degree resembling Laennec's ægophony. Nor have I been singular in the failure of my attempts to obtain the effect spoken of by Laennec. Dr. Davies and other experimentalists in this country and on the Continent have arrived at the same results; and Skoda distinctly states,[1] that under the conditions alluded to, "the voice sounds just the same as through a piece of liver of the same thickness as the depth of water in the bladder." Further, and quite independently of experiment, extended observation and pathological research have shown that Laennec attached too much importance to the presence of fluid in the pleura as a cause of ægophonic resonance; for although they have confirmed to the fullest extent the frequent coexistence of pleuritic effusion and ægophonic resonance; nay more, though they have proved that in a pure, intensely developed and persistent form ægophonic resonance is never met with without the presence of fluid in the pleura, yet they have also shown that effusion may exist without this peculiar vocal resonance;

[1] Loc. cit., p. 69.

and conversely, that ægophonic resonance may and does sometimes occur to a greater or less degree in pneumonic or tuberculous consolidation without any fluid in the pleura. Thus it has become evident that, however frequent the coexistence of the two phenomena, they cannot be regarded as necessarily in the relation of cause and effect.

I have already stated that ægophonic resonance will cease at any given spot as soon as a large quantity of fluid has accumulated, and the level of the fluid has risen above the spot at which the peculiar sound is heard; and further, that the surface over which it is heard is very limited in extent; but exceptional instances are sometimes met with which seem to negative both these statements. When adhesions exist between the lung and the costal pleura, so that the lung is retained in partial apposition with the chest-walls, ægophonic resonance may continue for weeks, or even months, in spite of very copious accumulations of fluid in the pleural cavity; and under the same circumstances, if the adhesions are extensive, but partial only, it may be heard over a very large surface of the chest—sometimes, indeed, over the whole of the affected side. I have myself observed instances of both these occurrences, and instances have been put on record by various observers in this country and on the Continent.[1]

Further, the statement that ægophonic resonance may occur without the presence of any fluid in the pleura, requires a few words of explanation. Of the fact itself there cannot be a doubt. A peculiar resonance, not indeed so purely or so intensely ægophonic as that which accompanies pleuritic effusion, but still of a sufficiently shrill and tremulous character to deserve the title of ægophonic, is often heard in cases of pulmonary hepatization or tuberculous infiltration, and, according to Skoda,[2] is met with occasionally in the interscapular region of women and children, even when the lungs are perfectly sound. I cannot say that I have ever met with such a sound in any woman or child whose lungs were healthy; nor do I think that with ordinary care a sound arising under such circumstances could be mistaken for or confounded with true ægophonic resonance; but the occurrence of such a sound over solidified lung is not very uncommon. It is most strongly marked when the ordinary voice of the patient is of a shrill, tremulous, or nasal character; and sometimes it so closely resembles the ægophonic

[1] See Laennec, "Traité de l'Auscultation Médiate et des Maladies des Poumons et du Cœur;" Andral's "Clin Méd.," t. ii, obs. xxi; Walshe, loc. cit., p. 140.

[2] Skoda, loc. cit., p. 68.

resonance of pleurisy that it is impossible by the sound alone to determine its cause. But careful and oft-repeated examinations will enable us to discriminate between the two cases. When the sound is dependent on pleuritic effusion, its seat is usually very limited, and often varies with the position of the patient; it rises higher and higher in the chest as day by day the effusion increases, and is not heard where the percussion-sound is dullest, or in other words, at the base of the lung, where the accumulation of fluid is necessarily greatest; whereas, when it arises in connection with pulmonary consolidation, it invariably occurs where the lung-tissue is most condensed and the percussion-sound dullest; is usually heard over a somewhat extended surface; is not affected by change of posture, and though seldom enduring above a few hours, may persist in the same spot for a considerable period in spite of continued extension of the area of dulness.

Skoda[1] has suggested that ægophony may "possibly be produced occasionally by a portion of mucus, &c., partially closing the mouth of the bronchial tube, imitating the thin tongue in the mouthpiece of tongued instruments," but that "probably in most cases the walls of the bronchial tubes within which the air consonates, react by impact on the air contained within them, and so give rise to the tremulous sound." The latter part of this theory, on which in fact Skoda rests the occurrence of ægophony under ordinary circumstances, is not only inexplicable, but utterly inconsistent with acknowledged facts; for if the bronchi always vibrate, as he admits they do when the air within them consonates, it is obvious that ægophony ought to be a constant accompaniment of the bronchophonic resonance which attends pneumonic hepatization, and which, according to Skoda, results from consonance of the voice in the bronchi; whereas the very opposite is the fact, its occurrence under such circumstances being quite exceptional.

What, then, is the cause of ægophonic resonance? Setting aside those spurious cases in which an ægophonic character is said to attach to the vocal resonance in healthy women and children, and in which this peculiarity, if indeed it exists, must be attributable to the shrill, tremulous character of the natural voice—I say, setting these aside, there are two conditions which I believe to be invariable accompaniments of, and indeed essential to the production of ægophony—1st, a condition of lungs calculated to give rise to bronchophonic resonance; 2dly, the existence of some agency able to impart to that resonance a tremu-

[1] Loc cit., p. 72.

lous, bleating character. This must be one of two kinds: either some tenacious secretion vibrating in the bronchial tubes and producing an effect analogous to the vibrating tongue of reed instruments, which possibly may be, though I somewhat doubt it, an occasional cause of ægophonic resonance over hepatized lungs; or, which I believe to be its source in cases of pleuritic effusion, the impulse of the vibrating and partly solidified lung against the costal pleura, an effect—viz., the repeated impulse of one solid vibrating body upon another—exactly analogous to that which takes place in the schoolboy's trick of speaking upon thin paper placed over the teeth of a comb, or in that of speaking, as Punch and Judy's showmen do, with a thin disk of metal or ivory so placed in the mouth as to lie between the lips and the teeth, and so to obstruct the egress of air from the mouth, in which case an ægophonic character is imparted to every sound by the jarring vibrations excited by the repeated impulses of the disk against the teeth. Under ordinary circumstances solidified lung lies closely in contact with the costal pleura, and practically for all purposes of vibration may be considered as connected with it; consequently, unless some cause of jarring, tremulous vibration exist within the lung itself, as in the instance of a vibrating piece of mucus in a bronchus, the vocal resonance will be purely bronchophonic, and not ægophonic. But just at the surface of a pleuritic effusion there must be a point at which the lung is barely separated from the chest-walls, and in which the bronchophonic vibrations of the lung must lead to that light, jarring impulse of the visceral against the costal pleura, which analogy proves conclusively to be a frequent cause of a peculiarly tremulous, bleating sound—the sound which characterizes ægophony.

Skoda laughs at the idea of three or four ounces of fluid giving rise to ægophonic resonance, and asserts that ægophony "will not be produced unless the fluid in the pleura be sufficient to deprive completely of air, by compression, a portion of lung large enough to contain a cartilaginous bronchial tube." But I am satisfied that this is not the fact, and that Laennec was correct in his statement, that a very small quantity of fluid may occasion intense ægophonic resonance. I have so often traced in the dead-house of St. George's Hospital the coexistence of scanty pleuritic effusion and ægophonic resonance, that I cannot doubt as to their mutual relation; and according to my views as to the cause of ægophony, the scantiness of the fluid is of little importance, if indeed it be not actually conducive to the production of ægophony. In fact, I consider the fluid to have no further influence over the production of ægophony than that of separating the two surfaces of the pleura to such an extent as to admit of the necessary vibratile impulse between them, and further, of interposing an imperfect conducting medium between the ear and the seat of bronchophonic resonance, which has the effect of preventing the transmission of vocal resonance, except in those parts where the stratum of fluid is very thin, and even in those parts so far diffuses the sound as to deprive it of its concentrated bronchophonic character. Consequently, if a lung be hepatized, or be otherwise in a condition to emit bronchophonic resonance, the smallest quantity of fluid may impart to that resonance an ægophonic character. It is impossible to prove that Skoda's suggestion relative to the presence of a vibrating plug of mucus in the bronchi may not *possibly* hold good, but the rarity of ægophony in patients oppressed by a quantity of tenacious mucus in their air-tubes seems to render such an explanation highly improbable.

Practically, then, ægophonic resonance implies one of two conditions: either hepatized or otherwise solidified lung (or, according to its position, a large superficial bronchus or a cavity in the lungs), with tenacious mucus so placed in the air-passages as to vibrate in a particular manner; or else the same condition of lung with effusion into the pleural cavity, slightly separating the two surfaces of the membrane, but admitting of their coming so far in contact at the spot where the peculiar sound is heard as to vibrate the one against the other. The latter is its common, its usual source; and when the sound is well developed and persistent, I believe it to be its invariable source. In certain instances of pneumonia, and in cases of tumors pressing into the pleural cavity, the voice has somewhat of an ægophonic character; but if due care be taken in making the examination, and a conclusion be not arrived at until after two or three interviews, I believe with Laennec that well-developed ægophony may be depended on as proof of the existence of fluid in the pleural cavity.

Autophony, or the sound of the observer's own voice, as heard whilst his ear is applied to the chest, either directly or through the intervention of a stethoscope, affords another class of phenomena which have been called in aid of the diagnosis of thoracic disease. Few persons can have failed to remark that their own voice appears to them to sound or reverberate more strongly under certain conditions than it does under others. Thus, if the palms of the hands are placed over the ears whilst a person is speaking, the vibrations of the voice appear to the speaker to be confined within his own head, and to acquire an intensity very different from that of which he is conscious during ordinary speaking. The same effect is observed in a lesser degree when one ear only is covered. M. Hourmann went farther than this, and made a series of experiments of the same kind with substances of varying thickness, from which he concluded that the character and intensity of the resonance bear a definite proportion to the thickness and density of the substance with which the ear is in contact. From this followed the inference that autophonic resonance would vary with the density of the contents of the thoracic cavity, and that these variations would prove valuable auxiliaries in the diagnosis of thoracic disease.[1] Unfortunately, observation has not borne out the practical results anticipated by M. Hourmann. There cannot be a doubt that the intensity of autophonic resonance varies greatly under different circumstances, and that consolidation of the lung is in many instances accompanied by a marked increase in its intensity; but it is equally certain that in some instances no such augmentation takes place, and that when the amount of consolidation is small, the variations of intensity are hardly perceptible. Indeed, observation has led me to believe that the intensity of autophonic resonance is attributable to the reflecting power of the substance with which the ear is in contact, rather than to its mere thickness or density; and if this be so, it must be modified as much by the condition of the chest-walls as by the contents of the thoracic cavity. Be this as it may, the increase of resonance is certainly not proportioned to the mere density of the subjacent parts; and the numerous exceptions to the rule of there being any increase of it over consolidated lung, render autophony useless as a means of exploring the chest.[2]

Auscultation of the cough is another mode of examining the chest, and sometimes furnishes important information. In health, tussive resonance, as heard on the chest-walls, is a short, dull, and indistinct sound, evidently produced at a distance, diffused over the surface of the chest, and accompanied by some degree of vibration of its walls. Its intensity is proportioned to the force of the cough and the thinness and elasticity of the thoracic walls, and it is heard loudest, and of a somewhat bronchophonic character, over the larger bronchial tubes. Indeed, whether in health, or in disease, tussive resonance undergoes changes corresponding precisely with the alterations in the vocal and respiratory sounds under similar conditions of the thoracic organs. Thus, its intensity may be simply increased without material alteration in its character; or it may acquire a bronchophonic, pectoriloquous, amphoric, or even a quasi-ægophonic character;

[1] "Revue Médicale," 1839.

[2] Barth and Roger, Piorry, Skoda, and other authorities abroad, and Davies, Walshe, and others in this country, have arrived at the same conclusion.

and when pneumothorax exists, or large cavities are present in the lungs, it may be accompanied by metallic tinkling and amphoric echo. If fluid, mucus, or other matter be present in the air-passages, it may be attended by râles and rhonchi, which, just as with respiration, will vary, *cæteris paribus*, according to the sizes of the passages or spaces in which they take their origin. In short, tussive resonance affords little or no information which may not be derived from the voice and the respiration; and in an auscultatory point of view, cough acquires its importance, not from the peculiar resonance which it produces, but from its power of removing mucus or other matters by which the air-passages are apt to become obstructed. Cough consists of a sudden and forcible expiration, followed by a quick, yet deep inspiration. The suddenness and force of the expiratory effort is often such as to dislodge obstructions in the bronchi, which cannot be got rid of in any other way; and thus a passage is cleared for the free ingress of air during the deep inspiration, by which each paroxysm of cough is succeeded. By its means, therefore, we are often enabled to judge of the permeability or impermeability of portions of lungs over which no respiratory sounds could previously be heard; and may often discover a cavity which had remained uninfluenced by the voice and the respiratory efforts, in consequence of its approaches having been obstructed by mucus. By its means, again, we may sometimes discriminate between pleuritic friction, and those pseudo friction-sounds which are met with occasionally in chronic bronchitis, as the result of tenacious mucus in the bronchi. For in the one case the friction-sound continues uninfluenced by the act of coughing; whereas in the other it ceases as soon as the mucus by which it is caused is dislodged. In this manner cough becomes a valuable auxiliary in the exploration of the thoracic cavity.

The subjoined tables show at a glance the different varieties of vocal resonance in health and disease, their character, mode of production, and usual seat, and the diseases with which each morbid sound is usually associated:

TABLE

RESONANCE OF THE VOICE, AS HEARD OVER THE

Variety.	*Synonym.*	*Character of sound.*
Laryngophonic. Tracheophonic.	Laryngophony. Tracheophony.	The sound is loud and concentrated, in the highest degree, conveying the idea of being produced immediately beneath the stethoscope. and of passing through the instrument so as to strike directly and forcibly on the ear.
Bronchophonic.	Natural Bronchophony.	The sound is less loud, less concentrated, and less distinctly articulated than in tracheophony: it appears to be produced at a greater distance, to be more diffused, and to strike less distinctly against the ear, so that the words seem to be at the end of the instrument instead of passing through it into the ear.
Pulmonary.	Ordinary vocal resonance.	A diffused vibratory sound, amounting to little more than a humming or buzzing, and conveying the impression of distant origin. The articulation of the voice is not appreciable.

I.

LARYNX, TRACHEA, AND CHEST-WALLS, IN HEALTH.

How produced.	*Its usual seat.*
The vocal vibrations which pass directly down into the trachea, being concentrated in that narrow sound-reflecting tube, reverberate there with great intensity.	Over the larynx and trachea.
The vocal vibrations which pass down the trachea and bronchi, owing partly to their being weakened by diffusion, partly to the lesser sound-reflecting power of the bronchi, as compared with the trachea, and partly to the intervention of the spongy tissue of the lung, reverberate less loudly than in the former situation.	The upper part of the sternum and the posterior surface of the chest, between the spines of the scapulæ.
The vibrations of the voice which pass down the trachea and larger bronchi are obstructed, weakened, and diffused by passing through the subdivisions of the bronchi and the spongy tissue of the lung before reaching the surface of the chest.	Over the whole surface of the chest, except the upper part of the sternum and the space which lies between the spines of the scapulæ, where the larger air-tubes are superficial and near to the chest-walls.

RESONANCE OF THE VOICE, AS HEARD

Class.	*Variety.*	*Synonym.*	*Character of sound.*
Diminished.	Weak or feeble.	Feeble resonance.	The sound is simply less intense than it is over healthy lung tissue. Sometimes an indistinct humming only is heard.
	Suppressed or absent.	Absence of resonance.	No sound audible.
Increased.	Simple increase.	Increased resonance.	The voice-sounds, unaltered in quality, are heard louder than natural, but diffused over the surface of the chest, just as in health. This variety of increased vocal resonance gradually passes into the next variety.
	Bronchophonic resonance.	Bronchophony.	The sound is louder and clearer than in simple increased resonance, is not distinctly articulated, and does not appear to pass through the stethoscope into the ear, but is concentrated, as it were, beneath or at the end of the stethoscope. This bronchophonic resonance passes imperceptibly into the next variety.
	Pectoriloquous ditto.	Pectoriloquy.	The voice-sounds are distinctly articulated, and transmitted directly through the stethoscope into the ear.
	Amphoric ditto.	Amphoric resonance.	A ringing sound, of a metallic quality, not distinctly articulated, not transmitted forcibly through the stethoscope, and resembling the sound produced by speaking into an empty jar.
	Ægophonic ditto.	Ægophony.	A tremulous, bleating or quavering sound, forming a sort of Punchinello voice, synchronous with, but of a higher pitch than, the voice of the patient, or else following it like a feebly whispered echo, and rarely traversing the stethoscope.

THROUGH THE CHEST-WALLS, IN DISEASE.

How produced.	*Its usual seat.*	*Disease with which it is commonly associated.*
By the occurrence of some physical change in the parts beneath the stethoscope, calculated to interfere with the transmission of the vocal vibrations to the surface.	Not confined to any portion of the lung.	Bronchitis, with copious secretion obstructing the bronchi; aneurismal or other tumours pressing upon a main bronchus; extensive inelastic pulmonary consolidation; emphysema, and slight pleuritic effusion.
By the same causes as lead to feeble resonance, if proceeding beyond a certain limit.	Not confined to any portion of the lung.	Each and all of the diseases above specified, when proceeding beyond a certain limit; but most commonly by extensive pleuritic effusion, pneumothorax, or the presence of large tumors lying between the lung and the surface of the chest.
By slight consolidation of the lung-tissue around the air-tubes, whereby the sound-reflecting power of the tubes is increased, and the pulmonary parenchyma is rendered more homogeneous, and a better conductor of the vocal vibrations to the surface of the chest.	Not confined to any portion of the chest, but usually most marked and of the greatest significance towards the apices of the lungs in connection with the deposition of tubercle.	In various morbid states, but, as a persistent condition, most common in the early stages of tubercular phthisis.
Sometimes by an increase of the pulmonary condensation which gives rise to simple increase of vocal resonance, and sometimes by the formation of abnormal spaces in the lungs, capable of concentrating and reflecting sound to as great a degree as moderate-sized bronchi.	Not confined to any portion of the lungs, and inasmuch as it may be heard over the bronchi even in health, it is of clinical significance only when heard where bronchophony does not exist in health. Of great importance as suggestive of phthisis, when existing at the apex of the left lung.	Pneumonic hepatization of the lungs; tubercular, cavernous, and other infiltrations of the lung-tissue round moderate-sized bronchi; small vomicæ, as in phthisis; and dilated bronchi.
Sometimes by the condensation of lung-tissue around a large bronchus, whereby the transmission of the sound to the ear is facilitated; more generally by the formation of abnormal spaces or cavities possessing smooth, sound-reflecting walls.	Not confined to any portion of the lungs, but occurring most commonly at the apices and in the upper lobes.	Chiefly in phthisis, but occasionally in pneumonia and other diseases in which the necessary physical conditions may exist.
By the reverberation of the voice in a large hollow space, bounded by tense sound-reflecting walls.	The apices and upper portions of the lungs.	Phthisis pulmonalis.
The vibration of the visceral against the costal pleura, at the point where the two surfaces are barely separated from each other, whether by fluid or by any other cause.	Not confined to any portion of the lungs, but most persistent at the posterior surface of the chest, near the roots of the large bronchi, where the vocal vibrations are most intense.	Chiefly in pleurisy, but also in certain instances of pneumonia, and of tumors projecting into the pleural cavity.

CHAPTER IX.

ON THE ADVENTITIOUS SOUNDS PRODUCED WITHIN THE CHEST BY THE ACT OF BREATHING.

The varieties of abnormal breathing already described are marked by modifications of the natural sounds of respiration. Those to which I have now to direct attention are characterized by the presence of sounds which find no analogue in healthy breathing, but are new and adventitious phenomena superadded to or replacing the healthy respiratory murmur. Such sounds may arise either from the air-passages or cavities communicating with them, from the substance of the lung itself, or from the pleural cavities; and as they vary in character according to their seat and mode of production, it is necessary to understand the mechanism in which each takes its origin, before a proper estimate can be formed of their true significance.

First, then, in regard to those adventitious sounds which originate in the air-passages or in cavities communicating with them. It has been already stated that in a healthy condition of the respiratory organs the air-passages are free and open, whilst the mucous membrane which lines them is constantly moistened by a thin, watery exhalation which lubricates its surface and renders it smooth. Further, it has been shown that anything which checks the exhalation of this moisture, or produces unevenness on the surface of the bronchi, or gives rise to irregularity in or narrowing of their calibre, increases the friction encountered by the respired air, impedes its escape during expiration, and leads to a corresponding alteration in the breathing-sounds. The first effects perceived by auscultation are roughness and prolongation of the sounds, especially of the sound of expiration. If the local changes are dependent upon causes of a transient nature, the roughness may gradually subside, and give place to the natural respiratory murmur. But more commonly, when mischief has gone so far as to produce roughness of breathing with prolongation of the expiratory sound, it is followed by further local changes, with a fresh train of physical phenomena. Not only does the calibre of the air-passages undergo a still greater diminution, but the dryness of the bronchial mucous membrane is succeeded by morbidly increased secretion. Thus, if the pulmonary mischief be not speedily resolved—in which case the roughness of the breathing will gradually cease—it usually happens that simple rough or harsh respiration is after a time replaced or accompanied by sounds which are dependent on the presence of secretion, and vary with its amount and character and with the altered size and condition of the bronchial tubes.

These sounds have been termed indifferently râles or rhonchi; the former being the term originally employed by Laennec, the latter, derived from the Greek word ῥόγχος, being the term more generally employed in the present day.[1] Whatever they may be styled, however, these sounds are divisible into two species, which acquire their distinctive characters from the fact that in the one the sounds are occasioned by the vibrations into which the air is thrown by passing through channels irregularly contracted in their diameter; whilst in the other they arise from the bursting of bubbles caused by the air being forced through fluid, more or less viscid and tenacious, which has accumulated in and obstructs the air-passages. The former, therefore, have been termed "dry" sounds, as contrasted with the latter, which being connected with the presence of fluid, have been termed appropriately "moist" sounds.

But although these so-called "dry" sounds are invariably connected with narrowing of the channels through which the air passes, a very common element in their production is the presence of tough, viscid mucus or other semi-plastic material, which does not admit of the passage of air and the formation of bubbles, but which, without entirely obstructing the air-tubes, adheres to their sides, contracts their diameter, and gives rise to, or admits of abnormal vibration. Indeed, in some instances the secretion and local accumulation of such mucus appear to be the chief cause of the constriction of the air-passages and of the vibration on which the production of the "dry" sounds depends; and in all cases, whatever the primary cause of the constriction—

[1] Neither term, however, has any special signification, and each of them is inappropriate when applied to *all* the adventitious sounds originating within the air-passages; for the term râle or rattle can hardly be regarded as applicable to many of the snoring, burring, cooing, and grunting sounds which are often audible in the chest, and the term rhonchus, which signifies snoring, is clearly inapplicable to the minute crepitations of pneumonia and to the fine bubbling sounds of capillary bronchitis. It appears to me better, therefore, to simplify our phraseology by substituting the word "sound" for these unmeaning or inappropriate terms. If, however, for the sake of convenience, the terms "râle" and "rhonchus" are to be retained, the former should be restricted to the sounds of bubbling, the latter to those of vibration. A distinction would thus be drawn between sounds which, though not necessarily indicative of a very different condition of the pulmonary tissue, yet take their origin in a different mechanism.

whether thickening of the bronchial mucous membrane, the pressure of tubercle or other matter, or, as often happens, irregular spasmodic contraction of the circular fibres of the air-tubes—the secretion of viscid mucus at the seat of obstruction contributes greatly to the general result. This is evident from the frequent intermission of these sounds, and from their cessation for a time after the act of coughing—facts, which though quite incompatible with the theory of persistent contraction of the air-passages, are strictly in keeping with the supposition that the contraction is in great measure referable to the presence of tough, viscid mucus which admits of removal and change of position by the full blast of air which accompanies forcible expiration or the act of coughing. Therefore, although the impression conveyed to the ear by these so-called "dry" sounds justifies the term which has been applied to them, it must always be remembered that the use of the term does not imply a total absence of secretion, but only the absence of secretion of such a nature and in such quantity as shall lead, during the act of respiration, to the formation and bursting of air-bubbles, and the consequent production of "moist" sounds.

In this limited sense the term "dry" is applicable enough; but if it be attempted, as is often done, to classify all the adventitious sounds originating within the lungs as either "dry" or "moist," then does the term "dry" become highly objectionable, as being inappropriate and calculated to mislead. Many of the sounds arising from simple abnormal vibration are connected, as already stated, with the presence of viscid secretion in the bronchi, and some even of the sounds associated with the bubbling of air through fluid in the air-passages convey a distinct impression of "dryness." Such are the fine crepitations which are frequently met with in pneumonia, and the dry "clicking" which often accompanies the early stage of tubercular deposit. Moreover, a slight modification of the conditions which give rise to "dry" crepitation and "dry" clicking will occasion the setting up of "humid" crepitation and "moist" clicking; and thus sounds which pathologically acknowledge the same origin, and which differ only in the quantity and tenacity of the existing secretion, are separated and placed in contrast to each other, as if they were distinct and independent phenomena. The course, therefore, which appears to me simplest and least likely to lead to error and misapprehension is to divide all adventitious sounds originating in the air-passages, according to their mechanism, into "sounds of vibration" and "sounds of bubbling," and then the special character of each variety of vibration and each modification of bubbling may be described if necessary according to the notion it conveys to the ear.

Viewed in this light, and considered in reference to their practical significance, the adventitious sounds originating within the lungs may be arranged advantageously in the following manner:

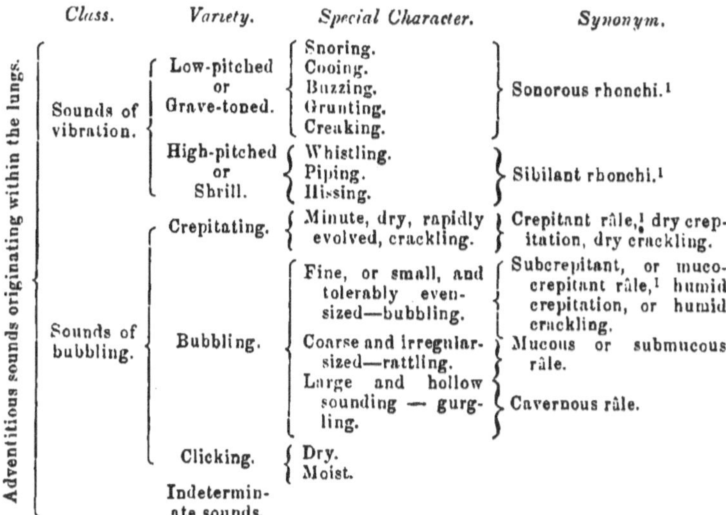

Adventitious sounds originating within the lungs.

Class.	Variety.	Special Character.	Synonym.
Sounds of vibration.	Low-pitched or Grave-toned.	Snoring. Cooing. Buzzing. Grunting. Creaking.	Sonorous rhonchi.[1]
	High-pitched or Shrill.	Whistling. Piping. Hissing.	Sibilant rhonchi.[1]
Sounds of bubbling.	Crepitating.	Minute, dry, rapidly evolved, crackling.	Crepitant râle,[1] dry crepitation, dry crackling.
	Bubbling.	Fine, or small, and tolerably even-sized—bubbling.	Subcrepitant, or muco-crepitant râle,[1] humid crepitation, or humid crackling.
		Coarse and irregular-sized—rattling.	Mucous or submucous râle.
		Large and hollow sounding — gurgling.	Cavernous râle.
	Clicking.	Dry. Moist.	
	Indeterminate sounds.		

The nomenclature above employed is extremely simple, and the classification itself is not only in accordance with the mechanism of the sounds, but includes all the varieties which have the least practical significance. Other sounds or subdivisions of sounds have been dwelt upon at length by many authors as diagnostic of particular forms of bronchial effusion or special forms of pulmonary mischief, but such distinctions have no real existence, and only tend to confuse the student and complicate the study he has in hand. Nothing can be more perplexing than a needless multiplication of terms, and therefore in this classification I have confined myself to those sounds which can be traced as having a constant relation to some particular condition of the pulmonary apparatus, and consequently are of real practical importance.

The precise character of the sounds of vibration varies with the force of the blast and the size, form, and condition of the

[1] These have been termed indifferently râles and rhonchi, but, as already stated, the term rhonchus is most appropriate as applied to the sound of vibration, and the term râle to the sound of bubbling.

channels through which the air has to pass. The stronger the current of air and the narrower the chinks or channels through which it is forced, the higher (*cæteris paribus*) the pitch, and the more whistling the character of the tone produced. And as every possible variety in the condition of the air-passages may exist in different portions of the chest, so every variety in the sounds of vibration, from the deep, low-pitched note of a large tube to the shrill, high-pitched whistle of a narrow chink or channel, may be heard sometimes in a single case. The grave, low-pitched tones are usually compared to the snoring of a person asleep, and do sometimes very closely resemble it; occasionally they more nearly approximate to the noise of grunting, buzzing, cooing, or creaking,[1] and now and then they assume a "rubbing character;" but whatever their nature, they may all be included under the general appellation of "sonorous" rhonchi; whilst those of a higher pitch, whether whistling, piping, or hissing, may be grouped under the title of "sibilant" or whistling rhonchi. Both classes of sounds are connected with the same mechanism, and the variations in their character are due to differences in the size of the air-passages in which each takes its origin. As a general rule, the low-pitched "sonorous" varieties originate in the large bronchi, and the high-pitched "sibilant" varieties in the smaller tubes; but inasmuch as the calibre of the large tubes may be constricted and rendered small by disease, it is obvious that no dependence can be placed on the mere discovery of sibilant rhonchus as evidence of an affection of the smaller order of bronchi. The grave, sonorous rhonchi are undoubtedly characteristic of an affection of the larger tubes, and an extensively diffused sibilant rhonchus may be relied upon as proof of an affection of the smaller bronchi; but when such a rhonchus is of limited extent, it may not impossibly proceed from a large bronchus which has undergone constriction.

[1] This is a point to which I would especially direct the attention of the student; for I have repeatedly seen this creaking and rubbing rhonchus mistaken for the friction-sound of pleurisy. Unfortunately the mistake is not confined to students. Several cases have fallen under my observation which have been wrongly treated by practitioners more or less acquainted with the use of the stethoscope, under the belief that their patient was suffering from pleurisy. The mere absence of ægophony, as will be shown hereafter, is not sufficient to mark the case as one of bronchitis, but a distinction exists between the sound in the two cases which is simple and conclusive. Cough and expectoration, by getting rid of the mucus, removes the cause of the rhonchus, and thus leads to its modification or complete temporary cessation; whilst if the sound is dependent on pleuritic friction, it continues uninfluenced by the act of expectoration. Thus, if care be exercised, a satisfactory conclusion can be arrived at, often at the first, and always in the course of two or three examinations.

It might be imagined, from what has been already stated relative to the source of these vibrating sounds, that they would necessarily accompany both acts of respiration. Very generally they do so to a greater or less extent, but sometimes they are heard during inspiration only, and at others are confined to the act of expiration. Varying greatly in intensity and duration, they are at one time slight, and at another so loud, as to mask or overpower the natural respiratory murmur. They may be audible not to the patient only, but to persons at some distance from him, and, when they are of a low-pitched sonorous character, they may convey a sensation of fremitus or vibration to the hand placed upon the chest-walls. They rarely accompany every act of respiration, but occur irregularly and interruptedly, persisting occasionally for some length of time at a particular spot, and then, suddenly and for a time at least, ceasing—the cause of their cessation in most cases being the removal of some tough mucus by the act of coughing. In many instances, however, the high-pitched sibilant varieties are less readily got rid of by coughing and expectoration than those which are grave or of a lower pitch—a circumstance which is explicable by the comparative weakness of the blast in the smaller bronchial ramifications whence the sibilant rhonchi usually originate, and by the greater difficulty which consequently attends the removal of the tenacious mucus.

According to the form of disease by which they are occasioned, these sounds of vibration may be confined to one spot, or one particular portion of the chest, or may be diffused generally over its surface. Probably, if caused by the pressure of aneurismal or other tumors, or by the lodgement of foreign bodies within the bronchi, they will be limited in extent, and more than usually persistent; if occurring as a consequence of bronchitis, connected with the deposit of tubercle or other matters, they not only will be persistent, but will abound in those parts especially which are prone to suffer from the disease in question; whereas, if they arise in consequence of bronchitis unconnected with any local cause of irritation, they will usually extend more generally over the chest, and will yield more readily to treatment. Moreover, if the disease does not speedily subside, they will be accompanied or followed, after no lengthened period, by the occurrence of bubbling sounds, resulting from the increased secretion which accompanies its further progress. Their most striking development is certainly met with when emphysema coexists with chronic bronchitis, in which case the expiratory sound is enormously prolonged, and the noises produced within the chest are of such variety as to defy description.

Thus, then, to sum up the practical information afforded by these sounds of vibration, it may be stated generally that their presence indicates constriction of the air-passages, and the probable presence of tenacious mucus. Their persistence for a lengthened period over a limited area, leads to a presumption of some mechanical local obstruction, such as the pressure of an aneurismal or other tumor, or the presence of a foreign body in the bronchus, or, when they are confined to the apices of the lung, the irritation of tubercular deposit. On the other hand, their existence over the entire chest, or over an extensive area, indicates the existence of wide-spread bronchitic irritation, which, if not arising from idiopathic causes, must be due to some generally diffused cause of local mischief, such as tuberculous or cancerous infiltrations. In either case their presence should lead us to expect the speedy occurrence of increased bronchial secretion, and the consequent production of bubbling sounds.

The low-pitched sonorous sounds which invariably originate in, and are altogether confined to the larger bronchi, are usually of less serious import, than the high-pitched sibilant rhonchi, which, when widely diffused throughout the chest, indicate the implication of a smaller order of air-tubes, and the existence of extensive affection of the respiratory apparatus, requiring immediate attention and active treatment.

The "bubbling" sounds, whatever their seat or position within the chest, are caused by the passage of the respired air through fluid more or less tenacious, which has accumulated in and obstructs the bronchial tubes. Their character, however, varies considerably, according to the consistence of the fluid, the size of the space in which it is collected, and the force with which the air is driven through it,—circumstances which determine the size and number of the bubbles, the compression to which the contained air is subjected, and consequently the sound which the bubbles cause on bursting.[1] The less tenacious the fluid, the smaller, *cæteris paribus*, will be the size of the bubbles; the less the resistance to their passage through it, the more rapid their succession, the more frequent their bursting. Again, the greater the energy of the respiratory movements, and the greater consequently the force with which the air is driven through the

[1] The influence exerted on the pitch and intensity of the sound, and on the readiness with which it is conveyed to the ear, by the nature of the space in which the bubbles burst, and by the state of the surrounding pulmonary tissue, will be fully discussed hereafter. My present object is simply to point out how the character of the bubbling and the sound it produces are modified, *cæteris paribus*, by the consistence of the fluid, the size of the space in which it is collected, and the force with which the air is driven through it.

fluid, the more numerous will be the bubbles in any given time, and the less distinct or separate the sound which each one produces in bursting. But the size of the tube or cavity containing the fluid exercises a further and most important modifying influence. The larger the space, the greater is the freedom with which the air passes, and the larger and *more unequal* is the size of the bubbles; so that the same fluid which in the smaller bronchial tubes or air-passages would give rise to a fine and tolerably uniform bubbling, will in the larger tubes produce a coarse and more irregular-sized bubbling,—sometimes resembling a rattling. In tubes or cavities of still larger size it will occasion a still more irregular and hollow-sounding bubbling, or, in other words, a gurgling, quite characteristic of air passing freely through liquid, and producing large unequally sized bubbles; so that according to the character of the sound we are enabled to form a tolerably correct estimate of the size of the tube or cavity from which it originates, and of the consistence of the fluid by which the bubbling is caused.

Crepitation.—If the bubbles are very minute, of uniform size, and burst in rapid succession, almost as in a volley, and emit a dry, crackling, or crepitating sound, rather than a liquid, bubbling sound on bursting, the air-passages in which the sound originates are extremely small. Indeed, such a bubbling indicates the entrance of air into the pulmonary vesicles, and proves that no disease exists of such a nature as to render the lung impermeable and prevent the expansion of the air-vesicles. This sound very closely resembles the noise one hears when a lock of dry hair is rubbed transversely and slowly between the finger and thumb close to one's own ear, or when common salt is thrown on the fire; and has been termed, not inaptly, crepitation. It is the crepitant râle or rhonchus of authors, and usually, but not invariably, accompanies the accession of pneumonia, and in some few instances is met with in connection with congestion of the lungs and œdema of the pulmonary tissue.[1] It is almost confined to the act of inspiration, which it sometimes accompanies from beginning to end, though it occurs more commonly towards its close; it is usually persistent for some length of time at the

[1] This fact is too often lost sight of in practice, and leads those who trust too exclusively to auscultation into grievously erroneous treatment. Instances of well-marked crepitation, not to be distinguished from the crepitation of pneumonia, are met with not unfrequently during the pulmonary congestion which often accompanies continued fever and other asthenic and hæmic disorders, as also during the early stage of pulmonary œdema arising in connection with acute anasarca. It is obvious that the active and depressing treatment which might be appropriate in combating inflammation of the lungs, would be productive of mischievous results in cases such as these.

spot where it is first evolved; is heard most distinctly after deep inspiration or after a fit of coughing, and is not checked by expectoration,—circumstances which usually serve to distinguish it from the fine bubbling sound, the variety next to be described, into which it gradually merges, and with which it is confounded not unfrequently.

Some persons have been unwilling to admit crepitation to be referable to bubbling, on the ground that if it were due to the passage of air through fluid in the pulmonary cells, it should accompany expiration as well as inspiration. Hence they have suggested that it is dependent upon the sudden unfolding of the walls of the air-cells, rendered dry and crackling by disease, or else upon the giving way of exudation-matter which has been effused between the vesicles in the substance of the lung. Neither of these theories, however, appears to me to be tenable. Observation in the dead-house proves conclusively that pneumonia does not produce dryness of the pulmonary tissue, such as could give rise to crepitation; and even were it otherwise, the dryness of the tissue should occasion crackling during the collapse of the lung almost as much as during the act of expansion. This may be proved experimentally, by inflating a dried bladder, and then compressing it so as to cause it to assume a folded condition. Nor does the second hypothesis rest on more stable ground. Indeed, in most cases, pathological research proves the absence of the cause to which the sound is attributed. Dissection of a portion of lung, in the first stage of pneumonia, shows congestion of the pulmonary capillaries, and effusion of a viscid sero-sanguinolent fluid into the air-vesicles; but in a vast majority of instances, it fails to discover the inter-vesicular or parenchymatous plastic exudation, to the giving way of which this theory refers. And if it be supposed that the sound is attributable to the movement of fluid in the areolar tissue during respiration, no greater advance is made towards an explanation; for obviously, under the conditions referred to, the sound should accompany the collapse of the lung during expiration, just as much as it does its expansion during inspiration. Indeed, I cannot but think that the objection raised to the bubbling theory of crepitation rests altogether upon a false assumption. It is quite intelligible, as stated by Dr. Davies, that "the air having forced its way through fluid into the interior of the lung-cells, may be expelled from those little cavities without being compelled to pass a second time through the secretion. At the end of an inspiration, the cells are fully expanded, and space is afforded for the presence of air and fluid. At the commencement of the expiration, the more elastic fluid, the air, is first driven out, finding a free passage from the vesicles into the terminal bronchial ramifications."[1] I am satisfied, therefore, that crepitation may be regarded as the result of air bubbling through fluid in spaces of very small calibre, such as the pulmonary air-cells, and that this is the only satisfactory explanation.

Fine or small even-sized bubbling.—If the bubbles be less delicate and minute, less uniform in size, less rapidly evolved, less regular in their occurrence, and, though distinctly crepitating, are less dry in their character, and convey the impression of being connected with the bubbling of air through fluid, the bronchi or air-passages in which they originate are still small, but the fluid is larger in quantity; for it is only in such tubes or spaces, and with such an amount of fluid, that fine even-sized bubbling can be produced. This is the "subcrepitant" or "muco-crepitant" rale of authors, and is the sound which accompanies capillary bronchitis, the first stage of œdema of the lungs, the effusion of blood or other fluids into the pulmonary tissue, and the resolution-stage of pneumonia. It is, in fact, only a modification of crepitation, but unlike the true crepitus of pneumonia, it is influenced, and for a time got rid of, by coughing, and accompanies expiration as well as inspiration, the fluid being in the terminal bronchi, and in such quantity that the air has to force a passage through it as well in expiration as in inspiration.

Coarse, irregular-sized bubbling; rattling.—If the bubbles be larger, of variable size, less regular in their occurrence, and distinctly liquid, the air-passages whence the sound originates are of larger size, admitting of more free and unrestrained bubbling. This is the "mucous" or "sub-mucous" râle of authors, and is the sort of bubbling which accompanies the secretion-stage of bronchitis, the occurrence of œdema, hemorrhage, or whatever leads to the presence of fluid, whether mucus, pus, blood, or serum, in air-tubes or passages of moderate size.[2] It accompanies both respiratory movements, and is modified by the acts of coughing and expectoration, whereby the fluid in the air-passages is removed, and the sounds for a time diminished or arrested. It must be remembered, however, that whatever the precise character of the bubbling, its amount does not necessarily afford an index to the quantity of fluid contained in the air-

[1] Davies, loc. cit., p. 175.

[2] The attempts which have been made by M. Fournet and others to draw distinctions between the sounds produced under these several circumstances are calculated to give rise to endless confusion. No difference exists sufficiently well marked and of such constant occurrence as to possess the slightest diagnostic value.

passages, for these may be so filled that little or no air can pass, and, therefore, little or no bubbling can be produced. In some such instances, all respiratory sound ceases, whilst in others slight bubbling sounds still continue. In the more advanced stages of bronchitis, and during the occurrence of œdema, this happens not unfrequently, and can be detected by the presence of more or less dulness on percussion; by the absence of respiratory murmur; by the want of proper expansion of the chest-walls; and, indeed, by the various signs which indicate the existence of impervious lung. But if the bubbling be small in amount, consisting of single or occasional bubbles, and the vesicular murmur remains distinctly audible, it is fair to conclude that little fluid exists in the air-passages;[1] whilst if the bubbling be extensively diffused, and be not accompanied by a distinct respiratory murmur, or be heard only in combination with a faint, indeterminate respiratory murmur, it is equally a legitimate conclusion that the bronchial tubes contain a large quantity of secretion.

Large, hollow-sounding bubbling; gurgling.—If, again, the bubbles be still larger and of unequal size, less numerous, and more irregular in their occurrence, often disappearing for a time, and conveying to the ear the impression of hollowness, the fluid is probably more uniform in its consistence than in the former instances, and the air-passages or spaces whence the sound originates are larger. This is the gurgling or so-called cavernous râle of authors, and is a modification or further development of the so-called mucous or sub-mucous râle which often merges into it. It is the sort of bubbling which occurs in the main divisions of the bronchi, in large dilatations of the bronchi, in tuberculous or other excavations in the lungs, and, in short, wherever there is an accumulation of fluid in large spaces having a free communication with the air-passages. It commonly exists with both inspiration and expiration; occurs at different periods of the expiratory act, according to the relative position of the contained fluid and the communicating bronchus; is greatly modified by cough and expectoration; and when the walls of the cavity are thin, tense, and elastic, has a peculiarly ringing, metallic character.

The size of the bubbles of which gurgling is made up varies greatly in different cases, and as the character of the sound produced is regulated to some extent by the size and number of the bubbles, some authors have applied the terms "amphoric,"

[1] Even under the conditions here supposed, fluid may be pent up in cavities undisturbed by the respiratory blast.

"cavernous," and "cavernulous" to different-sized gurglings, under the impression that the size and number of the bubbles form an index to the size of the cavity. Nothing, however, can be more erroneous than such a supposition. The character of the sound depends not only on the extent of the cavity, but on the nature of its walls, the freedom of its communication with the air-passages, the condition of the surrounding lung-tissue, the quantity and nature of the fluid contained in it, and the force with which the air is driven through that fluid. If the cavity, though large, be nearly full of fluid, and if that fluid be very viscid, and if the communication between the cavity and the air-passages be not free, the amount of bubbling must be very small, whatever the size of the cavity. Again, cavities bounded by thick, inelastic walls, and surrounded by solidified, uncollapsable lung, are incapable of contracting, and equally so therefore of expelling much air during expiration. Hence it follows that they cannot receive much air during inspiration, and consequently can admit of comparatively little gurgling. On the other hand, cavities bounded by thin and elastic walls, and surrounded by pervious lung-tissue, are greatly influenced by the respiratory movements, and expel a large portion of their contained air at each expiration. They are capable, therefore, of admitting a considerable quantity of air during inspiration, and this, if fluid exists in the cavity, produces a large amount of gurgling. Moreover, if the walls are thin, tense, and elastic, the sounds reverberate loudly and perfectly, and if the cavity is seated superficially they reach the ear without much difficulty; whereas if the walls are thick, spongy, and inelastic, the sounds are very imperfectly reflected, and are not readily transmitted to the ear. Thus it often happens that a small cavity with thin, elastic walls, produces more gurgling and splashing, and transmits a louder sound to the ear, than a much larger cavity bounded by thick, spongy, inelastic walls.

The statements already made show that anything which prevents the passage of air through fluid must necessarily check the production of gurgling. Hence the temporary cessation of the sound may be due to causes of several kinds—either to complete evacuation of the fluid; or to its diminution to such an extent as to bring its level below the opening of the tube or tubes communicating with the cavity; or to obstruction of the air-passages which lead to the cavity. When its cessation is dependent on the complete evacuation of the fluid, the gurgling is usually replaced by hollow, breathing-sound; when it is referable to the second-mentioned cause, the gurgling may be often reproduced by change of posture and the consequent bringing of one of the openings below the level of the fluid in the cavity; and when it depends upon the third named cause, coughing will often lead to its re-establishment by dislodging a plug of mucus or muco-purulent matter. When the obstruction is due to external pressure, cough has not much influence in restoring it, and its cessation is more permanent; and again when the space or cavity is completely full of fluid and does not admit the presence of air, gurgling cannot be produced.

Sometimes when a cavity is unusually large, and not only has very elastic walls, but contains a large quantity both of air and fluid, succussion of the patient, or the impulse against the cavity produced by the action of the heart, will give rise to an admixture of the air and fluid, and occasion a gurgling or splashing sound. I have had the opportunity of noting this fact on three occasions; and it is quite possible to conceive that if the communicating bronchi were obstructed, a gurgling or splashing sound might be thus produced, when no such sound was attendant upon respiration.

Another modification of bubbling is "clicking." Authors are not agreed as to the mode of origin of this sound, nor, indeed, have I ever met with a reasonable interpretation of it. Dr. Walshe says of dry clicking,[1] "The rhonchus, *though its mechanism is unexplained*, is of considerable diagnostic importance;" and again in reference to moist clicking,[1] "Its mechanism is almost as obscure as that of the dry crackling." But careful consideration of the conditions essential to its existence, and repeated experiments as to its mode of production, have convinced me that when it is met with in the lungs it is due to the sudden and forcible passage of air through a small bronchus, the sides of which, at one or more points, have been brought close together by external pressure, or have been agglutinated as it were, by tenacious mucous secretion. Thus its common cause is the presence of tubercle pressing here and there upon the walls of the smaller bronchi, and not only rendering them impervious, but exciting slight local irritation, with the consequent secretion of viscid, tenacious mucus. The bronchi are completely obstructed, and therefore do not admit of the production of the ordinary sounds either of bubbling or vibration; but now and then, perhaps two or three times in the course of an inspiration, their walls are separated for a moment under the pressure of the inspired air, and as the obstruction yields and the sides of the passage are forced asunder, the connecting mucus, which is drawn out into a sort of membrane, suddenly bursts,

[1] Loc. cit., p. 115.

and a sharp click is produced, which conveys an impression of dryness or moisture, according to the quantity and tenacity of the fluid.

Dry clicking, like dry crepitation, originates in connection with a very small quantity of viscid secretion, and is met with almost exclusively during inspiration; whereas humid clicking, like humid crepitation or fine bubbling, is connected with a somewhat larger quantity of fluid, and though most distinct and constant during inspiration, occurs not unfrequently during expiration. When dry clicking first makes its appearance, it is audible only with forced respiration, and even then does not accompany each inspiratory act; but after a time it becomes more persistent, and is heard accompanying ordinary respiration. The dry variety passes, after a time, into the moist variety, and the latter, as the pulmonary disease progresses and softening commences, is replaced by the ordinary sound of bubbling. This arises from a diminution in the amount of pressure on or obstruction of the bronchi whence the sound originates, and from an increase in the quantity of fluid in the air-passages, whereby it happens that the air no longer encounters mere films of mucus, but, by forcing its way through a more copious secretion, creates distinct bubbling.

Clicking, then—the "clicking" or "crackling" râle or rhonchus of authors—is commonly associated with the presence of tubercle, and the moist variety is regarded by some persons as quite characteristic of the early stage of tubercular softening. But careful observation at the bedside of the patient, borne out by the result of dissection after death, has convinced me that this opinion is erroneous. Not only have I met with dry clicking under circumstances in which the whole history and symptoms of the case, no less than the subsequent career of the patient, have satisfied me that no tubercles existed, but twice in the wards of St. George's Hospital I have met with its characters well marked, when subsequent *post mortem* examination has enabled me to establish the non-existence of tubercle.[1] Therefore, although it is perfectly true that tubercle is its ordinary cause, and that when clicking is met with, as it usually is, in the infra-clavicular, supra-clavicular, or supra-scapular regions, its presence justifies the gravest suspicion of tubercle, we must not conclude without further evidence that our patient is indeed consumptive. Anything, whether tubercle or other matter, which,

[1] In one of the cases alluded to the existing disease was chronic bronchitis, in the other sanguineous effusion into the lungs, as the result of valvular disease of the heart. The clicking, though well developed, was of temporary duration, and in neither instance was heard longer than five days.

in one or more of the smaller bronchi, surrounded by partially consolidated lung-tissue, shall produce occlusion of such a nature as will yield now and then to the force of the respiratory blast, may prove the cause of this peculiar sound. But it is usually in the early stage of phthisis, when tubercles are sparsely scattered through the pulmonary tissue, that clicking is most frequent; for it is only when the disease is not extensive, when consolidation is of small extent, and when considerable expansion of the lung still takes place during inspiration, enabling the air to force its way through an obstructed passage, that the conditions exist which are essential to its production. The pressure of a large mass, whether of tubercle or other matter, will usually obstruct the bronchi permanently, and by presenting an insuperable obstacle to the passage of air will prevent the occurrence of clicking.

If it be asked how the clicking which results from tubercle can be distinguished from that which sometimes, though rarely, accompanies other forms of pulmonary disease, I answer that they both take their origin in the same mechanism, and that by the character of the sound alone it is impossible to discriminate between them. But I am convinced that oft-repeated examinations of the chest will generally assist us to a correct diagnosis. In no single instance in which there were not good grounds to suspect the existence of tubercular disease have I found clicking a persistent condition; and if when dry clicking is first observed it continues for a considerable period, and is after a time followed by humid clicking, which, in its turn, is ultimately replaced by bubbling, there cannot be a doubt as to the existence of tubercle. So that this sign, so little insisted upon by some authors,[1] is, in fact, of great clinical significance.

I must not quit this part of our subject without alluding to a sound which has been described by Laennec under the title of "râle crépitant sec à grosses bulles, ou craquement." It is characteristic, he says, of emphysema, is heard only during inspiration, and resembles the noise made by the sudden inflation of a dried bladder. It is a sound of rare occurrence, and of little diagnostic value; and it would not be worthy of special remark had it not been described and first pointed out by the Father of auscultation, and adopted more recently by Skoda of Vienna. Even now, my chief object in mentioning it is to add my testimony to that of the many modern writers who refuse to assign it a place in the list of râles and rhonchi. It certainly

[1] Dr. H. Davies does not even allude to it in his chapter on "Râles and Rhonchi," nor is any mention made of it in Skoda's work.

does not possess a true bubbling character, as its name would imply, and probably is a sound of vibration connected with the extremely viscid and tenacious bronchial secretion which ordinarily accompanies vesicular emphysema. Certain it is that such a condition of secretion, exciting vibration in air-passages altered by emphysema, is quite sufficient to account for its existence, and is far more likely to give rise to its production than the condition of the lung-tissue, suggested by Laennec; for, with Dr. H. Davies, "I cannot conceive the pulmonary membrane which forms the lung-cells to be so dry in emphysema as to be capable of crackling on expansion like a dried pig's bladder."[1] Further, pathology does not countenance Laennec's hypothesis; and we are bound, therefore, to reject it. Whatever the precise cause of the sound, there cannot be a doubt that at present it possesses little practical value; and my advice, therefore, to the student would be, not to encumber himself with the term given to it by Laennec, but to regard the sound, whenever he may chance to meet with it, as a subject for careful investigation.

Skoda speaks of bubbling sounds which he terms "consonating." "Râles," he says,[2] "arising in the larynx, trachea, or either of the bronchi, may consonate within the thorax, just as the voice or the respiratory murmur consonates in the diseased states of the lung-tissue already referred to, and thus become distinctly audible throughout the thorax." They are "clear and high-pitched," he says,[3] "formed by unequal bubbles, and accompanied by a resonance which has neither an amphoric nor a metallic character." Thus, his description answers to that of ordinary, coarse, unequal-sized bubbling, taking place in tubes surrounded by lung-tissue in certain states of consolidation; and as Skoda himself admits that these so-called consonating sounds are accompanied by bronchial breathing and increased resonance of the voice, and "generally speaking, indicate the presence either of pneumonia or of tubercular infiltration," there can be little doubt that pulmonary consolidation is a condition essential to their existence.

What, then, is the peculiarity, and what the diagnostic value of these so-called "consonating râles?" In my opinion, they possess no single feature which entitles them to an independent place amongst recognized râles and rhonchi. It is universally admitted, and is strictly in accordance with the laws of acoustics, that sounds produced in any part of the air-passages may be propagated to the thoracic walls through air contained in

[1] Loc. cit., p. 181. [2] Loc. cit., p. 118. [3] Loc. cit., p. 131.

bronchi capable of reflecting sound, and through lung-tissue in certain states of homogeneity and consolidation; and it is equally certain that in peculiar states of the bronchi and pulmonary tissues, all sounds generated therein will be of a high-pitched character. But, if due care be exercised in conducting the examination, there is little chance of râles propagated from other parts of the chest being mistaken for râles of local origin. The history and progress of the case, the condition of the respiration at the spot where the sounds are heard, and the presence or absence of dulness or unusual resonance on percussion, will generally enable us to determine the question. Even if it were not so, and admitting, for the sake of argument, what really is not the fact, viz., that consonance is a principle of general application in the intensification of sounds propagated through the chest, still the sounds alluded to by Skoda would not possess any special significance; for Skoda himself fails to point out any certain signs by which to recognize the fact of their being generated at a distance, and reinforced and propagated by consonance. The only features which he indicates as calculated to excite suspicion as to the true nature of the case—viz., the clearness and high pitch of the sounds, and the dulness or unusual resonance on percussion—are one and all compatible with the local production of the sounds, and their direct transmission by conduction. Therefore, my advice to the student would be to discard "consonating râles" from his vocabulary, as having at the most a problematical existence, and as being undistinguishable from other rales, and therefore possessing little diagnostic value. In discarding the term, however, he should recollect the fact, that large irregular-sized, high-pitched bubbling sounds heard over solidified lung do not necessarily imply the existence of a cavity immediately beneath the surface of the part auscultated, but may be sounds generated in the larynx or trachea, in a distant cavity, or in large bronchial tubes, and thence transmitted along the air in the bronchi, and through consolidated lung-tissue, to the point at which they are heard. When once alive to the possibility of such occurrence, he is not likely to mistake their nature, or to err as to their true clinical significance. Sounds are rarely, if ever, propagated from a distance, unless extensive alteration has occurred in the physical condition of the lung, in which case they are heard over an extended area, and, with ordinary care, may be traced back to their source.

Various sounds have been described, I believe erroneously, as emanating from the lung-substance. Such, for instance, is the fine crepitation accompanying inspiration, which is heard sometimes at the base of the lungs, posteriorly, even in healthy persons, who have long been reclining on the back, when suddenly made to respire deeply—a sound which disappears after a few acts of respiration. This has been referred by Dr. Walshe[1] to the "forced unfolding of air-cells which are unaffected by the calm breathing habitual to the individual." I am satisfied, however, that it does not emanate from the lung-substance, but is true crepitation, due to the presence of a small quantity of secretion in the pulmonary vesicles. It is but natural to suppose that a small quantity of secretion should gradually accumulate in those portions of the lungs which are not brought into play during ordinary calm respiration, and in which, therefore, the air-passages are not exposed to the drying influence of the atmosphere; and this supposition is quite consistent with the speedy cessation of the crepitus as soon as respiration commences in the part; whereas the notion that the mere unfolding of healthy lung-tissue should give rise to crepitation, is utterly at variance with all that is known, experimentally and practically, of the wonderful delicacy, softness, and pliability of the pulmonary apparatus, and of its smooth and noiseless action, the only perceptible sound having its origin in the friction encountered by the inspired air.

So, also, in regard to the creaking sound heard not unfrequently towards the apices of the lungs in phthisical patients during the act of inspiration; a sound which M. Fournet described as a variety of "pulmonary crumpling," and which Dr. Walshe attributes to "the unfolding of induration-matter in the lung."[2] It is difficult to conceive the mechanism by which the presence of any solid or semi-solid matter in the lung-tissue should give rise to the production of such a sound; and the frequent existence of every variety of deposit in the pulmonary tissue without its occurrence, seems to negative the idea of its being so produced. Indeed, close and repeated observation has convinced me that the sound in question is a sound of vibration analogous to the creaking sometimes met with in the other parts of the chest during the existence of chronic bronchitis, and is due to a contracted condition of the bronchial tubes, and to the presence of viscid secretion in them; for, like the creaking sometimes heard in bronchitis, it is usually removable by cough and expectoration —a circumstance which serves to distinguish it from a very similar sound which sometimes results from local pleurisy.

So, again, in respect to the sound described by M. Fournet as resembling the crumpling of tissue-paper. This, I feel con-

[1] Loc. cit., p. 119. [2] Loc. cit., p. 120.

vinced, does not arise from the pulmonary tissue, as suggested by some authors, but is attributable to pleuritic friction. Certain it is that it is heard only at the apices of the lungs, where, in phthisical patients, pleuritic inflammation is very frequent, and is met with only during the early stage of phthisis, when it is probable that no adhesions exist, and whilst there still remains sufficient mobility of the apices of the lungs to produce the sound of friction. Further, it is persistent, and is not removable by coughing, as it would be if it were a variety of bronchial vibration. Such being the case, and as we know of no condition of the pulmonary tissue peculiar to the upper part of the lungs, where alone this sound is heard, it is fair to conclude that it does not emanate from "crumpling" of the lung-tissue, as suggested by Fournet, but is referable to friction, resulting from congestion and local dryness and roughness of the pleural membrane, the peculiarity of its character being referable to its position in relation to surrounding parts.

Indeed, the only sound which in my opinion can be strictly referred to the lung-substance is a peculiar variety of crepitation very rarely met with, and which hitherto, as far as I am aware, has never been recognized or described as distinct from ordinary crepitation. Its source, I believe to be, the presence of serosity in the areolar tissue of the lung; and its peculiarity, as contrasted with ordinary crepitation, is, that the crepitations are less numerous, and less regularly and less rapidly evolved, and that they are heard just as strikingly during the act of expiration as they are during the act of inspiration. The period of its occurrence and its duration correspond with the fine bubbling sound of capillary bronchitis—the subcrepitant râle of many authors; and this fact, when viewed in connection with the minuteness of the crepitations, the peculiar dryness of their character, and the fact of their being unaccompanied by expectoration, serves to distinguish them from any sound known to arise from the presence of fluid in the air-passages, and points to the existence of serosity in the lung-tissue as the only satisfactory explanation of the phenomenon. With fluid in the areolar tissue surrounding each lobule, crepitation could not fail to be produced by every movement of the lung, whether in inspiration or expiration.

A somewhat similar sound, and referable to the same cause, viz., the movement of fluid through cellular tissue, is heard sometimes over the sternum, when the mediastinum is infiltrated with either air or serosity. The crepitations, just as in the former case, accompany expiration as well as inspiration, and may vary considerably in tone, dryness of character, and number. They may be heard during ordinary tranquil breathing, or may be audible only during forced respiration; they may be constant, or only of occasional occurrence; and may endure for a considerable period, or may cease after a very few respirations. I have heard them strongly developed in several cases in which the mediastinal cellular tissue was infiltrated with serosity in connection with inflammation of the pericardium, and have also observed them in connection with the presence of air, as the result of accidental perforation of the chest. Their clinical importance arises from the possibility of their being mistaken, as I have known them to be, for the crepitations of pneumonia. It is strange, however, that such mistakes should be made; for the peculiarity of their character, the limitation of the area over which they are heard, and the period of the respiratory act at which they occur, together form a safe criterion of their true origin.

The adventitious sounds which arise in the pleural cavity require a more extended notice. In a state of health the two layers of the pleural membrane being smooth, and slightly humid, glide upon each other evenly and noiselessly; but anything which roughens one or both of them, or interferes with their lubricity by drying their surface, increases the friction between them, and thus gives rise to the production of sound. Hence the sounds so produced, whatever their precise character, are commonly termed "pleuritic friction-sounds." At first they are often soft and of a rustling or grazing character, and, as first suggested by Messrs. Barth and Roger, very closely resemble the sound produced when the back of one hand placed over the ear is rubbed lightly in one direction with a finger of the other hand. But as mischief advances, the sounds speedily lose this light, grazing character, and assume either a harsh, rubbing, grating, or crumpling character, or else become distinctly creaking. Very commonly they are more or less jerking or interrupted, as if the motion of the roughened surfaces over each other was momentarily suspended, and then suddenly recommenced. The precise character of the sound, however, affords no clue to the nature or extent of the pleural exudation; indeed, it appears doubtful, whether in some cases, at least, pleuritic friction-sound may not be caused by mere dryness and increased vascularity of the pleural membrane. Moreover, the precise condition of the pleural membrane and the character of the exudation on its surface have less to do with the intensity of the sound than have the permeability of the lung-substance and the force of the respiratory efforts, inasmuch as free expansion of the lung is necessary to produce that friction between the two opposed surfaces of the pleura to which the sound is attributable.

The soft, grazing variety is that which occurs at the first onset of the disease; and is probably confined to the dry period of pleurisy, for its duration is very short, and as soon as dulness on percussion and other symptoms of effusion occur, it is replaced by the rubbing or creaking varieties. It is dry in character, and superficial in situation, is heard only during inspiration, does not accompany each inspiratory act, is usually confined to a very limited area, and commonly passes away, or else is replaced within twenty-four hours by some other variety of pleuritic friction-sound. When it arises in connection with idiopathic pleurisy it soon disappears, and the patient is seldom seen until it is replaced by the rubbing, grating, or creaking variety. Indeed, so generally does this hold good, that in most instances when it is met with in auscultation, it indicates the circumscribed pleural inflammation, which accompanies the irritation of tuberculous deposit, and often serves to establish the true character of the pains by which that inflammation is attended.

The rubbing, grating, crumpling, and creaking varieties are the forms in which pleuritic friction-sound most commonly presents itself. Whichever its precise character, and whether loud, or whether faint and barely audible, it is always superficial in its seat, conveys the impression of dryness, and, though often confined to a limited area, is usually heard over a more extended surface than the grazing variety. Sometimes it is a continuous sound, the movement of the two surfaces of the pleura over each other being even and uninterrupted, but more commonly the motion of the pleura is somewhat interrupted, and consequently the sound is made up of a series of jerks, rarely exceeding four or five in number, which in strongly marked cases occasion vibrations so distinct and forcible as to be perceptible to the patient, and to be felt even by the hand placed over the seat of their production. This is the case especially in regard to the rubbing variety. Sometimes it is confined to the act of inspiration, but more commonly accompanies both inspiration and expiration; in most instances it is audible during ordinary respiration, but occasionally is developed only after deep inspiration. Its duration in any one locality seldom exceeds a few days, inasmuch as when effusion takes place into the pleural cavity, the fluid separates the two layers of membrane, and thus puts an end to the cause of the sound; but in some few cases where partial adhesions exist, keeping the two surfaces of the pleura more or less in apposition, and in the comparatively dry circumscribed pleurisies which often result from tubercular deposit at the apices of the lungs, I have known it continue for many weeks.

Laennec asserted that pleuritic friction-sound is a common accompaniment of interlobular emphysema, and one of its most characteristic physical signs; and although later and more extended observation has served to negative this statement, if taken in its broad and general sense, it still remains the opinion of some, that emphysema may prove an occasional cause of it.[1] But I quite agree with Dr. Stokes, that it is only when the surfaces are rendered dry by an arrest of secretion, or are roughened by the effusion of lymph, that their motions produce sounds perceptible to the ear. On several occasions, whilst auscultating lungs extensively emphysematous, I have detected sounds very greatly resembling the friction-sound of pleurisy; and, in some instances, the death of the patient has enabled me to determine, by post-mortem examination, that they did not proceed from the friction of an inflamed and lymph-covered membrane. But it does not follow that they are therefore to be attributed to the friction of subpleural emphysematous vesicles. In two only out of the instances just alluded to, did any emphysematous projections exist at the spot over which the sounds had been heard during life; and it is quite certain that in *most* cases of emphysema, however strongly developed, no sounds are heard at all resembling the friction-sound of pleurisy. If it were possible for irregularity or unevenness of the visceral pleura, resulting from emphysema, to occasion pleuritic friction-sounds, these sounds ought to be present whenever such a lesion exists, provided only that there remains sufficient mobility of the lung to permit of a free gliding of the visceral on the costal pleura. But their existence under these circumstances is notoriously quite exceptional. Such being the case, and as sounds not to be distinguished by their mere character from the friction-sound of pleurisy are sometimes heard during chronic bronchitis quite independently of emphysema, and equally so of pleuritic inflammation, I cannot doubt that the cause of the sound in the two cases is identical, viz., the vibration of viscid mucus in the air-passages under certain conditions of bronchial constriction and surrounding pulmonary solidification.

Unfortunately this intra-pulmonary source of sounds which resemble friction-sounds is often overlooked, or rather is not fully recognized. But I have met with too many examples of these bronchial creaking sounds, and after death in such cases

[1] Thus, for instance, Dr. Walshe.—See loc. cit., pp. 123–4.

have had too many opportunities of verifying the absence of pleuritic friction, to entertain the slightest doubt as to their true source. Unfortunately, too, I have more than once seen lamentable effects produced by erroneous treatment founded on this mistaken diagnosis. I am anxious, therefore, to express my conviction that, with due care, the sounds alluded to ought never to be mistaken for pleuritic friction-sounds. They are always, I believe, associated with a certain degree of bronchitic irritation, and the presence of viscid mucus in the air-tubes; they rarely convey the impression of being very superficial in their situation, or of being so dry as true pleural friction-sounds; they are usually loudest during expiration, and are always of short duration, intermittent in their occurrence, and removable by long-continued coughing and expectoration. Indeed, it is a safe rule when creaking or other sounds resembling friction-sounds are heard—especially if there is any coexistent cough and expectoration—to make the patients inspire deeply, and give two or three hearty coughs, before deciding upon the nature of the mischief. If this precaution were always taken, mistakes would seldom be made respecting the source and significance of these quasi friction-sounds.

Before quitting this part of our subject, it may be well to advert to certain adventitious sounds originating in the parietes of the chest, which are apt to simulate, and, in many instances, have been mistaken for morbid intrathoracic sounds. Two of these, connected with the pressure of the stethoscope, I have alluded to in a previous chapter, namely, the pseudo crepitation produced when the instrument is placed over hair on the surface, and when, again, the subcutaneous cellular tissue is infiltrated with either air or fluid. But there are other sources of morbid sounds which have not as yet been mentioned. The crepitus of fractured ribs has been mistaken for pneumonic crepitation, and the creaking often heard over the dry costal cartilages of elderly persons, and occasionally also over the scapula, arising probably from some morbid condition of the cellular tissue between that bone and the chest-walls, very closely resembles the creaking sound of pleural friction. The former may be recognized by its position, by its long persistence without extension of the area over which it is heard, and by its being unaccompanied by any of the other symptoms of pleuritic inflammation; whilst the latter may be got rid of without difficulty by a few brisk rotations of the arm.

The subjoined table exhibits at a glance the character of the adventitious sounds originating within the chest, the period of respiration at which they are usually developed, their mode of production, their ordinary seat, and the diseases of which they are commonly the index. Many, very many exceptions occur; but when the character of the sounds is strongly marked, the facts indicated by this table will generally hold good.

MORBID SOUNDS EVOLVED WITHIN THE

Class.	*Variety.*	*Synonym.*	*Character of the sound.*
Sounds of Vibration.	Low-pitched or grave-toned.	Sonorous Rhonchi.	Deep or grave-toned sounds of a snoring, cooing, buzzing, grunting, or creaking character, attended by fremitus, communicated more or less forcibly and extensively to the chest-walls.
	High-pitched or Shrill.	Sibilant Rhonchi.	High-pitched sounds of variable duration and intensity, and of irregular occurrence, of a whistling, piping, or hissing character.
Sounds of Bubbling.	Crepitating.	Crepitant râle. Crepitation, or dry crepitation.	A dry crackling or crepitating sound, conveying an impression of the bursting in rapid succession, almost as in a volley, of minute bubbles of uniform size. It closely resembles the sound one hears when a lock of dry hair is rubbed transversely and slowly between the finger and thumb close to one's own ear.
	Fine bubbling.	Subcrepitant râle, or humid crepitation.	A sound resembling the bursting of small bubbles of tolerably uniform size
	Coarse bubbling.	Mucous, or sub-mucous râle.	A coarse bubbling sound, conveying the impression of the bursting of somewhat large bubbles of unequal size.
	Large bubbling.	Cavernous râle, or gurgling.	A hollow, gurgling sound, often of a metallic character, conveying the impression of very large bubbles bursting in a large space.
	Dry clicking.	Dry mucous râle, or dry clicking.	A succession of three or four short crackling or clicking sounds, conveying an impression of dryness.
	Moist clicking.		A succession of three or four clicking sounds less dry, and less distinctly crackling than the sounds of dry clicking.

MORBID SOUNDS ASSOCIATED

Friction-Sounds.	Rustling or grazing. Rubbing. Creaking. Grating. Crumpling.	Pleural friction-sounds.	A sound sometimes continuous, but often abrupt, jerking, or interrupted, conveying an impression of the friction of two surfaces. It may or may not be attended with perceptible fremitus, and may be of a rustling, grazing, rubbing, creaking, grating, or crumpling character.

LUNGS DURING THE ACT OF RESPIRATION.

Period of respiration at which it occurs. Inspiration.	Expiration.	How produced.	Its usual seat.	Disease by which it is usually caused.
——	——	Caused by the vibration excited by the passage of air through the larger bronchi, irregularly narrowed either by spasmodic contraction of their circular fibres, or by the adhesion of viscid mucus to their walls.	Not peculiar to any portion of the lungs.	Bronchitis, especially when in a subacute or chronic form, or associated with emphysema.
——	——	Engendered by the same cause as the sonorous rhonchi, but originating either in the smaller air-passages, or in those of a larger size, which are temporarily contracted and reduced in calibre by reason of the disease.	Not peculiar to any portion of the lungs.	Bronchitis.
——		Due to the passage of air through fluid in the minute air-vesicles of the lungs.	Base of one or both lungs.	Pneumonia.
——	——	By the bubbling of air through fluid, more or less viscid, in the minute or capillary bronchial tubes.	Middle or base of one or both lungs.	Capillary bronchitis or the resolution of pneumonia.
——	——	Caused by the bubbling of air through liquid in bronchial tubes of varying size.	Middle portion of the lungs.	Bronchitis after secretion has been thoroughly established.
——	——	Referable to the bursting of large bubbles in a space of considerable size; the gurgling and metallic quality of the sound being more pronounced the smoother and more tense and sound-reflecting the walls of the space in which the bubbling takes place.	Apices of the lungs when due to— or, in other parts when referable to—	A cavity the result of softened tubercle. Dilated bronchi, inflammatory abscess, or the presence of pus in the pleural cavity, between which and the bronchi there exists a fistulous opening.
- - -		Caused by the forcible passage of air through a film of mucus, at a spot where the sides of a small bronchus have been brought together by external pressure, and agglutinated by tenacious secretion.	Apices of the lungs.	First stage of tubercular deposit.
- - -	- -	Produced in the same way as dry clicking, but associated with a more copious and less tenacious secretion.	Apices of the lungs.	Commencing softening of tubercle.

WITH THE MOTIONS OF THE PLEURA.

Period of respiration	How produced.	Its usual seat.	Disease by which it is usually caused.
The first variety is confined to inspiration; the others are sometimes confined to inspiration, but more commonly accompany both inspiration and expiration.	By the rubbing together of the two opposed surfaces of the pleura rendered dry or roughened by inflammation.	Not peculiar to any portion of the chest.	Pleurisy.

CHAPTER X.

ON SUCCUSSION, AND ON CERTAIN PECULIAR PHENOMENA CONNECTED WITH THE RESPIRATORY SOUNDS, THE BUBBLING RALES, AND THE RESONANCE OF THE VOICE AND COUGH—AMPHORIC RESONANCE; METALLIC TINKLING.

There still remains one other method by which we may sometimes obtain important information respecting the condition of the thoracic organs; I refer to succussion. So long ago as the time of Hippocrates, it was observed that if patients laboring under certain forms of thoracic disease are shaken by the shoulder whilst the ear of the observer is applied to their chests, a peculiar splashing sound may be heard on the affected side.[1] This sound, which closely resembles that obtained when a nearly empty cask is shaken, and which in fact is produced by the splashing of fluid in a large, air-containing cavity, is known by the name of the "succussion-sound." Its tone varies with the size of the cavity, and the condition of its walls, the density of the liquid it contains, and the relative proportions of air and liquid. The larger the relative quantity of air and the thinner the fluid, the more readily does splashing occur, and the more easily, therefore, is this sound produced;[2] whilst the smoother and more tense the walls of the cavity in which the splashing takes place, and the thinner the fluid it contains, the more metallic is the character of the sound. It may be produced by the least movement of the patient, by the act of coughing, or even by the action of the heart; and when not so producible, it may be readily excited by gently but suddenly pushing the patient backwards and forwards, or from side to side, so as to agitate the fluid. It is often perceptible to the patient himself, and may be audible by bystanders at a distance from the chest, but more commonly it is heard only when the ear of the observer is applied to the chest-walls. Its duration is variable; sometimes it lasts for a few hours only, more commonly for several days, but in some instances it has continued for many

[1] "Hippocrates de Morbis," lib. 11, sec. 45.

[2] Hippocrates observed that "those in whom much sound is heard have a smaller quantity of pus than those in whom the sound is less;" and although he does not appear to have understood the cause of this, his observation was perfectly correct.

months, and has been known to persist even for years. In these chronic cases, which are very rare, the sound is perceived by the patient whenever he shifts his position suddenly, as in running, jumping, walking down stairs, or riding on horseback. It ceases immediately the pleural cavity becomes full of fluid, and also when the fluid has been absorbed or evacuated, inasmuch, as already stated, it is never heard unless air and fluid coexist in the cavity.

From what has been already stated respecting the cause of this peculiar sound, it is obvious that it must have its origin in one of two conditions: either when an enormous excavation, partially filled with fluid, exists in the lung-substance, or else when air and fluid coexist in the pleura. The latter is its most frequent, and as some assert its invariable cause, but at St. George's Hospital I have met with it well marked on three occasions in which post-mortem investigation revealed an enormous tuberculous cavity but no pleuritic effusion, and I am therefore enabled to state positively that it may occur, though of necessity very rarely, in connection with excavations in the lung-substance. The extent of surface over which it is heard depends in great measure upon the nature of the lesion by which it has been occasioned. If, as is commonly the case, it results from the presence of air and fluid in the pleural cavity, and no adhesions exist between the two layers of the pleura, it will be heard over the whole of the affected side; but if the lung has contracted adhesions to the thoracic parietes, it will be audible at those parts only where air intervenes between the lung and the chest-walls. If it originates in a large pulmonary excavation, the surface over which it is heard will be proportionately limited in extent. I apprehend, however, that there need be little difficulty in distinguishing a succussion-sound arising in a pulmonary excavation from one originating in the pleural cavity; for the limitation of the area over which it is heard is sufficient to arouse suspicion; and in pulmonary excavations the resonance rarely acquires a metallic quality at all equal in intensity to that which usually accompanies pleural succussion-sound. Indeed, in the only three cases in which I have heard the sound in connection with pulmonary excavations, it had a very slightly metallic character, though it was strongly marked as a splashing sound.

Thus, then, it would appear that the forms of disease in connection with which a succussion-sound may be heard are large pulmonary excavations with tense and indurated sound-reflecting walls, and pneumothorax, whether arising from an ulcerated opening between the lung and the pleural cavity, or from an accidental perforation of the pleural membrane from without. Theoretically, it is possible for pneumonic abscess and other forms of pulmonary excavations to give rise to it, but I have never met with a case in point, nor can I find a record of such a case in the experience of others. Practically, therefore, phthisis pulmonalis may be regarded as almost invariably the cause of the succussion-sound; for pneumothorax, when arising in the first-mentioned manner, is generally referable to tuberculous ulceration, and so also are the enormous pulmonary excavations from which the sound sometimes emanates.

Under much the same conditions as those which obtain when a succussion-sound is audible, viz., during the existence of a large, air-containing cavity, certain other phenomena are sometimes observed which deserve special notice. I refer to the sounds known as amphoric echo[1] and metallic tinkling. Every person must be aware that on blowing, speaking, or coughing into a large empty jar with a narrow neck, the sound emitted by the observer is accompanied, or rather followed by a peculiar hollow sounding echo or reverberation, quite distinct, and sometimes even of a different pitch from the original sound. Very commonly this echo is of a humming or buzzing character, and hence was termed by Laennec, "bourdonnement amphorique," but sometimes it is high-pitched, and of an intensely metallic quality, its precise tone and character being dependent on the form, size, and reflecting power of the jar, and on the nature and strength of the vibrations by which the sound is caused. The same holds good in regard to the chest. Amphoric echo may be produced in the chest just as in the jar, but only under analogous conditions, viz., when a large air-containing cavity exists, with tense and good sound-reflecting walls. And just as in the jar, so in the chest; the more tense, and smooth, and sound-reflecting the walls, and the more distinct and forcible the original sonorous vibrations, the greater the probability of the echo being of a ringing, metallic character. The ringing quality which it ordinarily acquires under such circumstances resembles that produced when an empty cask is stricken forcibly, or when a falling stone strikes the bottom of a deep well, but under certain conditions a peculiarly distinct clear tinkling sound is heard, which has been termed, not inaptly, metallic tinkling.

[1] This is sometimes styled "amphoric resonance," but the term is objectionable, inasmuch as "amphoric resonance" is a designation applied to a peculiar sound, elicited on percussion; and if the same term is applied to the two phenomena, which have no relation to or connection with each other, it is apt to lead to confusion. Amphoric echo is a term applicable only to the phenomenon we are discussing.

Its distinctive characteristics, as contrasted with ordinary amphoric echo of a metallic quality, are, that it is clearer and of a higher pitch, of short duration, more purely metallic, and remarkably tinkling in its character, and recurs at distinct but irregular intervals. Laennec compared it to the sound excited by dropping grains of sand into a hollow metallic vessel; and a close imitation of it may be heard by applying the ear to a glass decanter, or a thin, hollow metallic vessel, containing a small quantity of fluid, whilst water is poured into it drop by drop.

The nature and cause of amphoric echo do not admit of doubt. It is merely the echo or reverberation, in a large cavity more or less adapted to reflect and concentrate sound, of certain noises excited either within the cavity itself or in its immediate vicinity by the respiration, by the râles or rhonchi, by the voice, or by coughing, by ordinary succussion, or by the impulse occasioned by the heart's action.[1] Its metallic quality is attributable to and varies with the form and size of the cavity, the sound-reflecting power of its walls, and the force and character of the original sonorous vibrations. The only questions which have been raised repecting it are, as to how small a cavity is capable of producing it, and as to whether a fistulous communication with the bronchi is essential to its existence. The size of the cavity has not been determined; indeed the peculiarity of the resonance appears to depend less upon the mere size of the cavity than upon the sound-reflecting power of its walls. But in regard to the question as to the necessity for the existence of a direct communication between the bronchial tubes and the cavity in which this sound occurs, much difference of opinion has been expressed. Laennec and his followers maintained that such a connection is indispensable, and even in the present day some persons profess to entertain the same opinion. But Skoda,[2] Barth and Roger,[3] and other foreign pathologists, and most English authorities on auscultation, are opposed to this view, and I am disposed to enrol myself amongst their number. Rarely is a case of pneumothorax met with in which amphoric echo is not producible, yet it seldom happens that the fistulous opening is not covered with effused lymph, or the access to it obstructed. And experiment leads us in the same direction; for as was pointed out by Skoda, if, whilst one "person speaks into a stethoscope placed

[1] I have noted this last-named source of amphoric echo on four occasions; and a case in which the phenomenon was well marked has been recorded by my late colleague, Dr. Bence Jones, in vol. viii, p. 61, of the "Transactions of the Pathological Society."

[2] Skoda, loc cit., p. 134.

[3] "Traité pratique d'Ausculation," p. 213.

on a stomach filled with air," another person listens through a stethoscope placed on its surface, he will hear amphoric echo sounding in the stomach, although of course no direct communication exists between the air within the cavity and that external to it. Therefore, although when this sound is heard in connection with pulmonary cavities, it is usually referable to vibrations transmitted to the cavity directly through an opening from the bronchial tubes, I do not hesitate to express the opinion that such a communication is not essential to its production; but that if the air in the cavity be separated from a moderate-sized bronchus merely by a thin layer of pulmonary tissue, the sounds existing in that bronchus may pass into the air of the cavity with force sufficient to excite sonorous vibrations therein accompanied by amphoric resonance. The same holds good in respect to the cavity of the pleura; and in all cases of pneumothorax in which there is not a free communication with the bronchi, the transmission of sonorous vibrations through the walls of the cavity affords the only feasible explanation of the phenomenon.

The cause of metallic tinkling in the chest has been a subject of even greater dispute. Laennec attributed it to the reverberation produced by the falling of drops of fluid into liquid effused in the pleural cavity: Fournet, Dance, and others, supposing it to occur only when a fistulous opening exists between the pleural cavity and some bronchial tube, have referred it to the bubbling of air through liquid in the pleura, and most modern authors have connected it in one way or another with the bubbling of air through fluid, or the falling of drops of liquid into fluid in the pleura. On the other hand, Skoda[1] maintains that neither the presence of fluid nor the existence of a fistulous opening into the bronchi is at all essential to its existence; that it may be often heard in cases in which no opening into the bronchi can be detected after death, and therefore, that if even an opening did exist during life, it must have been far too small to transmit vibrations capable of producing metallic tinkling. He further appeals to the result of experiment, and states,[2] that "if a person speak into a stethoscope placed on a stomach filled with air, both metallic tinkling and amphoric echo may be heard sounding within the stomach, and this whether the stomach be partly filled with, or contain not a drop of fluid." In short, he refers this phenomenon to consonance of the voice, or of the respiratory sounds, or of abnormal sounds in the pleura or some other cavity, and asserts that it may be produced in pulmonary cavities just as in the pleura. My own conviction is, that metallic

[1] Loc. cit., p. 138.

[2] Loc cit., p. 134.

tinkling is a mere reverberation or echo of certain sounds reflected by the tense and indurated walls of a large hollow space within the chest; that it may occur without any direct communication with the bronchi; and that its peculiar character is referable chiefly to the nature and tension of the walls of the cavity and the character of the original sonorous vibrations, but in part also to circumstances tending to modify the concentration, quality, and pitch of the sound. Theoretically, therefore, it ought to be producible in pulmonary cavities just as is ordinary amphoric echo; but practically, for reasons hereafter to be mentioned, its occurrence under such circumstances is extremely rare. Amongst the thousands of pulmonary excavations which I have examined, I have only met with it fully developed in a single instance; and in the case alluded to the cavity was of unusual size, considerably larger than a full-sized orange, and was adherent anteriorly to the parietes of the chest. Further, it was bounded by smooth tense walls, surrounded on all other sides by consolidated lung-tissue, and contained remarkably thin homogeneous pus.

I have already referred to Skoda's opinion, that tinkling may occur without the presence of fluid, and have detailed the experiment by which he endeavors to support that opinion. But any one may convince himself by repeating this experiment, that although an intensely metallic-sounding resonance is developed, true metallic tinkling cannot be thus excited. Hence it would appear that the Viennese professor does not confine the term metallic tinkling to the sound described by Laennec as tinkling, but makes it comprise the metallic sound by which amphoric echo is usually characterized—a sound which derives its peculiarity simply from the hardness and tension of the walls within which the vibration occurs. If, then, this wide interpretation of the term is to be taken, Skoda's assertion is undoubtedly correct; but if Laennec's definition is to be followed, and a distinction is to be made between ordinary high-pitched amphoric echo and true metallic tinkling—the sound as of grains of sand falling one by one into a thin metallic vessel—I am satisfied that he is in error, and that fluid is essential to the production of the phenomenon.

But experiment leads me to go further than this, and to assert that, with a view to the production of true metallic tinkling, it is necessary not only that fluid be present, but that it be thrown into active vibration. Dr. Walshe has shown that if a little water be placed in a glass decanter in which amphoric echo is occurring, no tinkling will be produced so long as the fluid remains at rest, any more than if no fluid were present;

but that when the fluid is set in motion, as it may be, by letting drops of water fall slowly on its surface, the echo which results is the purest metallic tinkling. Let the experiment be varied by substituting gruel or some other viscid fluid for the water, and it will be found that by gradually increasing the viscidity of the liquid, a point will be reached at which it will be difficult to say whether the sound emitted deserves the title of "tinkling," and that on rendering it still more viscid, the tinkling sound will cease altogether. In short, it would appear, that the vibrations which give rise to the true metallic tinkling must be very free, and cannot originate in very viscid fluid. Hence, doubtless, the rarity of the phenomenon in its purely developed form, even in cases of pneumothorax, and its almost entire absence in pulmonary cavities, even when metallic echo is well marked; the matter which is secreted being not sufficiently homogeneous or else too viscid to admit of the free bubbling, or the active vibration, which is necessary to the production of pure metallic tinkling.

So also it would appear that the vibrations which are productive of this peculiar sound must be excited by some cause of almost momentary duration, as if tinkling were the echo of a single note intensified and concentrated to the utmost by reflection from the walls of the cavity in which it is heard, and liable to be interfered with by any vibrations set up concurrently with or shortly after it. The result of experiment seems decisive on this point. Thus, if in a decanter containing a little water a few drops of fluid be let fall slowly, drop by drop, on the surface of the water, metallic tinkling is the constant result; but pour water in a continuous stream into the decanter, and the pure tinkle is replaced by ordinary amphoric echo, oftentimes of a low-pitched, buzzing character.[1] Thus, if care be taken in gradually increasing the rapidity with which the water is dropped, a good approximation may be arrived at in regard to the frequency with which loud vibrations may be excited without so far interfering with each other as to destroy the true tinkle.

From what has been stated, it is obvious that in cavities in which no fluid exists amphoric echo alone can be heard, and that in cavities which contain both air and fluid, amphoric echo or metallic tinkling will occur, according as the original sonorous impulses are clear and separate, and do or do not produce free vibration of the fluid. The vibrations calculated to give rise to metallic tinkling may be produced by the falling of drops of

[1] See Walshe, loc. cit., p. 158.

fluid from the sides of the pleura into the liquid contained in the pleural cavity, or by the formation and bursting of bubbles by the entrance of air into the pleura. If the sonorous impulses thus produced are sharp and distinct, following each other slowly, so as not to interfere with and destroy each other, and if at the same time they give rise to active vibration of the fluid, metallic tinkling is the result; if, though sharp and distinct, and following each other slowly, they do not produce agitation of the fluid, amphoric echo of a metallic quality occurs, but not metallic tinkling; and if again they follow in quick succession, so as to interfere with or obstruct each other, the resulting echo is of lower pitch and only slightly metallic in quality, however sharp and forcible the original sonorous impulses, and however great the agitation of the fluid. Thus it often happens that metallic tinkling may be traced passing gradually into amphoric buzzing echo, the conditions upon which the change depends being the gradual disappearance of the fluid, or an alteration in the character of the fluid, or an increased rapidity in the occurrence of the sonorous impulses, the vibrations thereby excited becoming gradually less distinct, and of a lower pitch.

Thus, to sum up my opinion respecting the nature of amphoric echo and metallic tinkling, I would say that they are both essentially the same, viz., the echoes of a sound reverberating in a large cavity with tense, smooth, sound-reflecting walls; that they may both occur without any direct communication between the cavity and the bronchi, though the existence of such a communication is by no means unusual, and increases the probability of their occurrence; that the former—amphoric echo—whether low-pitched or of an intensely high-pitched and metallic quality, may be produced by the rales and rhonchi in the chest, by the voice, and by the act of coughing, and may be excited by such noises, whether they are set up in the cavity itself or in a bronchus or other contiguous cavity, and whether the cavity itself be filled with air alone, or contain both air and fluid; but that the latter—metallic tinkling—cannot occur in the chest without the presence of a homogeneous fluid which admits of free vibration, and cannot be produced except by vibrations excited by an agency capable of imparting only a momentary impulse to that fluid.

In a practical point of view the facts stand thus: If in a case of pneumothorax, with a fistulous communication between the pleura and the bronchi, the fistulous opening be free and above the level of the fluid, amphoric echo is usually heard alone. When with the increase of effusion the fluid rises above the level of the opening, or when again, from change of posture, the opening is brought temporarily below the level of the fluid, amphoric echo is produced if the air bubbles freely and rapidly through the liquid, but metallic tinkling if the bubbles escape slowly and intermittently, one by one. Thus, if the opening be large and free, amphoric echo is heard, whereas if it be small and partially covered or obstructed by false membrane or other secretion, metallic tinkling is the result. The acts of coughing, speaking, or moving, by producing agitation of the fluid, may give rise to tinkling, under the conditions above mentioned, even when there is no opening into the bronchi; and so possibly may the echo of rales and rhonchi in adjacent bronchial tubes, providing they have force sufficient to occasion such agitation of the fluid as is requisite for the production of metallic tinkling. Sometimes when tinkling is not producible, either by coughing, speaking, or forcible breathing, it may be distinctly heard following succussion of the patient, or as the result of a sudden change from the recumbent to an erect posture. Indeed this is perhaps the most certain method of producing it, and its cause in such cases is doubtless the fall of drops of fluid from the upper part of the cavity into the liquid below, or the bursting of air bubbles resulting from the agitation of the fluid or from the passages of air through the liquid in the chest, or through a thin film only of the liquid which has been brought by change of posture over the fistulous opening.

Thus, then, amphoric echo indicates the existence of a cavity in the chest, and its most frequent source is an air-distended pleura; but occasionally it is met with as a consequence of an enormous excavation in the lungs. Metallic tinkling very rarely occurs except in connection with an air-distended pleura containing some amount of fluid, which is commonly serous or seropurulent rather than pure pus. Both the phenomena may, of course, arise as consequences of gangrene of the lungs, the bursting of a pneumonic abscess, or the giving way of an emphysematous air-vesicle; these occurrences, however, are so rare, that practically amphoric sounds, of whatever nature, may be regarded as almost certain signs of far-advanced tuberculous disease. They may be heard in every part of the chest, but usually are most audible about half way up posteriorly, or in the lateral regions. Amphoric echo may accompany both sounds of respiration, or may be limited to either, the limitation being generally to the inspiratory sound. Metallic tinkling may occur at any period of respiration; but when it is caused by the act of breathing, it is usually, if not always, produced during inspiration, though it is sometimes prolonged into the succeeding expiration. It never occurs where amphoric echo is not producible, and generally alternates with, or occurs occasionally during the continuance of that resonance. Oftentimes it is audible for a few days only; is rarely of long continuance, and seldom accompanies many successive respirations. Amphoric echo always indicates the existence of a cavity; metallic tinkling the presence of fluid in the cavity.[1]

[1] It has been objected to this supposition by Skoda, Davies, and others, that it is altogether opposed to physical laws, inasmuch as an opening which would admit air from a bronchus into the pleural cavity would also serve to let out the fluid from the pleura into the bronchus. But they appear to have overlooked an occurrence which often takes place, and which renders the explanation perfectly feasible, viz., that the opening is covered with a false membrane, which may serve as a valve to prevent the escape of fluid from the cavity, though it may not effectually bar the entrance to a few bubbles of air from the lungs.

PART II.

CHAPTER I.

IN the present division of our subject I purpose describing the various diseases of the respiratory organs, and shall endeavor to point out the general symptoms by which they are each accompanied, the principal morbid changes induced by their actions, the physical signs to which they give rise, and the treatment best calculated to give relief. By so doing, I hope to enable the student to interpret the signification of different physical signs under the varying conditions of disease, and thus to utilize the knowledge he has acquired of the principles on which the diagnosis of chest disease is founded.

PLEURISY.

Pleurisy, or inflammation of the investing membrane of the lung, is one of the commonest, and not unfrequently one of the most striking of serous inflammations. Characterized from the first by chilliness and fever, acute, catching pain on one side of the chest, oppression of the breathing, decubitus on the affected side, a short, dry cough, and a quick, hard pulse, it seems unlikely to be confounded with any other disorder. But its general features, though usually well marked, are by no means distinctive. They are one and all met with from time to time associated with other forms of thoracic disease, and the merest tyro in medicine must have observed how frequently one or more of them is absent, even in severe examples of the disease, and how those which are present are apt to mislead. Indeed there is no affection of the chest on which physical diagnosis throws more light than on pleurisy, and none in which its aid is more needed by the practitioner.

It may be well to premise, as bearing upon the physical

signs of the disorder, that the pleura is not only a serous membrane, but is also a shut sac, one portion of which is attached to the chest-walls, whilst the other is reflected over and attached to the lungs, forming their serous envelope. Hence, on the one hand, the morbid products met with in pleurisy are, serum, lymph, blood, and pus—the usual products of serous inflammation; and, on the other, there arise phenomena resulting from the attrition of the two inflamed and roughened surfaces of the pleura, as also, it may be, from their complete or partial adhesion as a consequence of plastic exudation on their surface. Further, it may be stated that, just as with inflammation of other serous membranes, the general symptoms vary greatly under different conditions of health and constitution. In one case the catching pain in the side and other well-marked features of the complaint leave little room for doubt as to the nature of the malady. In another the pain in the side, the difficulty of breathing, and other characteristic symptoms may be absent, notwithstanding the rapid progress of mischief. These differences are doubtless attributable to causes the nature of which it is difficult to fathom; but, apart from innate constitutional peculiarities, there is reason to believe that age, sex, and the previous habits of life exercise a powerful modifying influence. Be this as it may, the fact must be constantly borne in mind whilst considering the following details of the disease.

For convenience of description, the course of acute pleurisy may be divided into three stages, characterized by well-marked anatomical differences, and corresponding to striking alterations in the physical signs and general symptoms of the disease. These are—1st. The dry stage, or stage of congestion. 2d. The stage of effusion. 3d. The stage of absorption and resolution. Each of these varies in its general symptoms, and also, more or less, in the results which it produces, according to the age, strength, and constitutional peculiarities of the patient; but the structural changes and physical signs to which an attack of acute pleurisy *ordinarily* gives rise in a healthy person, will be seen by reference to the following description in which the different stages of the disease are each considered separately. The varieties of the complaint, as it is met with in the weakly and unhealthy, and the complications which often arise during its progress, require separate consideration.

Morbid Anatomy of Pleurisy.

First Stage, or Stage of Hyperæmia.

The pleural membrane is much drier than natural; its normal transparency is lessened; its smoothness impaired, and its surface covered with a delicate network of finely injected blood-vessels, or else studded with tufts of them in streaks or patches.

Second Stage, or Stage of Effusion.

The sub-pleural cellular tissue is injected and œdematous; the pleural surface is extremely vascular, as in the dry stage, and the pleural sac contains lymph and serum in varying proportions. Sometimes the effusion consists of pale straw-colored serum, with a few flakes only of albuminous matter floating in it; but more generally, in addition to this, a thick layer of lymph coats and roughens both surfaces of the pleura, either universally or in patches. If the congestion of the pleural membrane be excessive, a few blood-globules may also escape, and the serum may be more or less distinctly blood-tinged. The quantity of fluid poured out may vary from a few drachms to many ounces, or even to several pints; and according to its amount so is the effect produced on the lung. Whatever its quantity, compression of the lung-tissue is the natural consequence. Gravitating as it does to the lowest portion of the chest, it usually gives rise, if small in amount, to partial compression of the lower lobe, which is pushed upwards, backwards, and inwards towards the root of the lung; but if it be more copious, it produces more extensive compression of the lung as it rises higher and higher in the chest, and, if no adhesions already exist, forces the entire lung backwards against the spinal column. As soon as the pleural cavity is thoroughly filled, and the lung has been compressed into the smallest compass, the continued effusion of fluid causes the sac to stretch, and to encroach in all directions on the surrounding structures. The walls of the chest yield to the outward pressure, and enlargement of the entire side takes place, or bulging occurs at its inferior part; the diaphragm is thrust down, and a considerable prominence is often perceived in one or other of the hypochondria; the liver also is forced down; the mediastina are encroached upon, the heart is often displaced, and when effusion exists on the left side of the chest, is pushed over to the opposite side. After a time, if the effusion is persistent, the intercostal depressions are effaced, and, yielding to the outward pressure, the spaces may even bulge.

The lung, of course, is reduced in size, in proportion to the amount of pressure to which it is subjected. At first it is only partially emptied of air, and is simply tougher and less crepitant than a healthy lung, but it yields by degrees to the pressure of the fluid, and the gradual contraction of the plastic exudation-matter which coats it, and parting with all its aeriform contents lies flattened against the spine, a small, carnified mass, wrinkled perhaps on its surface, tough and leathery to the touch, and of density sufficient to sink instantly in water. In this case it is perfectly non-crepitant, but may be healthy in structure, and capable of re-expansion. On section it exhibits a smooth homogeneous surface, of a slaty, gray color.

Third Stage, or Stage of Absorption and Resolution.

The sero-plastic portion of the exudation is gradually absorbed, and ultimately the two surfaces of the pleura, covered with plastic material, come into apposition. Adhesion then takes place between them at one or more points, by means of the interposed lymph. Vessels are developed in this lymph, which rapidly becomes organized, and forms either a dense false membrane of fibro-cellular structure, or else mere fibro-cellular bands. In the one case the two surfaces of the pleura are firmly and universally agglutinated; in the other they are connected at a few points only.

If the effusion has been copious, and more especially if it has been of long duration, the lung is often so much compressed, and so firmly bound down by adhesions, that it is incapable of re-expansion, and becomes permanently impervious to air. In this case, as the fluid is absorbed the chest-walls necessarily fall in under the effect of atmospheric pressure, and considerable distortion of the chest ensues. The affected side is diminished in every diameter, and its surface becomes irregular and uneven: in front it is often concave; the intercostal spaces decrease in width; the ribs undergo distortion, the lower ribs especially approximating and overlapping each other; the dorsal spine curves laterally, the convexity being sometimes towards the diseased side, but more generally towards the sound side, whilst a curve in the opposite direction occurs in the lumbar portion of the column; the intercostal muscles waste, in consequence of the entire loss of motion in the chest-walls on the affected side; the diaphragm and the subjacent viscera are often drawn upwards, and sometimes the heart is more or less displaced. These symptoms may occur when the exudation consists of lymph and serum, but they receive their most striking development in cases in which absorption fails to take place; and the fluid, whether sero-purulent or purulent, is drawn off artificially, or escapes from the pleural sac spontaneously through a fistulous opening.

Physical Signs of Pleurisy.

First Stage, or Stage of Hyperæmia.

Inspection shows a diminution in the respiratory movements, both of elevation and expansion, in the affected side, consequent on the pain which is felt on inspiration.

Mensuration gives similar results.

Palpation yields only negative results.

Percussion-sound not materially altered.

Auscultation sometimes makes us aware of a slight, interrupted grazing friction-sound on the affected side, referable to the dryness of the membrane and the congested state of the vessels; more commonly, it only tells us of a jerking unevenness in the rhythm of respiration, and of weakness or indistinctness of the respiratory murmur, consequent on the imperfect and irregular expansion of the lung. There is exaggerated respiratory murmur on the unaffected side.

Second Stage, or Stage of Effusion.

Inspection detects increased motion on the unaffected side of the chest and great deficiency in the respiratory movements, especially of the expansion-movement, on the affected side; after a time, perfect immobility of the chest-walls is perceived, with general enlargement of the side, and flattening or total obliteration of the intercostal spaces. Sometimes no general enlargement of the side is noted, but there is bulging of the inferior portion.

Palpation informs us of great diminution in the intensity of vocal fremitus if the amount of effusion be small, and of its absence if the effusion be copious; sometimes, though rarely, of a vibration due to friction of the roughened surfaces of the pleura; frequently, after a time, of displacement of the heart, and of the presence of fluid in the pleural cavity, as indicated by fluctuation in the intercostal spaces. The surface of the chest on the affected side is often œdematous, and is felt to be unnaturally smooth.

Mensuration proves the existence of deficiency in the movements of the chest, especially in that of expansion, and after a time of enlargement of the affected side, both in its circumference and in its antero posterior diameter. The enlargement is most evident over the false ribs.

Percussion ascertains the existence of dulness, and an increased sense of resistance, most marked in the inferior portions of the chest, and generally terminating rather abruptly above. The line which marks the area of dulness is not altered by respiration, but sometimes may be made to shift by changing the patient's posture. Loud, hollow resonance, of a shallow character, is often met with in the upper part of the affected side, and sometimes, though rarely, a *quasi* cracked-pot sound may be detected there.[1] After a time, as effusion progresses, the entire side becomes dull on percussion, and the area of dulness extends far beyond the natural limits of the lung.

Auscultation, at an early stage, may discover pleural friction-sound of a rubbing, creaking, or grating character, which ordinarily disappears, with the increase of effusion, but may possibly continue throughout the attack; total absence of the respiratory murmur, where the effusion is most abundant; weakness, or diminution of the sound, when the fluid is less copious; harshness or hollowness of breathing above the level of the effusion. In the interscapular region, at the root of the main bronchi, the respiration remains loud, hollow, and of a blowing character. The vocal resonance, in the earlier stages, is greatly increased, and in the posterior and lateral regions, especially at the root of the lung, it is often, but not invariably ægophonic. In the latter stages, when the lung is much compressed, and air ceases to enter, the respiratory sound and vocal resonance are altogether absent, and ægophony is never heard. If the heart is thrust out of its natural position, it will be heard pulsating to the right of the sternum, or possibly far beyond its normal bounds to the left. The respiration in the healthy lung is exaggerated, or compensatory.

Third Stage, or Stage of Absorption and Resolution.

Inspection informs us that the enlargement of the affected side is disappearing; that the intercostal spaces are regaining their normal condition, and the mobility of the chest-walls is returning.

Palpation gives us notice of returning vocal vibration and friction fremitus.

[1] I have met with two or three marked instances of this.

Mensuration proves a gradual return to the normal admeasurements of the chest.

Percussion gradually, though slowly, yields a clearer sound, the dulness being most persistent in the inferior portions of the chest, where the compression of the lung, and the accumulation of solid plastic material is often so great, that the percussion-note never regains its normal clearness.

Auscultation.—The respiratory sounds are again heard, at first weak and distant; then possibly harsh; subsequently of a normal character. Sometimes, as absorption progresses, and the two surfaces of the pleura come again into apposition, a friction-sound reappears for a time, but ceases as soon as adhesion takes place. Occasionally pseudo râles are caused by serous infiltration of the subpleural cellular tissue. The vocal resonance, if at any time ægophonic, speedily loses this character, and becomes, at first, simply bronchophonic, and ultimately normal. The heart, if at first displaced by the effused fluid, regains its normal position, and is felt and heard pulsating under the left nipple. If, as in certain instances, the lung remains permanently impervious to air, there is entire loss of motion on the affected side; the respiratory sounds and the vocal resonance do not return, and the dulness on percussion is persistent. If, on the other hand, a portion of the lung—usually the upper portion—becomes partially pervious to air, percussion may elicit a hollow, but shallow resonance over it; whilst, in the same situation, the vocal resonance will be loud, and the respiration coarse and blowing.

These, then, are the morbid changes and physical signs which an uncomplicated attack of pleurisy ordinarily produces, whether in an acute or chronic form. A few remarks, however, must be added, before we pass on to the varieties of the disease, and the complications often met with during its course.

Reference has been made to the friction of the two inflamed and roughened surfaces of the pleura; and it might be inferred, from the description of the morbid appearances, that friction-fremitus and friction-sound must be felt and heard in most cases of pleurisy. This, however, is not so. Experience proves that friction-sound and friction-fremitus are exceptional phenomena in pleurisy, owing, probably, in the early stage of the complaint, to the rapidity with which effusion takes place, and to our failing to make our examination during the continuance of friction; and in the later stage—the stage of absorption—partly to the solidification the lung has undergone, whereby expansion of the chest and consequent friction of the two surfaces of the pleura is in great measure prevented, and partly also to the rapidity with which the thickly lymph-coated pleural surfaces become agglutinated. And as more motion of the lung takes place during forced than during gentle breathing, it happens sometimes that, although friction-sound is inaudible during ordinary breathing, it may be heard when the patient is directed to take a deep breath.

Again, it has been stated, that the line which marks the area of dulness may be *sometimes* made to shift its position by changing the patient's posture. Now, this is strictly and literally correct; but many authorities have asserted broadly, that, provided no adhesions exist between the two surfaces of the pleura, and that the quantity of the fluid in the pleural sac be not sufficient to compress the entire lung, the position of the dulness or resonance on percussion will vary according to the attitude in which the patient is placed; the lowest and most depending part—the part to which fluid gravitates—being always dull; and the part which, for the time being, happens to be uppermost, being always resonant on percussion. And the legitimate inference has been drawn from this statement, that when dulness is met with which is not attributable to circumscribed pleurisy, and which, nevertheless, does not shift its position with the varying attitude of the patient, it cannot be due to pleurisy, but must be referable to an inter-thoracic tumor, or to a lung solidified by pneumonic, tuberculous or other infiltration, which does not admit of change of position, but necessarily occupies the same part of the chest. Nothing, however, can be more erroneous. True it is that percussion-dulness referable to solidification of the lung or to the presence of an intrathoracic tumor, does not shift its position with the varying posture of the patient; and that when the seat of dulness is found to vary with change of posture, the dulness must be attributable to the presence of fluid. But it is not true, as the unreserved statement above referred to would lead students to believe, that when no adhesions exist to bind down the lung, or to circumscribe the effusion, the line of demarcation between dulness and normal resonance on percussion will always be found to shift its position with the varying posture of the patient. Case after case has come under my observation in St. George's Hospital, in which the contrary has been noted; indeed, in my experience, it seldom is observed to any considerable extent, except when the examination is made within a few hours after the effusion has occurred, and before the lung has suffered from compression. Even under these conditions it is not always well marked; for when, as commonly happens, the products of inflammation consist not only of serum, but of solid albuminous and fibrinous matters in large quantity, the parietes of the chest, covered with the soft, pulpy, inelastic material, yield a dull sound on percussion, in whatever attitude the patient may be placed. All that can be fairly stated is, that the percussion-note is not so dull as prior to the change in the posture of the patient; it rarely yields the good, clear resonance of healthy lung-tissue, nor that which is observed under similar circumstances in hydrothorax.

Another point of some importance relative to the dulness on percussion in pleurisy is the rapidity of its occurrence. From the moment that effusion commences the percussion-sound in the lower part of the chest becomes comparatively dull; and as serous fluid is often poured out rapidly and early in the attack, pleuritic dulness may be clearly established within a few hours from the onset of the disease. I have known one side of the chest completely dull as the result of pleuritic effusion within thirty-six hours after the first accession of pain in the side—a phenomenon rarely if ever witnessed in pneumonia, which produces hepatization of the lung slowly and gradually, and is, therefore, comparatively, a long time in occasioning extensive dulness on percussion.

The intensity and extent of the dulness are other points of considerable interest, as bearing on the diagnosis between pleurisy and pneumonia. The effusion of pleurisy may push aside the heart, encroach upon the mediastinum, and extend itself under the whole of the sternum, which, thereupon, yields a dull sound on percussion; whereas a consolidated lung never passes beyond its natural boundary, and therefore does not give rise to dulness over more than one-half of the sternum. Moreover, the dulness of pleuritic effusion is hardly to be confounded with the dulness produced by pneumonic or other solidification of the lung. It is far duller—a shorter, flatter sound, resembling the short, dead tap which results from percussing the thigh; and the resistance offered to the percussing finger is beyond comparison greater than that which is met with in pulmonary consolidation from whatever cause arising.

Again, reference has been made to enlargement of the side as one of the later symptoms of pleuritic effusion. Hence, without further explanation, it might be inferred that the chest-walls will only yield to long-continued outward pressure of the fluid, and that enlargement, therefore, can only take place in chronic cases after effusion has existed for a considerable space of time. In most cases, undoubtedly, this inference would be correct; but instances are met with in which distinct enlargement of the side, as shown by admeasurement, occurs at a very early period of the attack. I have observed more or less local bulging in the lower part of the chest before the end of the third day of the attack; and cases are on record in which the chest has been considerably enlarged before the end of the fifth day. In my experience, however, this early enlargement has occurred only in childhood, when the chest-walls are unusually yielding; and in those children only in whom the costal pleura has suffered severely, and in whom the intercostal muscles being paralyzed, and the subserous structures softened and infiltrated with serum, the chest-walls yield to the outward pressure with unusual readiness. It may possibly occur at an equally early period in adults, but I have not chanced to meet with it.

So again it has been stated that the intercostal spaces are apt to be effaced under the outward pressure of the fluid; and the statement, if unexplained, might lead to the conclusion that this is the common, if not the invariable result of pleuritic effusion. But it should be clearly understood that this is not the fact. Commonly, indeed, a filling up of the intercostal spaces is observed whenever effusion into the pleural sac is considerable in amount; but instances not unfrequently occur in which the side is considerably enlarged, and in which, nevertheless, there is neither bulging of the interspaces, nor very marked deficiency in the action of the intercostal muscles during inspiration. The difference is explicable, I believe, by reference to the state of the costal pleura and the subjacent structures. When that portion of the pleural membrane is intensely inflamed, the subjacent structures become infiltrated with serum, and otherwise involved in the mischief; their contractile power is lost, and their resistance to the outward pressure of the fluid is greatly lessened. On the other hand, when the inflammation is less intense, or when the chief fury of the attack has been directed on the pulmonary pleura, the intercostal muscles and adjacent structures are not paralyzed, as in the former case, but continue in action, and offer great resistance to obliteration of the interspaces. The idea formerly entertained, that bulging occurs only when the pleura is distended with pus, has been clearly proved inconsistent with fact. It occurs more frequently under these circumstances than when the effusion consists of serum, for the simple reason that the costal pleura and subjacent structures are commonly involved in such cases, but there is no necessary connection between bulging and empyema.

The opposite condition of the chest, or that produced by retraction or falling in of its walls, consequent on compression of the lung, absorption of the fluid, and contraction of the sero-plastic exudation, also requires further elucidation. The difference observed in the size of the two sides is referable not only to the falling in of the diseased side, which becomes narrower and smaller than the sound side, but to positive enlargement of the sound side, consequent on actual compensatory hypertrophy of the sound lung, on which all the work of the body devolves.

Again, it should be understood that retraction of the chest-walls, and the distortion which results from it, are not in every instance persistent phenomena. The lung, as already stated, is compressed, but not necessarily diseased. Consequently, when absorption of the fluid takes place, its structure will admit of inflation; and if the plastic exudation which surrounds it and the adhesions which bind it down are not so firm as to render its expansion impossible, they may gradually yield to the inspiratory efforts, and the lung may again expand. In this case, the chest will enlarge under the influence of the outward pressure which inspiration exerts, and the distortion of the chest will gradually pass off. Such instances are naturally rare, and although I have seen many in which considerable recovery has taken place, I have never met with one in which complete re-expansion of the side has occurred; but a remarkable case in point has been put on record by Sir Thomas Watson, which serves to establish the possibility of complete recovery.[1]

Thus far we have discussed the phenomena usually attendant upon an uncomplicated attack of pleurisy occurring in previously healthy persons of vigorous constitution. In such persons the exudation-matter is commonly very plastic, and of high vitality, readily becoming organized; so much so that anastomosing red lines have been found traversing the newly-formed false membranes, even when death has taken place after only a few days' illness. But, unfortunately, the disease is prone to arise in the old, the scrofulous, the intemperate, and the cachectic, in whom the effused lymph is almost necessarily of low vitality, and is found curdy, ill-concocted, and perfectly unorganized even after months of pleuritic suffering. Under these circumstances, the effused matters often tend to become puriform, and thus the case becomes seriously complicated. Either the chest remains permanently enlarged, or the purulent fluid which it contains seeks to escape from the cavity in which it is pent up, by perforating the pleura and discharging itself through the lungs, the thoracic walls, or the diaphragm. If, which rarely happens, the pus discharges itself through the diaphragm, peritonitis is the probable result; if it perforates the lung, symptoms of pneumonia will probably occur, and the moment of complete perforation will be marked by severe paroxysms of cough, resulting in a copious

[1] See "Watson's Lectures," ed. 1, vol. ii, p. 126.

discharge of pus, or sero-purulent matter through the bronchi; but if it perforates the costal pleura, and finds an outlet through the thoracic walls, one or more soft, inelastic tumors, in which fluctuation is perceptible to the touch, will be perceived on some portion of the thoracic parietes before the skin gives way. In either of the last two cases air will find its way into the chest, as soon as perforation occurs, and pneumothorax will be added to the existing mischief. The physical signs of this complication must be reserved for separate consideration.

Another point which requires notice in regard to empyema is, that when the pus makes its way externally, and perforates the costal pleura, the subcutaneous swellings to which its escape gives rise, may increase and decrease in size with expiration and inspiration, and if situated on the left side of the chest, may even pulsate synchronously with the heart. I have witnessed both these phenomena on two occasions, and instances have been put on record by other observers. They are of no practical importance; but the latter acquires an interest which would not otherwise attach to it, from the fact that the student or careless practitioner might possibly mistake its real nature, and refer the pulsation to aneurism. The absence of aneurismal thrill and murmur, and of symptoms indicative of pressure on the trachea, the œsophagus, the larger veins, and the recurrent nerve, and further the presence of the physical signs and general symptoms of empyema, and not unfrequently the situation of the pulsation, will suffice to establish its true character.

Sometimes, again, when empyema exists on the left side of the chest, pulsation may be perceived in the infra-clavicular region, and over the arch of the aorta, even when no circumscribed tumor exists. I have noted this transmitted impulse to a slight extent in several instances, but never to a degree at all likely to mislead even a student; therefore I should not have considered it worthy of mention, had not Dr. Walshe referred to two cases in which, while the side was generally enlarged, gentle local bulging was manifest in the site of the pulsation, which was forcible enough "to jog the head at the end of the stethoscope," and in which, therefore, the question of aneurism might possibly have arisen. Such cases must be exceedingly rare; for, on inquiry amongst my friends, I cannot hear of any one who has met with an instance in point, nor can I find a record of any others than those above alluded to. Practically, too, they cannot be very important; for the presence of well-marked symptoms of empyema, backed by the feebleness of the heart's sounds over the seat of pulsation, by the absence of aneurismal thrill and murmur, by the equality in the radial pulse on the two sides, and by the absence of symptoms denoting pressure on the spine, the trachea, the œsophagus, the larger veins, and the recurrent nerve, ought at once to remove all doubt as to the real nature of the mischief.

Other modifications in the physical signs of pleurisy are occasioned by the attack being set up in persons whose lungs have previously contracted adhesions, which keep them forcibly in apposition with the costal pleura and the parietes of the chest. In these cases the physical signs of the disease are so much altered, that he who fails to bear the fact in mind will be often misled in his diagnosis. Thus, when the lung is universally adherent to the posterior walls of the chest, it is almost impossible for any amount of effusion to stop the vocal resonance, or annihilate the respiratory sounds over the surface with which the lung is in contact, or to cause entire dulness on percussion. Even the vocal fremitus will not cease until a large amount of fluid has been poured out. A striking case in point has recently occurred to me in St. George's Hospital. George Godfrey, aged twelve, was admitted under my care into the Hope Ward, suffering from empyema of the left side of sixteen days' duration. The whole side was considerably enlarged, and measured an inch and a half in circumference more than the sound side; it was perfectly dull on percussion anteriorly, but posteriorly did not yield such a thoroughly dull sound; and over the inner portion of the whole posterior surface of the chest diffused blowing respiration was audible, vocal fremitus could be distinguished, and vocal resonance was everywhere present, and of a hollow bronchophonic character. The pus was evacuated by an artificial opening; and enormous retraction and distortion of the chest, with curvature of the spine, ensued.

Adhesions give rise to another difficulty in the diagnosis of the disease. When the two surfaces of the pleura have become adherent at various parts by means of adventitious membrane, the effused fluid may be circumscribed, or, in other words, may not be contained in the general cavity of the pleura, but in a sac or sacs formed by the adventitious membrane. These sacs may be independent of, or may communicate with each other; but in either case the fluid is confined within their boundaries, and does not shift its seat as fluid ordinarily does, with change of the patient's posture. Hence there may be perfect dulness on percussion over the seat of effusion, with entire absence of vocal fremitus, vocal resonance, and respiration; and there may be even local bulging; and these symptoms, though due to pleurisy, may remain unaffected by change of posture. Thus, these circumscribed collections of matter may simulate a solid tumor in the pleura or in the lung, and may lead the careless and inexperienced to an erroneous diagnosis. But if due caution be observed, a mistake of this kind can seldom arise; for on the confines of the sac or sacs where the lung, compressed by the fluid, is adherent to the walls of the chest, there is usually found some dulness on percussion, and diffused hollow blowing respiration, with considerable increase of vocal resonance,—conditions which are rarely met with in cases of intrathoracic tumor, unless the adjacent lung be solidified by pneumonia, or infiltrated by tubercle or other matter, in either of which cases there would probably be other evidence as to the nature of the mischief. Moreover, in these cases of pleuritic effusion, there is entire absence of symptoms denoting centripetal pressure—of pressure on the spine, the trachea, and larger bronchi, the œsophagus, the larger veins, and the recurrent nerve—and yet these symptoms are common attendants on intrathoracic tumors.

Other modifications in the physical signs attendant upon pleurisy are occasioned by the occurrence of pleuro-pneumonia, or in other words by the coexistence of pleurisy and pneumonia. It might naturally be supposed that inflammation affecting the investing membrane of the lung would be likely to spread to the lung-structure, and, conversely, that inflammation affecting the tissue of the lung would probably extend to its investing membrane. And observation in this instance is consistent with hypothesis; for it is found that the two inflammations frequently coexist, though the one is apt to predominate over the other. Pleurisy, indeed, is usually met with unassociated with pneumonia; and though inflammation of the lung may occur coincidently with pleurisy, it is rarely, very rarely, set up in sequel of that disease. But it is otherwise when pneumonia is the primary disorder; for acute inflammation of the tissue of the lung rarely runs its course without the supervention of pleurisy. Hence arise modifications in the physical signs of pleurisy, which vary, according to the condition of the lung at the time the pleuritic symptoms commence, and to the extent to which effusion had proceeded before the lung became implicated. Ordinarily, if the lung has undergone compression from pleuritic effusion before inflammation of its tissues commences, the supervention of that disease does not seriously modify the physical signs of pleurisy; but if, on the other hand, the lung is solidified by pneumonic inflammation before the commencement of pleuritic effusion, it will resist to a great extent the compression of the fluid, and will continue to occupy the larger part of the pleural cavity, whilst at the same time the larger bronchi will remain pervious, and will be surrounded by a good sound-reflecting and sound-conducting structure. The result of this will be loud and widely diffused tubular breathing, such as is never met with in simple pleurisy; a more than commonly intense and persistent vocal fremitus and vocal resonance; a more than ordinarily pronounced and persistent ægophony; and a very rapid extension of pleuritic dulness on percussion;—signs which will receive a development commensurate with the rapidity with which the two diseases respectively run their course.

It should be added, that acute pleurisy is often associated with, if it be not the cause of, circumscribed pericarditis and peritonitis, leading to limited plastic exudation; and that in the former case, especially, the unwary student might mistake the rough sound of pericardial friction for the friction-sound of pleurisy. The cardiac rhythm of the friction-sound in the one case, and its respiratory rhythm in the other, which can be readily determined by directing the patient to hold his breath, ought, in most instances, to resolve any doubt on the subject.

Having now completed the investigation of the physical signs attendant on pleurisy, we will pass on to a consideration of the general symptoms, and then discuss the indications for treatment.

Acute pleurisy is ushered in by symptoms resembling those which mark the accession of pneumonia and pericarditis. Chilliness sometimes amounting to actual shivering, followed by fever, with heat of skin, acute catching pain in the side, dyspnœa, a short, dry cough, a hard pulse, and a difficulty or impossibility of lying on the affected side, are the symptoms usually observed at the outset of pleurisy. But they are not all present in every instance, and some of them at least are by no means characteristic.

Shivering is the symptom which is least constantly present, and possesses perhaps least of a distinctive character. Seldom equal in severity to the shivering which marks the accession of pneumonia, it frequently amounts to little more than mere chilliness; and in some severe attacks even the sense of chilliness is altogether absent. It usually precedes the "stitch in the side" by some hours or even days, but occasionally it is not observed until a later period of the attack, or after the characteristic pain has been developed. And as with the shivering, so with the fever which follows it. Sometimes, nay, generally, it is tolerably well marked, but it is rarely characterized by much intensity. Thus the skin, though hot, is often perspiring, and is seldom dry and burning, as in pneumonia; indeed the temperature rarely rises to 103°, and more commonly ranges between 100° and 102° 5″.

The pain in the side, which is frequently called "a stitch in the side," is described as of a catching or stabbing character, and is undoubtedly the most striking and characteristic symptom of pleurisy. Varying in severity from the slightest "stitch" to a feeling of positive agony, it generally shows itself at the very outset of the attack, and though often fugitive at first, very soon becomes constant and localized. Most commonly it is felt on a level with or immediately beneath one or other of the breasts, at a spot corresponding to the antero-lateral attachments of the diaphragm, but sometimes it is met with under the scapula, in the axilla, or beneath the clavicle; sometimes along the borders of the false ribs; sometimes even in the abdomen itself, especially in the right hypochondrium; whilst at others, it fixes upon the non-inflamed side, or extends over the whole of the inflamed side. This last, however, is an exceptional occurrence; for usually, no matter how extensive the area of inflammation, the pain is limited to a particular spot, and shows no disposition to shift its quarters. If the strain of the disease has fallen principally on the visceral pleura the pain is usually moderate in degree, and though felt on deep inspiration, on coughing, or on sneezing, is scarcely increased by pressure or percussion; but if the costal pleura is much inflamed the pain is generally acute, is greatly augmented even by gentle intercostal pressure and by percussion, by lying on the affected side, by various movements of the body, and by cough; and is so greatly aggravated by inspiration, that the patient feels as if he were being stabbed each time that he attempts to inspire freely. Sometimes cases will run their course unaccompanied by pain from first to last; sometimes severe pain is persistent throughout the attack, or proves irregularly intermittent, but more commonly the pain, whatever its character, is found after a time to remit in severity, and ultimately to cease altogether.

The cause of pleuritic pain has long been a fertile topic of discussion and conjecture. Some persons have attributed it to the friction encountered by the inflamed and roughened surfaces of the pleura; and in support of their opinion, they have referred to the frequency of its occurrence at the lower portion of the chest, where thoracic motion is greatest, and where, consequently, pleuritic friction is most intense. But cases sometimes present themselves in which pleurisy runs through its various stages, accompanied by intense pleuritic friction, and in which, nevertheless, no pain is perceived from first to last. Sometimes, again, the most intense pain exists, and yet the physical signs prove beyond dispute that the two surfaces of the pleura are separated by fluid, and that no attrition of the costal and pulmonary pleura can take place. I have met with several examples of both these phenomena. It is obvious, therefore, that some other explanation is needed; and having regard to the facts, that the pain is generally aggravated by pressure, that it is sometimes intermittent, and that it is not always felt at the seat of inflammation, it seems probable that, in some instances at least, it is purely neuralgic, and in others a true intercostal myalgia, as suggested by Dr. Inman.

Hurried breathing, followed by dyspnœa, is, perhaps, the most constant, though not the most distinctive feature of acute pleurisy. At the outset of the disease the respiration is accelerated, in consequence of the pain which inspiration excites—pain which induces the patient to endeavor to compensate by the frequency of the breathing for its shallow and imperfect character. But the breathing is not only short and hurried, it is jerking or irregularly interrupted; for the catching pain does not always occur at precisely the same period of each respiratory act, and consequently the thoracic movements are of unequal duration, and of a jerking or interrupted rhythm. Thus, for some time after the invasion of the disease, the breathing, though short and quick and jerking, hardly deserves the title of dyspnœa. Its increased frequency is due simply to the pain which a full inspiration excites, and not to any material encroachment on the breathing apparatus. But when effusion has occurred, the lung is encroached upon, and real difficulty of breathing sets in. If the pain persists, the respiratory movements will still continue jerking in rhythm, and will be more hurried even than before; whereas if, as often happens, the pain subsides, or decreases greatly after effusion has taken place, the breathing, though still hurried, will be no longer jerking or interrupted. Sometimes the dyspnœa is most distressing to the patient, and continues so throughout the attack; sometimes it is severe at the outset of the complaint but soon diminishes in intensity; sometimes it amounts to little more than shortness of breath, which annoys, but does not oppress; and cases are not wanting in which, though the respiration is hurried, the patient makes no complaint of difficulty of breathing, and seems hardly conscious of its existence. Indeed, instances must have occurred to many practitioners, in which it has been difficult to persuade the patient that serious mischief existed in the chest. In some of these cases the other subjective symptoms are also absent, and then the disease is said to be latent. Of course the frequency of the respiration, when the pain has subsided, is proportioned, *cæteris paribus*, to the amount and rapidity of the effusion, and to the condition of the opposite side of the chest. The larger the quantity of the fluid poured out, and the greater the rapidity of its outpouring, the more intense will be the dyspnœa. So again, if the other lung happens to be inflamed or otherwise diseased, the equilibrium between the quantity of air and the quantity of blood will be more seriously disturbed, and the difficulty of breathing will be greater than if such disease had not existed. Seldom, however, is the breathing very hurried, nor is the ratio naturally existing between the pulse and the breathing so much perverted as in pneumonia. In pleurisy the pulse rarely bears a smaller ratio to the breathing than 3 to 1, the respiration being 40 when the pulse is 120 per minute, but it often happens in pneumonia that the respirations will number 60 or 64 when the pulse is only 100 or 108. Indeed, putting aside every other consideration, the amount of perversion in the ratio naturally subsisting between the pulse and the respiration will almost serve to distinguish pleurisy from pneumonia. Dr. Walshe asserts that the ratio naturally subsisting between the pulse and the breathing is more perverted in the "sitting than in the lying posture;" and, in proof of this, he cites a case, in which the ratio was as 3.39 to 1 in a recumbent posture, and as 2.93 to 1 in an erect position. My own experience, however, does not accord with his in this respect; and I cannot help thinking that the result he obtained must be regarded as exceptional. Possibly it may have been due to the observation being made soon after the patient had assumed an erect posture, and before he had recovered from the disturbance of his respiration consequent on the change of position. Certain it is that in eight cases which I examined specially with a view to this inquiry, I obtained a similar result when the examination was made soon after a change of posture; whereas in six out of the eight the result was totally different when the examination was deferred until after the patient had been kept in an upright posture for half an hour before the observations were made. In the six cases referred to, the ratio was less perverted in an erect than in a recumbent posture.

Cough is another symptom which usually accompanies acute pleurisy. Sometimes occurring as a prominent feature of the complaint, and sometimes altogether absent, even when pleuritic inflammation is intense and effusion into the pleura extensive, it is not to be depended on as a diagnostic sign. When it does occur, it is a short, half-stifled cough, dry, or accompanied only by a scanty mucous expectoration. The student, however, must not be misled by the existence of copious frothy or rusty-colored sputa. He must bear in mind that copious expectoration does not militate against the presence of pleurisy; it only shows that if pleurisy exists, it is complicated by bronchitis or pneumonia, the signs of which will be discoverable on examination, in addition to the signs of pleurisy.

The position of the patient is a point of some importance in the diagnosis of pleurisy. Some persons have asserted that the patient lies on the affected side, some that he lies only on the sound side, whilst others have maintained that he may lie indifferently on either side, but commonly reclines on his back or in a diagonal posture between the back and the side, with his body inclined towards the affected side. The truth appears to be that, at the outset of the complaint, the pain in the side, which is aggravated by pressure, prevents the patient lying on the affected side, and he therefore sits up in bed, or else lies on the sound side, or on his back with his body inclined to the sound side. After effusion has occurred, other influences come into operation, and lead him to alter his position. The fluid in the pleura would press upon and interfere with the action of the sound side, and so would cause dyspnœa if he continued to lie on that side, and as the pain usually ceases, or else diminishes in intensity as soon as effusion has taken place, he has no longer any reason for not lying on the affected side. He therefore assumes the posture which allows the sound side greatest freedom of action, and gives greatest relief to the breathing. Thus he sits up in bed, or lies on the affected side, or reclines on his back with his head somewhat raised and his body inclined to the side on which effusion exists. This, however, though generally the case, must not be regarded as an invariable rule. Instances are met with in which the patient is unable to lie down throughout the attack; others in which, from the very outset of the attack, he lies on the affected side, as the easiest position; and others again in which, through the pleura is full of fluid, he reclines indifferently on the back or on either side without augmenting his distress or increasing the difficulty of breathing. Examples in point are not very common, but they are met with often enough to prove that the decubitus in pleurisy, though sufficiently characteristic, is not to be depended on as a sign of the disease.

The general febrile symptoms attendant upon acute pleurisy do not present any characteristic features. The face is seldom flushed as in pneumonia, the skin is hot and moist, rarely dry and burning; the pulse is quick and often hard, contrasting forcibly with the soft pulse of pneumonia; the urine is high-colored, usually clear, of high specific gravity, and occasionally albuminous, as in other acute affections,[1] and there is seldom much delirium. The relative proportion of fibrine in the blood is augmented, as in other inflammatory disorders, but seldom to the same extent as in pneumonia.[2]

These, then, are the general symptoms of acute pleurisy. They may all be present in a marked degree, or one or more of them may be absent. In the former case it is difficult to mistake the nature of the complaint, even without reference to the physical signs; in the latter the aid of the physical signs is essential to a correct diagnosis. Nay more, cases are sometimes met with in which the pain and all the other subjective symptoms are wanting, and in which it is impossible without the assistance which physical diagnosis affords, to form even a conjecture as to the nature of the patient's malady. Several instances of this sort, in which patients have been admitted with a chest full of fluid, and in which the existence of pleurisy had been entirely overlooked, have come before me in the wards of St. George's Hospital.

The course of pleurisy may be acute or chronic. Ordinarily, when pleuritic inflammation is set up it runs on rapidly to the production of copious exudation into the pleural cavity. But there is reason to believe that in some instances, at least, its progress may be at once arrested if vigorous and appropriate measures are adopted early. Thus, a few hours may witness the access of inflammation, the occurrence of plastic exudation on some portion of the pleura, and the entire subjugation of the disease, with adhesion of the two surfaces of the membrane. Several cases which have presented symptoms indicative of this train of events have occurred under my own observation, and some have been put on record by other practitioners. But these instances cannot be regarded as types of an ordinary attack of pleurisy. More generally, a considerable amount of fluid is effused, the lung is pressed upon and the physical signs and general symptoms already described as accompanying pleurisy are met with in a more or less fully developed form. In such cases, if appropriate treatment is adopted, the morbid action is gradually controlled, the fluid portions of the exudation are absorbed, the intercostal spaces cease to bulge, the two surfaces of the pleura which the fluid had separated again come into apposition, and become more or less generally adherent by means of the interposed lymph, the lung re-expands as the pressure of the fluid

[1] See a paper by Dr. Parkes in "Med. Times and Gazette," for January 1st, 1859.

[2] See Simon's "Chemistry."

is removed, and for all practical purposes, the chest regains its former condition. Neither spirometry nor external measurement can after a time detect any enlargement of the thorax, or any interference with the action of the lung resulting from the pleuritic seizure. And whilst these favorable changes are taking place in the physical condition of the chest, a corresponding improvement is observed in the condition of the patient. The feverish symptoms decrease and ultimately disappear; the cough and dyspnœa gradually subside; the patient gets to lie indifferently on either side, he gains flesh, and his general health and strength improve.

But in other cases, although no untoward complications arise, recovery is not so rapid or so complete. The feverish symptoms gradually subside, absorption of the fluid takes place, but the parts do not recover their former condition. The lung remains compressed by the adventitious membrane with which it has been coated during the progress of inflammation, or is bound down by firm adhesions. In either case it is incapable of expanding as the fluid is absorbed, and the necessary consequence is, that the other lung becomes hypertrophied or emphysematous, and pushes over towards the affected side of the chest, the diaphragm on that side is drawn up, the clavicle is depressed, the chest-walls fall in, the thorax becomes shrunken and distorted, and the patient, deprived of the use of one lung, remains permanently short-breathed. If, as usually happens, the false membranes which form the obstacle to the expansion of the lung are thick and firm, the lung never can recover from the effect of the attack, and the distortion of the chest is irremediable. If the membranes are not so unyielding, the lung may gradually expand again, under the dilating influence of inspiration, and when it comes in contact with the walls of the chest, adhesion of the two surfaces of the pleura may take place. Sometimes, indeed, as proved by a case recorded by Sir Thomas Watson,[1] complete recovery of the parts may thus gradually occur; but more commonly, the density of the adventitious membrane is sufficient to prevent the full inflation of the lung, and though the side may gradually enlarge, and the distortion disappear to a considerable extent, yet some evidence of former mischief will last for life. This has been the case in every instance which has fallen under my own observation.

In some instances, again, our utmost efforts fail to arrest effusion and produce absorption. The febrile symptoms become mitigated, but day by day the side enlarges, the intercostal

[1] Loc. cit., pp. 106–7.

spaces gradually become obliterated, or even bulge, the lung on the affected side is rendered impervious to air, and the fluid, encroaching upon the mediastinum, interferes with the action of the healthy side. In these chronic cases the patient becomes pale and cachectic in appearance; his lips are livid, his face is puffy, and the thoracic, and even the abdominal walls on the affected side become more or less œdematous. Yet, with all these evidences of obstructed circulation, there is no enlargement of the superficial veins of the chest, and none of the signs of centripetal pressure observed in cases of aneurismal or other intrathoracic tumors. The patient lies on the affected side or else on his back, with the head somewhat raised, and the body inclined towards the side on which the effusion exists; his breath is short, but often not so as to cause distress; the voice is weak, and there is frequent cough, which is dry in cases of idiopathic pleurisy, but accompanied by muco-purulent expectoration if there be any coexisting affection of the lungs; the skin is dry, and usually rather hot, the pulse is frequent and often weak, and though the patient seldom experiences pain in the affected side, he is unable to sleep, suffers from hectic, and becomes gradually thinner and weaker. Indeed, his principal, if not his sole complaint, is of debility and shortness of breath.

Now, in these cases, if the patient be not relieved by having the chest-walls punctured, and does not sink from exhaustion consequent on the oppression of the breathing, the effused matter after a time escapes by perforating the membrane which confines it. Sometimes the patient comes under notice before the process of perforation has commenced; and we are thus enabled to trace the establishment of a fistulous opening between the pleural sac and the external air. At others, one or more fistulous communications have been effected before the patient comes under observation. Most commonly the opening takes place through the costal pleura and the parietes of the chest; sometimes through the pulmonary pleura and the lung; more rarely, in both directions at the same time, or in other words, through the chest-walls and the lung; and still more rarely, the matter perforates the diaphragm, and escapes into the cavity of the peritoneum. When the latter event occurs, peritonitis is set up, and runs on rapidly to a fatal issue; when the fluid forms a fistulous opening through the pulmonary pleura, there is usually some antecedent bronchitis and pneumonia, and after a time, a sudden paroxysm of cough and dyspnœa occurs, and results in the expectoration of a large quantity—a pint or more—of purulent or sero-purulent matter, which threatens suffocation whilst it is being ejected, and leaves the patient very much exhausted. The cough and expectoration continue to a diminished extent for many days or even months; but if the case progresses favorably, they ultimately cease, the difficulty of breathing passes off, the patient regains his strength, and when he recovers, has usually less distortion of his chest than when the matter has produced complete and permanent compression of the lung, and has found its way out through the parietes of the chest. Of course, as soon as perforation occurs, air is admitted into the pleural sac, and the side, which continues enlarged and rounded, presents a variety of pathognomonic signs, which will be discussed under the head of Pneumothorax. Suffice it at present to say, that if the case terminates favorably, the air is gradually absorbed, the two surfaces of the pleura come again into apposition, and the progress of the case towards recovery is marked by symptoms similar to those observed when no communication has taken place between the cavity of the pleura and the external air.

When the matter perforates the costal pleura, and forces a passage through the parietes of the chest, one or more soft inelastic tumors, in which fluctuation is perceptible to the touch, will be visible on the chest-walls for some days before the skin gives way and an opening is established. Sometimes, however, patients may not come under observation until one or more openings have been formed between the pleura and the external air. In either case, the student or unwary practitioner might be misled as to the real nature of the mischief; for neither the tumors nor the openings in which they result are necessarily situated in the most depending parts of the chest. but appear to form in almost any part of the thoracic walls. Moreover, they may not communicate with the pleural sac directly, but by means of long tortuous sinuses. The discharge may be large or small in amount, according to the size of the opening, the freedom of its communication with the pleural sac, the quantity of effusion present, and the rapidity with which pus continues to be secreted; but it is often abundant, and, by the drain which it creates, gives rise to emaciation and exhaustion. Indeed, not unfrequently the patient sinks in consequence of exhaustion so produced; but if he is able to bear the drain, the discharge gradually decreases, and, after a longer or shorter period—a few weeks or, posssibly, some years—it ceases altogether, the opening in the parietes closes, the cough subsides, the difficulty of breathing passes off, the health and strength improve, and, in its ultimate progress, the case resembles those in which no communication has occurred between the pleura and the external air. The only difference is that, as the compression of the lung has been usually more complete, and the solid exudation-matter larger in amount, the collapse of the chest-walls and the consequent distortion is generally greater than when no fistulous communication has existed. In this instance, as in the last, air finds its way into the chest as soon as the opening is established. The consideration of the physical signs of this complication must be postponed until the subject of pneumothorax is discussed.

In what proportion of cases acute pleurisy passes on into a chronic stage has not yet been determined; and I believe that if the proportion were established, it would prove a very uncertain or fallacious guide in respect to the prognosis of any particular case. Age, sex, habits of life, constitutional peculiarities, the plan of treatment adopted, and the period of the attack at which it was commenced, are all circumstances which exercise a remarkable influence over the course of this disease, and render deductions from any number of cases taken indiscriminately inapplicable to individual instances. Experience, however, justifies the statement, that chronic pleurisy bears a small numerical proportion to the attacks of the acute disease; that it is seen most commonly in males, and amongst the very young and the aged, in persons of a feeble or unhealthy constitution and intemperate habits, and in those cases in which appropriate treatment is not adopted, or is deferred until a late period of the attack.

The remarks just made in respect to chronic pleurisy hold good almost equally in respect to empyema.

Krause[1] refers to 137 cases, which he collected from various authors; and amongst these, 96 occurred in males, 23 in children, and 18 only in women. Dr. Walshe[2] records three cases only in females, and 19 in males; and in like manner, Heyfelder,[3] who observed 20 cases of this disease, met with only one example of it in a female. My own experience of empyema, excluding instances connected with tubercular phthisis, extends to 43 cases, of which 29 occurred in men, 3 in children under fifteen years of age, and 11 in women. The proportion in women is larger than that reported by other observers, but not larger, I believe, than would be found to exist in the records of any large hospital. Even acute pleurisy is more common in males than in females, but not to the extent indicated by em-

[1] Krause, "Des Empyema," p. 106.
[2] Walshe, loc. cit., p. 371.
[3] Heyfelder, "Archives de Médecine," 3ème serie, tom. v, p. 59.

pyema; and I believe that the extraordinary tendency which the disease exhibits, when occurring in men, to degenerate into empyema, is attributable principally to their intemperate habits of life, and to their neglecting to seek relief in the early stages of the complaint.

The causes of pleurisy undoubtedly exercise an important influence over its course, and a due appreciation of their relative bearings is essential to a correct prognosis and treatment. Exposure to cold, the irritation excited by tubercular and other deposits in the tissues, the extension of inflammation from the lung or the pericardium, and mechanical violence, such as blows on the chest or fracture of a rib, are the principal causes which are commonly supposed to excite an uncomplicated attack of pleurisy; while penetrating wounds of the chest, the laceration of the pulmonary pleura which is caused by the splintered ends of a fractured rib, and perforation of the pulmonary pleura resulting from inflammatory ulceration of the pleural membrane, or from the extension of a tuberculous excavation, are the most frequent of those which not only excite pleurisy, but lead to the admission of air into the pleural cavity, and give rise to pneumothorax.[1] Of the agency of the first-named cause, viz., exposure to cold, which is generally said to be the most common source of pleurisy, I feel bound to express my entire disbelief. Nothing, I think, admits of clearer proof, than that cold, however intense, and however applied will not produce pleurisy in a healthy person; and that when exposure to cold is followed by inflammation of the pleura, the disease is due to some morbid condition of the blood, and not merely to cold. Possibly the patient might have escaped an attack if the exposure had not taken place, for his general health might have improved, and the blood might have regained its natural condition, without the supervention of pleuritic inflammation. But strictly speaking, cold is a mere accessory cause of the disease—a predisposing or exciting cause, but not the proximate or essential cause. Over-fatigue, anxiety of mind, undue excitement, or any of the thousand causes which tend to lower the vital energy and disturb the various functions of the body, might, equally with cold, have served to overcome the patient's power of resistance, and thus to excite an immediate attack. But popular feeling and casual observation combine to assign to cold the discredit of the seizure. Shivering precedes or accompanies the disease, and it

[1] Suppurative disease in the abdominal viscera has been known to produce perforation of the diaphragm, and to excite pleurisy, and hydatids have given rise to the same result; but such cases are extremely rare, and only deserve notice as medical curiosities.

is not surprising that the public should regard it as the result of exposure to cold. But the physician should be aware that something more is needed than mere exposure to cold, or to atmospheric vicissitudes, in order to induce inflammation of a membrane which is protected on all sides from the influence of cold.[1] The blood must be out of order, and probably contains some noxious, irritating matter, which acts directly on the pleural membrane, and excites inflammation, or else interferes with the nerve-supply of the part, and so disturbs the circulation and nutrition of the tissues. Experience has long since proved that a tuberculous, scrofulous, or cancerous diathesis, and the cachectic condition of the blood which results from Bright's disease of the kidneys, intemperate habits, syphilis, pyæmia, gout, and rheumatism, are amongst the more active provocatives of pleurisy, and equally so of other serous inflammation,[2] and we should at once suspect and search for one of these causes of poisoned blood or deranged nerve-supply when called to a case of pleurisy in which "cold" is the only apparent cause of the attack.

With regard to the other causes of the disease, it need only be observed, that those which are productive of uncomplicated pleurisy are less formidable in their consequences than those which lead to the admission of air into the pleural cavity. Pleurisy, when traceable to any of the first-named causes, excepting "cold," is usually limited in extent, easily controlled, and productive of adhesions, more or less partial, between the two layers of the pleura—an event which does not materially interfere with the function of respiration, and which in the case of tubercular deposits, protects the patient against perforation of the pulmonary pleura, and the admission of air into the pleural cavity, with all the untoward symptoms consequent thereupon; whereas, in cases in which air finds its way into the pleural sac, and goes on accumulating there, not only is the lung very forcibly compressed, but the entire surface of the pleura is sure to be inflamed, and the products of inflammation assume a puriform character. As a consequence of these events, the immediate danger incurred is far greater, and the risk of permanent injury more probable, than in the former cases. Even under the most favorable circumstances, the lung usually fails to re-expand to its former size; and as the air and fluid are

[1] For a detailed examination of the popular fallacy respecting cold, see the Introduction to the third edition of my work on "Rheumatism, Rheumatic Gout, and Sciatica," where the subject is fully discussed.

[2] See a paper, by Dr. Habershon, on the "Etiology of Peritonitis," in vol. xliii of the "Med.-Chir. Trans."

gradually got rid of, the chest-walls fall in, distortion of the thorax, if not of the spine, ensues, and the patient remains short-breathed for life.

Idiopathic pleurisy, occurring in a person whose lungs are not chronically diseased, seldom terminates fatally in its acute stage; so seldom that Dr. Walshe asserts that he has never lost a patient himself, and has never "known of an occurrence of the kind in the practice of others." I wish I could indorse this favorable report; but it happens that I have seen several instances of a fatal termination. Had the assertion been limited to cases in which appropriate remedies are had recourse to early in the attack, it would, I believe, have been strictly correct; but patients often neglect themselves at the outset of the disease, and do not send for a medical man until the disease has been committing its ravages for a week or ten days; and in this case death is not an unusual event. Day by day they get weaker and more exhausted, hectic sets in, the tongue becomes dry, the skin is covered with a clammy perspiration, and thus they gradually sink. During the year 1860 two such cases occurred at St. George's Hospital.

But death is not the ordinary result of acute pleurisy; indeed, recovery may be almost regarded as the rule. The course of the disease, however, differs remarkably in different cases. Sometimes the patient is seized with acute, catching pain in the side, and all the ordinary general symptoms of pleurisy, and loud but circumscribed friction sound is heard over the seat of pain. Under the effect of treatment these symptoms rapidly subside; no sign of other than circumscribed plastic exudation can be obtained, and in a few days the patient is quite well again, the two surfaces of the pleura having become adherent at the spot where the lymph was poured out. These attacks are designated attacks of "dry pleurisy;" they are often met with in cases of consumption, and give rise to the partial and often very limited adhesions which are found in the pleural cavity after death. But more frequently the disease goes on to effusion, and then, if recovery takes place, it occurs in one of two ways: either the fluid is gradually absorbed, or, absorption failing to occur, the effusion becomes chronic. In the former case recovery generally takes place, though the chest-walls on the affected side often become retracted; in the latter the effusion gradually becomes purulent and a considerable proportion of the cases prove fatal. When recovery takes place, the chest-walls are almost invariably retracted, the precise degree of retraction being proportioned, as in the former case, to the extent to which the lung has been compressed by adventitious membrane, or bound down by adhesions, which prevent it from re-expanding and filling up the void created by the escape of the fluid. When, as sometimes happens, pneumothorax is added to the evils incident to pleurisy, the proportion of fatal cases is even larger than in ordinary chronic pleurisy. Air in the pleura, if not admitted through a fistulous opening in the chest-walls caused by the escape of fluid from the pleural sac, has usually found ingress through a wound in the pulmonary pleura, resulting from organic disease of the lung, which is a fearful though not necessarily a fatal complication. The precise proportion of cases in which death speedily follows the admission of air in this way, it is difficult, if not impossible, to ascertain; that the proportion is large is a matter of notoriety, but it is susceptible of proof that temporary recovery often takes place. I have met with several well-marked examples of it, and one, in which the recovery was most remarkable, occurred under my care in St. George's Hospital no later than January 12, 1859.[1] Sometimes, indeed, perforation of the pleura and the admission of air is attended with little more than temporary inconvenience, a slight catching pain in the affected side and acceleration of the pulse being the only symptoms calculated to arrest attention. A girl, Mary Ann R—, æt. 16, admitted under my care in the Crayle Ward of St. George's Hospital on April 22, 1865, was an excellent illustration of this remarkable fact, for she had very little pain when the perforation occurred, and she left the hospital apparently in perfect health within three weeks of the time at which tympanitic resonance of the chest-walls, a splashing sound on succussion, and metallic tinkling had made it manifest that a communication existed between the bronchi and the cavity of the pleura.

In practice it is often important to discriminate between pleuritic effusion and pneumonic consolidation; therefore, before passing on to the treatment of pleurisy, I will endeavor to place in juxtaposition the principal points of distinction between these two disorders.

Pleuritic Effusion.	*Pneumonic Consolidation.*
History and general symptoms.—Shivering is seldom severe, and often absent. The attack commences with a sharp catching pain in the side. The cough is dry and rarely frequent; sometimes there is little or no cough. There is seldom much burning heat	*History and general symptoms.*—Shivering, usually severe, is almost a constant precursor of the attack. The pain is not so acute as in pleurisy, nor does it catch the breath in a corresponding degree. The skin is hot and burning. The cough, which is severe

[1] See "Hospital Case-Book," xxxviii, p. 178.

Pleuritic Effusion.	*Pneumonic Consolidation.*
of skin. The pulse is hard, and the ratio naturally subsisting between the pulse and the respiration scarcely ever falls below three to one. The patient lies on or inclined towards the affected side.	and frequent, is accompanied by copious, and often rusty-colored expectoration. The pulse may be full, but is frequently soft; rarely hard as in pleurisy; and the ratio subsisting between the pulse and the respiration often falls two to one. The patient lies indifferently on either side.
Inspection and Mensuration.—The chest is almost motionless over the seat of effusion and is enlarged, rounded, and smooth; the ribs are widely separated, the intercostal spaces are more or less obliterated, and occasionally the integuments are œdematous. Sometimes there is protrusion of the hypochondrium, and if the heart be displaced, as it often is, it may be seen pulsating out of its normal position.	*Inspection and Mensuration.*—There is somewhat diminished mobility of the affected side; but its dimensions and appearance, the state of the intercostal spaces, and the position of the heart, remain unaltered.
Palpation.—Vocal vibration is imperceptible on the walls of the chest, except at the root of the lung. Friction-fremitus is sometimes though rarely perceptible.	*Palpation.*—Vocal fremitus is increased, except when the feebleness or high pitch of the voice renders the vibration slight, and therefore imperceptible on either side.
Percussion.—Dulness is always intense, often extending beyond the middle of the sternum; sometimes but not invariably, it may be altered in position by changing the posture of the patient. Under the clavicle percussion often yields a clear but shallow resonance; it shows displacement of the viscera by eliciting their characteristic percussion-sounds.	*Percussion.*—Dulness is rarely so intense as in pleurisy; it never extends beyond the natural limits of the lung, and is not altered in position by change in the patient's posture.
Auscultation informs us that the respiratory sound and the vocal resonance are sometimes extremely weak and distant, more generally altogether absent, except at the root of the lung, where diffused hollow breathing and obscure bronchophonic or ægophonic resonance of voice may be heard. There is entire absence of fine crepitation. These signs, however, may be modified by adhesion of the lung to the walls of the chest.	*Auscultation* reveals loud, hollow, tubular breathing, and intense bronchophonic resonance of the voice over that portion of the chest which corresponds to the consolidated lung, and fine crepitation over the adjacent portions, if inflammation is still extending.

In some instances dulness on percussion occasioned by an enlarged liver or spleen has been mistaken for the dulness of pleuritic effusion, but ordinary attention to the general symptoms and physical signs, and to the facts referred to in the first portion of this work, relative to the effect of full inspiration in increasing the area of vocal fremitus and of clear resonance on percussion, will enable the careful practitioner to distinguish the dulness thus caused from that which is produced by pleuritic effusion, and which, as already stated, is not affected by respiration. Dulness referable to simple hydrothorax, intrathoracic tumors, and tuberculous or other infiltration of the lung, will be discussed in connection with those several diseases.

With ordinary care it is impossible to confound the pain in the side and the short, interrupted breathing which result from pleurisy with the symptoms occasioned by pleurodynia, intercostal neuralgia, and other affections of the thoracic parietes. Nevertheless, such a mistake has been often made by practitioners who, in full reliance on general symptoms, have neglected to appeal to physical signs; and it may be advisable, therefore, to point out the characteristic marks of the several disorders.

The pain of pleurodynia, or rheumatism in the walls of the chest, though often undistinguishable in other respects from the pain of pleurisy, may be recognized by the fact that it is affected to a far greater degree than the pain of pleurisy by pressure on the ribs and by movement of the trunk and arm. If this be not sufficient to establish its true character, the absence of fever and of the physical signs of pleurisy will usually serve to clear up the mystery. The only cases in which any doubt can be entertained are those in which the patient is seen soon after the accession of pain, and in which the pain coexists with bronchitis, accompanied by febrile action. Even here the delay of a few hours will serve to set the question at rest, inasmuch as it will admit of pleuritic exudation, and the consequent production of percussion dulness, pleuritic friction-sound, and ægophonic resonance of the voice.

The pain of intercostal neuralgia may be recognized by its paroxysmal character, and by its following the course of the affected nerve. Like pleurodynia, it simulates pleurisy, in so far that it gives rise to diminished freedom in the respiratory movements on the affected side, and to weak respiration, with jerking irregularity in its rhythm; but, like pleurodynia, it is distinguishable from pleurisy by the absence of percussion dulness, friction-sound, ægophony, and other characteristic physical signs of that disease.

Having, then, determined that the patient is indeed suffering from pleurisy, what measures does it behoove us to adopt for his relief?

If we see him in the stage of hyperæmia or congestion—in other words, in the first stage of the disease—the question of practising venesection may be fairly entertained, provided only that he be of sound constitution and that his strength be good, and his symptoms of a sthenic character. In most cases bloodletting is unnecessary, but in some, in which the stitch in the side is severe and the pulse hard, it unquestionably proves serviceable if it be not carried beyond the amount required by the exigencies of the case. The points to be obtained are, relief of pain, and moderation in the force and frequency of the pulse. The dangers to be avoided are, constitutional depression,—which so often leads to the production of suppurative instead of adhesive inflammation,—and anæmia, with its consequent protracted convalescence. The best method of making sure of the former without running undue risk of the latter is, to bleed the patient in an upright posture, and allow blood to flow in a full stream until he can take a deep breath freely, or else feels faint and exhausted. In an adult of ordinary vigor, the subject of acute pleurisy, the loss of from ten to twenty ounces of blood will usually suffice to produce one or other of these results, and a repetition of the bleeding will not be needed. If the pain continue, or the urgency of the symptoms be such that a further loss of blood is considered desirable, local depletion is usually preferable to venesection. Leeches should be applied to the painful side; and when they have done their duty, the leech-bites should be covered with a bread-and-water poultice, which effects the double object of fomenting the seat of pain and promoting the bleeding. Some persons recommend the employment of cupping in preference to leeches; but, on practical as well as on theoretical grounds, I would strongly urge the application of leeches. The inflamed membrane lies so superficially, that leeches are quite efficient in their action, and do their work without giving pain or uneasiness to the patient; whereas cupping over the painful part is often a severe infliction, and not only aggravates existing pain, but in some instances, I believe, aggravates the mischief it is intended to remove. Certain it is that on several occasions I have seen pain in the side increase after the use of the cupping-glasses—an event which has never occurred after the application of leeches; and my firm conviction is, that leeches are as much preferable to cupping in the treatment of a superficial disease like pleurisy, as cupping is to leeches in the treatment of deepseated mischief like pneumonia. If it be deemed necessary to employ bloodletting and the attack of pleurisy be not very severe, leeches may be employed at the very outset, to the exclusion of general venesection.

Following on, and in aid of bloodletting, it is generally expedient to administer a brisk purgative; but when once the bowels have been freely relieved, nothing proves so serviceable as mercury in small doses combined with opium. Tartar emetic, which is invaluable when the substance of the lung is inflamed, has little or no control over the course of pleurisy; whereas mercury combined with opium exercises a markedly beneficial influence. It appears not only to check the effusion of inflammatory products, but to promote their absorption. The system, however, must be brought under its influence before these effects are fully developed, and therefore its administration should be commenced from the first, with a view to obtain its constitutional action as speedily as possible. The common practice is to give two or three grains of calomel every three hours, in combination with a sixth or a quarter of a grain of opium, and to continue its exhibition until the mouth is affected. But I believe that more immediate and more satisfactory results may be arrived at by applying mercury externally, and making larger use of opium internally. My usual practice is to give half-grain or grain doses of opium every three or four hours, in combination with a grain of calomel and half a grain of digitalis, and to have the whole side covered with a piece of linen spread with mercurial ointment. Over this is placed a poultice covered with oiled silk. In this way the action of the mercury is induced more rapidly, and apparently with greater relief to the symptoms, and less subsequent distress to the patient, than when the drug is taken wholly or principally by the mouth. Of course the patient must be carefully watched, and as soon as the slightest symptom of ptyalism is induced, the ointment must be omitted, and the internal administration of calomel suspended, or the quantity of the drug or the frequency of its repetition diminished; the object being to obtain its constitutional effect without aggravating the patient's sufferings by salivation.

But though the mercury be omitted, the use of opium and digitalis should be continued, and with them may be combined two grains of squills, which acts beneficially as a diuretic. If fever runs high, and the urine is loaded, an ammoniated saline draught, with the addition of a few grains of the carbonate and acetate of potash, is not only grateful to the patient, but assists in maintaining the action of the skin and kidneys; and if the tension of the system is great and the skin proves inactive, a small dose of tartar emetic may be advantageously added, either to the pills or the mixture. If pneumonia coexists with pleurisy, the administration of tartar emetic will be urgently required. Meanwhile turpentine stupes should be applied to the chest; and when they can be borne no longer, the whole of the affected side should be kept incased in a jacket-poultice covered with oiled silk, or in a piece of spongio-piline wetted with hot water and sprinkled with laudanum. Such appliances, I am satisfied, do something more than assuage pain and afford comfort to the patient; they contribute largely to relieve local congestion and maintain the circulation in the tissues, and so much relief is often observed to follow their use that it is difficult to conceive that they can do otherwise than operate in arresting or modifying the course of the disease.

Some persons recommend the application of blisters to the seat of pain, and urge their employment not only as the most effectual method of producing counter-irritation, but as affording an absorbent surface to which mercurial ointment can be advantageously applied. Experience, however, induces me to agree with Sir Thomas Watson and others in recommending that vesication be not employed during the active stage of the disease, or that, if employed, the blister be not placed over the seat of pain. The close proximity of a blister to any part where pain is most acute, and where intense inflammation is going on, is calculated to increase the local irritation; and in several instances which have come under my notice the sudden disappearance of pleuritic friction coincidentally with the occurrence of percussion dulness, has told, in terms not easily to be mistaken, the mischief which a blister so applied has occasioned. The same objection cannot be urged against blisters applied at a distance from the affected part; but even in this way, they afford little benefit during the more active stage of the disorder. They do not appear to check the morbid action, and the vesication which they produce may prevent their employment at a later period when their influence may be needed. Indeed, their curative action is not displayed until the pain in the side and the fever are subsiding, or, in other words, until the first activity of the disease is overpast, and the patient is suffering principally from the accumulation of fluid in the pleural sac. Then they prove our most valuable allies, and operate probably by disgorging the vessels and stimulating absorption. Certain it is that, when they are thus employed, any lingering pain yields rapidly to their influence, and percussion and auscultation often mark the rapid disappearance of effusion.

Sometimes, however, effusion continues to take place in spite of blisters and mercurials, and sometimes even when the progress of effusion is checked the treatment fails in producing absorption of the fluid already effused; in short, the case passes into a chronic stage. Under these circumstances, tartar emetic ointment, croton oil liniments, moxas, setons, and even the actual cautery, have been had recourse to by certain practitioners. But I cannot help regarding most of these applications as little less than barbarous engines of scientific torture. When blisters fail, the other agents are seldom of any service, and they are always productive of needless suffering. But much good may be effected by a different plan of treatment. Hitherto inflammatory fever and pain have prevailed, and the diet has been necessarily somewhat low; but the case is altered now that the symptoms of active inflammation have subsided. If the patient, at this stage, be kept too low, or be unduly depressed, it will not only be impossible to induce absorption of the fluid, but there will be great danger of its becoming sero-purulent in character. A more generous diet must therefore be given, and the general health sustained by quinine and other tonics. Meanwhile it is expedient to make full and steady trial of diuretics and absorbents. Even when a succession of blisters has failed in relieving the patient, ioduretted ointments, or ioduretted lotions kept constantly applied to the whole surface of the affected side,[1] backed by the internal administration of cinchona with tincture of iodine, iodide of potassium, and small doses of bichloride of mercury, nitre, acetate of potash, squills, digitalis, and cantharides, have sometimes effected the desired object.[2] The fluid has been gradually reabsorbed, and recovery has ensued. Case after case has come under my observation, in which this treatment, steadily pursued, has produced the most satisfactory results; and I am inclined to think that, in many instances, the unfavorable issue of pleurisy in its chronic stage is attributable to a want of tone in the system, caused by injudicious treatment. The patient is kept too low, or is overmuch purged, or in some other way is unduly depressed, so that the reparative power is not duly exercised. In several instances, both in this country and in the Hôtel Dieu at Paris, I have seen patients progressing favorably until some lowering treatment has been had recourse to, and from that time the symptoms have put on an untoward aspect. Therefore, it is not without sufficient reason that I urge the necessity of a generous diet and tonic medicines

[1] Subjoined are formulæ I often employ:

℞. Hyd. Bichloridi, gr. iv;
Tr. Iodi., ʒvj—℥j;
Glycerini, ℥iij;
Aquæ destillatæ, ℥ivss. Ft. Lotio.

℞. Hyd. Bichloridi, gr. iv—v;
Ung. Iodi. co., ʒiv—vj;
Adipis, ʒiv—℥j. Ft. ung.

or,

℞. Hyd. Bichlor., gr. iv—v;
Potassii Iodidi, ʒij;
Aquæ destillatæ, qs. ut solventur.
Hyd. Bichlor. et Potassii Iod.
Adipis, ℥j. Ft. ung.

[2] A favorite diuretic is one contained in the Pharmacopœia of St. George's Hospital. It is a pill composed of digitalis, squills, and the pil. Hydrargyri.

in aid of any diuretics and absorbents which may be employed in the chronic stage of pleurisy.

But if all ordinary means fail, and the pleura continues distended with fluid, so that the breathing is seriously oppressed, the general health undermined, and life jeopardized, a question arises as to the propriety of relieving the patient by tapping, or, in other words, by puncturing the chest and letting out the fluid. The operation itself is simple enough, and neither difficult of performance nor formidable in its immediate results. Therefore, the sole question to be decided is as to the propriety of having recourse to its aid.

Some persons have advocated the practice of tapping the chest even during the acute stage of the disorder, on the ground that if accumulation of fluid in the pleura is prevented, compression of the lung, and the various evils connected therewith, will also be obviated. But theoretical considerations are decidedly opposed to puncturing the chest, and admitting air to the inflamed pleural membrane; and practice justifies the doubts which theory suggests as to the expediency of so doing. On the one hand, it has shown that the admission of air into an inflamed serous cavity is apt to be followed by suppurative inflammation, even when the effused matters had previously consisted of mere lymph and serum; and that fatal results very commonly ensue under these circumstances;[1] on the other, it has proved that the mortality from idiopathic pleurisy is very small, and that, however extensive the mischief in the pleura, and however profuse the fluid effused, it may be generally got rid of by the influence of remedies as soon as active inflammation has been subdued. Indeed, it may be taken as thoroughly established that, during the acute stage of the idiopathic disorder, the operation of tapping is not curative in its action, and is almost certainly productive of mischief, and that it is only when pleurisy is associated with organic disease of the lungs, or with inflammation of the opposite lung, that we are justified in entertaining the question of its performance in the early stage of the disorder. In some such cases, where death has appeared imminent from suffocation, it has afforded temporary relief, and has prolonged life, and, therefore, as a last resource, may be resorted to under the circumstances mentioned. But it must not be regarded as a curative agent—it is simply palliative—a mere expedient for giving temporary relief; and even in this view it often fails; for, although in a few instances it has afforded the patient a respite, it has much more frequently

[1] "Sédillot de l'Empyème," p. 127.

done little else than complicate his disorder, and lessen his chances of even perfect recovery.

But the question assumes a very different aspect when the operation is proposed as a means of relieving a chest in which active inflammation no longer exists, but which, nevertheless, is distended with fluid which remedies have failed to get rid of. In many of these cases, the fluid in the pleura interferes with the vital functions, and jeopardizes the patient's life; and we *must* interfere if life is to be prolonged. It matters not whether the disease commenced originally in an acute or in a chronic form; in other words, whether it has reached its present point rapidly or slowly; our duty is to obviate the tendency to death which the retention of the fluid in the chest occasions. The only questions which can arise are, as to the precise moment when the operation should be performed, and as to the mode of its performance, and the spot at which the opening should be made.

The first question is one on which a variety of views have been entertained, and on which, nevertheless, it is difficult to conceive how any difference of opinion can have existed. Some persons have advocated the performance of the operation even during the acute stage of the disorder; others have insisted on the propriety of postponing it until a very advanced period of the attack; some have been unwilling to recommend it under any circumstances; whilst others, fixing arbitrarily upon the fifteenth or some other given day of the disorder, have asserted that it may be undertaken on or after that day, but not until that day has arrived. Few arguments are required to refute each of these doctrines. The first is the only one which requires serious consideration, and that has been already disposed of. The objections to the others are obvious and unanswerable. A refusal to operate under any circumstances could not be maintained in the present day, in the face of the numerous recoveries which have taken place after all hope of amendment, except through the medium of the operation, has been necessarily abandoned; and it needs no special experience of pleurisy, to understand that the indications for the operation must be as marked in one case at the end of a fortnight, as they are in another after the lapse of a month. And with regard to the proposition for postponing the operation until a very late period of the attack, observation has long since furnished materials for an opinion adverse to that practice. Worn out and exhausted by the long-continued irritation in his chest, and by the oppression of the breathing, resulting from the large accumulation of fluid in the pleura, the patient is ill able to bear the operation, or to withstand the drain which subsequently ensues; while the lung, bound down and compressed by false membranes which have gradually become thicker and firmer, is irretrievably damaged for all purposes of respiration, and is incapable of re-expanding when the fluid is withdrawn. The result is just what might have been anticipated, namely, that the operation performed under these circumstances is commonly followed by fatal consequences, and that, even when recovery does take place, great and permanent distortion of the chest ensues.

The only practical tests as to when the operation should be had recourse to are the condition of the patient's health and respiration, and the absorption or non-absorption of the effused fluid. As long as the breathing is not seriously embarrassed, and the general health does not decline, so long we are justified in making full trial of our remedies, in the hope that absorption of the fluid may be brought about. But, as soon as distress of breathing or lividity and anxiety of the countenance denotes serious interference with the functions of life, delay is no longer justifiable; it becomes our duty at once to give our patient the chance which the operation affords. Even in the absence of any obvious oppression of the breathing, if day by day we note failure of the health and strength, and are not encouraged by the results of auscultation and percussion to believe that absorption is taking place, we are bound to hasten to his relief, and perform the operation of tapping. Sometimes, when neither deep oppression of the breathing nor rapidly failing strength proclaims the necessity for operative interference, the gradually increasing size of the chest, or the evidence of non-absorption of the fluid, may warrant or imperatively call for the operation; but, in this latter case, the indications as to the precise moment for its performance are not so unequivocal as in the two former instances. Experience has proved, that by steady perseverance in a given course of treatment, the fluid is sometimes absorbed after all hope of getting rid of it by natural means had been almost abandoned. Therefore, a proportional increase of caution should be observed in such cases before the operation is undertaken. Not only should repeated blisters have been employed, and mercurial action cautiously induced, but the patient's strength should have been thoroughly upheld by a generous diet, and tonics should have been administered perseveringly in combination with iodine and other diuretics and absorbents. Even if no evidence of absorption of the fluid can be obtained after some weeks' trial of the remedies, the practitioner should still hesitate to perform the operation if the general health is not suffering, and the respiration is not materially oppressed. In this condition of affairs, he may feel doubtful as to the possibility of inducing absorption; but whatever doubts he may entertain, his right course is, not to have immediate recourse to the operation, but rather to make a tentative exploration of the chest by plunging a grooved needle into the affected side. If simple serum follows its introduction, the needle may be at once withdrawn, and the effect of remedies tried for a further period; for the operation should not be undertaken until it becomes apparent that it is impossible to get rid of the effused fluid by natural means; whereas, if the fluid appears to be pus, the chest should be tapped without delay. Every day which is suffered to pass after pus has been ascertained to exist in the pleura must necessarily expose the patient to risk by lessening his rallying powers, and increasing the impediments to the re-expansion of the lung.

I have hitherto made only slight allusion to the oft-mooted question as to whether the propriety of performing the operation can be affected by the serous character of the fluid to be removed. But, after what has been already insisted on, it is only necessary to repeat that the state of the patient's health and respiration, and the absorption or non-absorption of the effused fluid, are the only points which must be allowed to guide the physician's practice. The precise character of the fluid may influence his opinion as to the prognosis of the case, but it cannot militate against the propriety of the operation. If the fluid be pus, it is manifest that the operation should not be delayed, whatever the condition of the patient may be; and if the fluid be only serum, the operation is not the less expedient and necessary if the functions of life are seriously interfered with. The utmost that can be said in such a case is, that if the general health be not declining, other remedies should be fairly and perseveringly tried before recourse is had to the operation.

The introduction of a grooved needle into the suspected side, which was originally suggested by the late Dr. Thomas Davies, is at all times desirable as preliminary to the operation of tapping. Not only does it serve to corroborate our diagnosis and enable us to determine with certainty the existence of pleuritic effusion, but it also informs us of its character. If the fluid be serous, it will flow along the groove of the needle, and trickle down the patient's side; if it be puriform, it will not flow so freely, and probably not a single drop will escape until the needle is withdrawn, and then a drop will make its appearance at the external orifice; or possibly none may be seen at the external orifice, but a small quantity may be visible in the groove of the needle. If no fluid escapes, and the groove of the needle is found not to contain pus, the practitioner will be made aware that no fluid is collected at the spot where he had imagined it to exist, and where, accordingly, he had introduced the needle. In this case he will have the satisfaction of feeling that the puncture has given little pain, and has done no harm to the patient, whilst it has been the means of preventing the introduction of the trocar or a lancet at a spot where its introduction might have been followed by dangerous consequences. But, although no evidence of effusion may have been obtained by the needle, the practitioner may feel convinced that fluid is contained in the pleural cavity, and that the existence of old adhesions of the pleura, or the presence of a thick coating of coagulable lymph at the spot where the needle was introduced, has alone prevented his obtaining evidence of the fact. In this case, so harmless is the puncture of the needle, that he may venture to make a tentative exploration at another part of the chest, and, if care be taken to employ a sharp needle, which shall penetrate the coating of lymph instead of driving it before it, he can hardly fail to succeed in his object. If then, on taking all necessary precautions, he fails in the second attempt to obtain proof of the presence of fluid, he will have reason to doubt the accuracy of his diagnosis, and may congratulate himself that by adopting the simple precaution of using the grooved needle, he has been saved from the dangerous error, to which his mistaken diagnosis would have led him, of tapping a chest in which no fluid existed.

Having satisfied himself, by careful examination, that the pleural cavity contains fluid, and that remedies are powerless to induce its absorption, and having had his views as to the existence of liquid corroborated by the introduction of a grooved needle, the practitioner, without further delay, should perform the operation of tapping. But how should he proceed in the matter? What instrument should be employed? Should any precautions be taken to prevent the entrance of air into the pleura? Where should the opening be made? Should all the liquid be let out at once? Should the wound be kept open or healed up? These are questions which force themselves on his attention, and ought to be solved before the operation is attempted.

The choice of an instrument must be regulated, in some measure, by the view which is entertained respecting the importance of preventing the entrance of air into the pleura. Some persons have laid great stress upon the non-admission of air as an element of success in puncture of the chest, and have suggested the use of canulæ, with valves and stop-cocks, as preventive of pneumothorax. Some of these instruments are very ingenious, and are calculated, as far as possible, to effect the object for which they were designed. But, in truth, they are found to be practically inoperative; for it is impossible to draw off any large quantity of liquid from the chest without admitting air into the pleura. The lung is seldom able to expand freely at once; and as the chest-walls will not yield beyond a certain point, air must be allowed to find ingress into the pleural cavity, or the fluid would not flow out. I have seen two of the most perfect of these stop-cock instruments employed; and so long as precautions were taken to exclude air from the pleura, so long they failed to draw off more than a few ounces of the fluid; directly air was admitted, the liquid flowed through them freely. Therefore, however desirable it may be, theoretically, to exclude air from the pleura, it is practically impossible to do so if we wish to relieve our patient. Fortunately in this advanced stage of the complaint when *acute* inflammatory action has subsided, the admission of air is not a matter of so much importance as theoretical considerations have appeared to some persons to suggest. It does not necessarily re-excite acute inflammation of the pleura, neither does it necessarily induce suppuration in a pleura already inflamed. This has been proved in many cases in which serous fluid has been drawn off by the canula without the subsequent occurrence of suppuration. Again, the existence of air in the pleura is not found to interfere with the re-expansion of the lung, in the cases now under discussion. The lung is already compressed by the fluid, and in most cases is slow in regaining its due expansion; and, as air is rapidly absorbed from the pleural cavity, its temporary admission is of little moment. I have known all traces of pneumothorax disappear within five days after the operation—a period within which it is improbable that the lung would have expanded fully even if no air had been admitted. Nevertheless, as the presence of air in the pleura is an abnormal condition, and is certainly conducive to suppurative inflammation, every precaution should be taken to prevent its admission, when the grooved needle has shown that the effused liquid is serous in character. In these cases probably the pleuritic inflammation has been less acute and not accompanied by the outpouring of so much lymph as in those in which the fluid is purulent, and the lung therefore being less bound down by adhesions, will be more capable of expanding than when pus is present. Under these circumstances, a stop-cock apparatus should be employed, and the opening in the chest should be carefully closed as soon as the instrument is withdrawn. When, on the contrary, the grooved needle has proved the existence of pus, I believe that the admission of air is not of the slightest importance; and that, even if it were, the patient could not be relieved of the fluid without its admission. In this instance, therefore, a large trocar, or even a lancet or a bistoury, may be employed; and, unless some special circumstances contraindicate such a course, the wound should be kept open, so as to admit of a constant discharge of matter. In the former class of cases, if the patient progresses favorably, no further operative interference will be necessary. But if shivering occurs, or if much constitutional disturbance arises, the fair inference is, that suppuration has commenced; and an appeal should again be made to the decision of a grooved needle. If this shows the conclusion to be correct, a free opening should be made at once into the pleura, just as in those cases in which the needle from the first revealed the existence of pus.

The selection of the spot at which the opening is to be made is a matter of some importance, and demands the exercise of care and discrimination. Of course, if the fluid has already begun to point externally, and has occasioned the appearance of a soft, elastic tumor on the chest-walls, there is no choice as to where the puncture is to be made, for nature herself has determined the question. The tumor must be opened, and a channel thus afforded for the egress of the fluid. But in most instances there is nothing of this sort to guide us, and our choice of a spot at which to introduce the trocar must be regulated by other considerations. The object to be attained is, to provide the easiest and most thorough vent for the liquid, without endangering the patient's safety. In some respects, the most dependent part of the antero-lateral portion of the chest is best fitted for the purpose; and the sixth intercostal space, a little in front of the digitations of the serratus magnus muscle, is the spot often recommended. But, on making a puncture so low in the chest, some risk is incurred of perforating the diaphragm, the liver, the spleen, and other of the abdominal viscera, and thus causing the speedy death of the patient. This is no fanciful danger; it is a mischance which has occurred on more than one occasion,[1] and which can hardly fail to occur if the diaphragm has been pushed upwards by enlargement of the abdominal organs. Prudence, therefore, suggests that the opening should not be made below the fifth interspace—the spot originally suggested by Laennec—and that if, on examination, the liver or spleen appears to be enlarged, and therefore probably has forced the diaphragm high up in the chest, the punc-

[1] See Watson, loc. cit., ed. i, vol. ii, p. 134.

ture should be made in the fourth instead of the fifth interspace. In a case of this sort, Laennec transfixed the diaphragm and perforated the liver by introducing the instrument in the fifth interspace; and it is well known that an enlarged liver or spleen will sometimes force itself up as high as the fifth rib. It is impossible, therefore, to take too much care in examining these points before deciding upon the spot for the operation; and as, when the fluid points spontaneously, the opening commonly occurs high up in the chest, there is reason to believe, as suggested by Dr. Stokes, that in selecting a somewhat higher level at which to make the opening, we should be consulting our patient's safety, and should not seriously compromise his chances of recovery.

Another necessary precaution is to determine that the lung is not bound down by adhesion to the part at which it is proposed to introduce the trocar. The existence of any respiratory sound, or of vocal resonance, vocal fremitus, or resonance on percussion, is quite sufficient to prove that the lung is in close proximity to the chest-walls at that particular spot, and should deter us from introducing an instrument there. So, also, in regard to the impulse of the heart, felt or seen at any spot. Further, it is important not to make the incision too near the margins of the ribs, lest the branches of the intercostal arteries be wounded. Again, considerable œdema of the integuments oftentimes exists, and some difficulty may occur in determining the precise spot at which to introduce the instrument. This may be overcome by exerting steady and continued pressure, which, by driving the fluid out of the cellular tissue beneath the fingers, discloses the ribs, and enables the operator to exercise his discretion in the matter.

When the opening has been made, and the instrument introduced into the pleural cavity, a question arises as to whether a portion only of the liquid should be drawn off, or whether the pleural sac should be emptied as far as possible. Some persons have maintained the propriety of the former mode of proceeding, and have adduced a variety of speculative reasons for the withdrawal of the liquid in successive portions; whilst others have advocated the latter method, and have supported their views by reference to practical results. In certain instances, success is known to have attended both modes of practice; but the weight of modern experience is greatly in favor of a full and free evacuation of the liquid, which affords the most complete and immediate relief. The only exceptions, perhaps, are cases in which the fluid has been ascertained to be serous, and those, again, in which the withdrawal of the fluid, whether serum or pus, occasions faintness. In the former, the removal of a small quantity of the liquid may relieve the tension of the vessels, enable the process of absorption to take place, and render further operative interference unnecessary; in the latter, the suspension of the operation is a matter of necessity, and not of choice, and the operation must be repeated as soon as the patient is in a condition to bear it. Some persons have proposed to make sure of emptying the chest by forcibly drawing off the fluid which accumulates below the level of the puncture. Laennec, for instance, suggested the use of a cupping-glass for the purpose; and syringes, syphons, and other instruments have been proposed by others; but common sense and practical observation are alike opposed to the practice. The lung, covered as it is with false membrane, cannot expand at once to fill up the space occupied by the fluid, and, therefore, if the liquid be withdrawn by artificial means, it must be replaced by air, or the chest-walls must fall in. In either case, the patient would not be in a better position than he was before the withdrawal of the fluid; whilst the operation of removing it would have to be repeated day after day, as fresh pus is secreted, and would, necessarily, be attended with much fatigue and considerable risk to the patient. With the existing knowledge on the subject, I doubt whether any one would recommend the operation; my own feeling is strongly opposed to it.

A plan, however, has been recently proposed, which accomplishes much the same object by natural means, and seems to promise very favorable results. I allude to what has been termed the system of "drainage." With a view to carry out the plan, "a firm, long iron probe, somewhat bent," is introduced through the opening, made in the usual place, as above described, and "is then directed towards the lower and back part of the pleural cavity, the lower the better. If the end of the probe be made to press against the side of the thoracic walls, it can be felt from the outside, through the intercostal spaces; though, perhaps, obscurely, owing to the thickness and toughness of the false membrane within. The lowest and most appropriate site in which the probe can be felt having been selected, an incision is made upon the end of the probe, which is then brought through the opening thus made. A strong piece of silk thread is passed into the eye of the probe, and drawn through the two openings, and the drainage-tube—an India-rubber tube, perforated at frequent intervals, in the way recommended by Chassaignac for the healing of sinuses—being firmly tied to one end, is then drawn through by means of the silk; the ends of the tube are tied together, and the operation is complete." The benefits derivable from the adoption of this plan are said to be, that "the openings in the chest-walls are always free; the matter is discharged drop by drop as it forms, so that, if the tube be suitably placed, there is never any collection of pus in the thorax; no time is given for decomposition, and the pus, therefore, is discharged in a healthy and pure state."[1] Now, it is obvious, from this description of the operation, that it involves a free admission of air into the pleural cavity, and therefore that it is not admissible in cases in which the effused fluid is of a serous character; but it is feasible enough, in cases of empyema, in which, as before stated, the admission of air is not productive of mischief. In some of these, as Dr. Goodfellow has shown, it may lead to the recovery of a patient in whom a single opening in the thoracic walls proves an inadequate channel for the escape of pus. I have not as yet had an opportunity of trying it; but the objects it effects are so thoroughly in accordance with my own views as to what is desirable in cases of empyema, that I shall certainly recommend its adoption when next I meet with a patient in whom operative interference is necessary.

Before quitting the subject of tapping, it may be well to refer somewhat more at length to the changes observed in the character of the secretion after the pleura has been punctured. Rarely does the newly-secreted liquid retain a serous character, even though the pleura contained simple serum before the operation was performed. The usual tendency is to the effusion of pus, or sero-purulent fluid, and sometimes a few hours suffice to complete the change. Before the operation the effused liquid is devoid of smell, provided no air has found its way into the pleura; but after air has been admitted, the effused liquid is often converted into puriform matter of a low type, and emits a fetid odor. Whether the smell is attributable to the character of the pus secreted, or to decomposition resulting from its detention in the pleura, has not yet been ascertained, but I am inclined to regard the latter as the primary and potential cause of the change, and to believe that the decomposing fluid reacts on the diseased pleural membrane, and leads to its secreting pus, which is of still lower vitality, and more readily yields to the putrefactive process. But whatever the cause of the fetor, there cannot be a doubt that when it occurs immediate steps should be taken to get rid of it, and nothing answers better than injecting the pleura with warm water, containing a weak solution of per-

[1] See a paper by Dr. Goodfellow and Mr. De Morgan in vol. xlii "Med.-Chir. Trans."

manganate of potash (ʒij of Condy's solution to Oj). At the same time the opening through the chest-walls should be enlarged, or another made in a more depending position, so as to afford a freer exit to the liquid, and prevent its accumulation in the pleura. The system of drainage by Chassaignac's tube, properly carried out, would probably accomplish this object.

The prognosis, in cases where the operation is had recourse to, is a subject involved in much uncertainty. Unfortunately there are not any data from which to obtain a general estimate as to the value of tapping, nor are there any trustworthy grounds for an opinion as to its issue in any particular instance. The result, however, bears some relation to the quantity and quality of the effused liquid, and to the length of time the operation is deferred. The larger the quantity of the fluid, and the longer the time for which the operation is postponed, the greater, *cæteris paribus*, is the displacement of the thoracic organs, the more serious the injury they will have suffered from the pressure to which they have been subjected, and the greater the amount of constitutional distress. As a consequence of this, the lungs and other organs return less readily to their normal condition, and the patient has less chance of rallying. Nevertheless, patients have often died when the effusion has been somewhat scanty, and others have recovered when it has been very abundant. In one of the cases treated by the system of drainage, no less than "eight quarts had escaped in twenty-four hours," and yet the patient recovered without a drawback.

The same uncertainty attaches to the character of the effused liquid as a foundation for our prognosis. In most instances, undoubtedly, mere serum in the pleura is less formidable than a sero-purulent or puriform fluid; but in many cases the liquid, which at first was serous, has subsequently become puriform, and the case has terminated fatally. On the other hand, cases in which the secretion has been purulent, even before the chest was punctured, have gone on steadily to recovery. In my own practice I have met with several examples in point, and many have been put on record by other observers. The most unfavorable cases are those in which when the chest is opened the fluid which escapes is sero-purulent and sanious, and possibly fetid. These are usually connected with the presence of dead bone, or malignant or other organic disease, and are almost invariably fatal. I know of no trustworthy statistics showing the ratio which the cases of recovery bear in the three sets of cases respectively, but my own observation leads me to believe that the proportion is largest when the effusion consists of serum only, and smallest when the liquid is sero-purulent and bloody. The cases in which the fluid which escapes is pure pus occupy a middle place, and in these the mortality is about one-third. So at least it has proved in cases under my own observation, for of these six out of seventeen died. When the fluid which escapes consists of pus from the first, the risk is less than when it is originally serum, and subsequently assumes a sero-purulent or purulent character.

After all, however, the general condition and the constitutional powers of the patient in most cases form a much more trustworthy ground for an opinion as to the issue of the operation than do the quantity and quality of the effused liquid. This, therefore, is a point to which attention should be directed, and the bearing of which should be carefully weighed, before the operation is undertaken. It is manifest that effusion is less likely to prove fatal when it results from simple pleurisy than when it is connected with a carious rib; it is less likely to have an unfavorable issue when the lungs are free from organic mischief, than when they are tuberculous, or otherwise diseased; and it will less probably run an untoward course when the patient is constitutionally sound, than when his kidneys or other organs are organically deranged, or when his health and strength are shattered by intemperance or by vicious habits. If the patient be young and vigorous, of sober habits and sound constitution, and if his digestive organs perform their functions properly, there will be far less risk of a fatal result than in persons advanced in years, or of an opposite condition of system. Indeed, judging from what I have observed in the cases which have come under my own observation, I am inclined to believe, that although the operation terminates fatally in a large proportion of those who have led an irregular or intemperate life, or are constitutionally or organically unsound, yet that it issues favorably in a large proportion of vigorous and temperate persons, provided only that it be not delayed too long, and that a free opening be made for the escape of the liquid.

In cases of double empyema, the prognosis, of course, is extremely unfavorable, and the operation is scarcely justifiable.

Pneumothorax and Hydropneumothorax.

It has been already stated that air or gas is sometimes found in the inflamed pleural cavity, and that its presence gives rise to pathological effects and physical signs which require careful consideration. But before entering into these details it may be advisable to define what is meant by the terms pneumothorax and hydropneumothorax, and to point out the various circumstances under which these affections may arise.

The term pneumothorax is used to denote a collection of gas or atmospheric air in the cavity of the pleura; and the term hydropneumothorax, if strictly applied, signifies the coexistence of air and fluid in the same cavity. But idiopathic pneumothorax rarely exists, and even when it does, is usually accompanied, in a few hours, by the effusion of fluid in the pleura; so that, practically, the terms pneumothorax and hydropneumothorax may be regarded as almost convertible. The various causes which may lead to the presence of air in the pleural cavity will be seen on inspection of the following table, in which they are classified under four separate heads:

1st. *When no communication exists between the pleura and the external air:*

α. Spontaneous evolution of gas from the decomposition of fluid in the pleura, or from gangrene of the pleural membrane.

β. The secretion or exhalation of air by the pleura. (?)

2dly. *When a communication takes place between the pleura and the alimentary canal:*

Resulting from inflammatory softening and perforation of the œsophagus or stomach.

3dly. *When a communication occurs between the pleura and the atmosphere through an opening in the chest-wall:*

The result of penetrating wounds of the thorax, or of parietal abscess.

4thly. *When a communication occurs between the pleura and the bronchi:*

α. The result of violence—rupture of the lung-substance, and tearing of the pulmonary pleura.

β. The result of disease. Perforation of the pulmonary pleura from without inwards, as by empyema; or from within outwards, as by ulceration excited by hydatids, or by tubercular, cancerous, pneumonic, gangrenous, metastatic and bronchial gland abscess, or by rupture of the pulmonary pleura in emphysema, and pulmonary apoplexy.

The first and second sets of cases are of extreme rarity, and the third belongs exclusively to the province of the surgeon; so that, practically, the fourth class comprises the causes to which our attention must be specially directed. Even these vary greatly in the frequency of their operation. Thus tubercular ulceration is the efficient cause of perforation of the pulmonary pleura in a vast majority of cases. Dr. Walshe has stated 90 per cent. as the proportion of cases in which, if traumatic cases are excluded, pneumothorax is referable to tuberculous ulceration, and my own experience leads me to believe that this is a close approximation to the truth. For although of 147 cases collected by M. Saussier, 81 only are reported to have occurred in phthisical patients, it is certain that these numbers do not furnish trustworthy data for an opinion as to the causation of pneumothorax. "Tuberculous perforation is an every-day affair, which passes unnoticed" and unrecorded; whereas "perforation from gangrene, vesicular emphysema, hydatids, pulmonary apoplexy, abscess, and other rare causes, are greedily caught hold of and registered." Hence it naturally follows that when examples are collected from different sources, the result, as in M. Saussier's cases, does not show a sufficiently large proportion of tubercular disease. In the absence of any trustworthy statistics bearing on the subject, I am constrained to refer to my own observation and to our combined experience at St. George's Hospital as evidence on the question at issue. My own experience relates to twenty-two cases, in eighteen of which tuberculous ulceration was the cause of perforation, in one pneumonic abscess, and in the remaining three empyema, which emptied itself through the lung.[1] The post-mortem records of St. George's Hospital for the ten years ending December 31st, 1850, tell of twenty-three cases, in twenty-one of which tuberculous ulceration was the cause of perforation, and in two pneumonic abscess.[2]

Thus, then, it is apparent that if traumatic cases are excluded, the presence of pneumothorax may be regarded as presumptive evidence of tuberculous ulceration of the pleura, and it may be satisfactory to inquire as to the usual seat of the perforation, the age and sex of the sufferers, and the period of the phthisical disorder at which this complication is prone to arise.

Of the 18 phthisical patients to whom my notes refer, 11 were males and 7 females. Of the males, 1 was from fifteen to twenty years of age, 6 were from twenty to thirty-five years of age, 3 were from thirty-five to fifty years of age, and 1 was fifty-four years of age. Of the 7 females, 1 was only eighteen years of age, 4 were from twenty to thirty-five years of age, 1 was thirty-nine years of age, and 1 was forty-two.

In most instances the patient was in an advanced stage of consumption. This, however, is not a necessary condition; for Louis has recorded a case in which perforation took place within

[1] These 16 cases do not include any of the instances which proved fatal in St. George's Hospital during the decennial period ending December 31, 1850.

[2] See St. George's Hospital Post-mortem and Case books; also "Decennium Pathologicum," cap. v, sec. v; by Dr. T. K. Chambers.

a fortnight after the phthisical symptoms first declared themselves, and experience has fully demonstrated the fact that it may occur at any period of the disease. Nevertheless, its occurrence at other than an advanced stage of the disease is quite exceptional.

In 10 of my 18 cases, the perforation took place on the right side, and in 8 on the left; but in 9 of the cases I am unable to state positively at what spot the perforation occurred, inasmuch as 5 of the patients passed from my observation before a fatal termination ensued; in 3 of the others there was no post-mortem examination, and in 1 case the firmness of the pleural adhesions, and the consequent tearing of the lung on removal, rendered it impossible to ascertain precisely the seat of perforation. In the remaining 9 cases the pleura gave way, as below stated, viz.:

On the right side—In the middle and outer part of the upper lobe, in 1 case.
" In the lower and posterior part of the upper lobe, in 3 cases.
" In the outer part of the middle lobe, in 2 cases.
On the left side—In the middle and posterior part of the upper lobe, in 2 cases.
" In the upper and posterior part of the lower lobe, in 1 case.

Of the 21 fatal cases which occurred in St. George's Hospital during the decennial period embraced in Dr. T. K. Chambers's "Decennium Pathologicum," "10 were males, and 11 females. The females were all between seventeen and thirty-one. Of the males, 7 were from fifteen to thirty, 2 from thirty to forty-five, and 1 forty-six years of age." In every instance, save one, the tubercular disease was far advanced. In 12 of the 21 cases the perforation had occurred on the right side, and in 9 only on the left—a fact which confirms my own observation, but is opposed to the statistics given by Dr. Walshe, relative to 74 cases collected from different authors, and to 10 observed by himself; of these 84 cases, we are told that 55 were examples of perforation on the left, and 29 only of its occurrence on the right. For obvious reasons, the results obtained from a number of cases taken indiscriminately at a large public institution are likely to be more trustworthy than those gleaned from selected cases, and I am, therefore, inclined to believe that there is probably little or no difference in the liability of the two sides of the chest to suffer from this complication of phthisis.

With respect to the precise seat of perforation, Dr. Walshe, without giving any statistics on the subject, asserts that "the pleura commonly gives way postero-laterally in the area comprised between the third and sixth ribs." I have already given the results of my own experience in the matter; but as I am disposed to give credence to the accurate statistics of a large public hospital, rather than to any generally expressed opinion, or to statistics derived from selected cases, I will again refer to our united experience at St. George's Hospital. Thus in the 21 cases already alluded to as having occurred during the decennial period ending December 31st, 1850, the fistulous openings from the bronchi by vomicæ into the pleura were as follows:

On the right side—In the middle of the upper lobe, in 3 cases.
" In the base of the upper lobe, in 1 case.
" In the middle lobe, in 2 cases.
" In the upper part of the lower lobe, in 1 case.
" In the lower edge of the lower lobe, in 1 case.
" Distributed through the upper part of the lung, in 3 cases.
" In another case the lung contained many tubercles, but none were softened at the time of death, and any solution of continuity in the pulmonary tissue, which was condensed by pressure of the air, seemed to have been obliterated.

On the left side—In the middle of the upper lobe, in 3 cases.
" In the anterior part of the lower lobe, in 2 cases.
" In the outer part of the lower lobe, in 1 case.
" Distributed through the lung, in 1 case.
" In 2 cases, though the lung was full of vomicæ, the exact position of the opening could not be ascertained.

These facts coincide very closely with those observed by myself, and with Dr. Walshe's statement, that perforation commonly occurs in the area comprised between the third and sixth ribs. The probable explanation of them is so well put by Dr. Chambers, that I am tempted to give it in his own words. He says,[1] "But how are the exposed parts," *i. e.*, the apex of the lung,

[1] Loc. cit., cap. v, sec. v, p. 65.

which is the ordinary habitat of vomicæ, and the base, which is the usual seat of abscess—how are these parts "guarded? I believe by the following provision: In breathing, the base, from the pressure of the diaphragm, and the apex, from being driven up into a corner, have less motion than the middle of the lungs, and consequently, when pleurisy occurs, they are most liable to become adherent to the thoracic walls. The adhesions thus formed not only prevent the perforations there from being detected, but they also prevent harm ensuing from them."

Thus, then, I think we are justified in stating that pneumothorax is not more prone to occur in one sex than in the other, nor on one side of the chest more than on the other; that it is most commonly met with between the ages of twenty and thirty-five, and is generally due to perforation in the middle of the chest on either side, for the reason that motion is greater there than in any other part of the chest, and that therefore the pleura in that position is less likely to be protected against the results of perforation, by adhesion of the two surfaces of the membrane.

The general symptoms of perforation of the pleura are the sudden accession of a sharp pain in the side, accompanied by a sensation as of something giving way, and followed by intense difficulty and distress of breathing. Sometimes the pain is agonizing in the extreme, and all the general symptoms are well marked; at others there is little or no pain in the side, and though the breathing is very hurried, there is no sensation of dyspnœa. In a case which I saw last year in consultation with Mr. Allen, of St. John's Wood, there was neither pain or distress of breathing; and in the case of Mary Ann R—, a young woman admitted under my care into the Crayle Ward of St. George's Hospital, on April 22d, 1865,[1] in whom a circumscribed empyema emptied itself through the lungs, not only was there very little pain or dyspnœa, but all traces of air had disappeared, and the patient left the hospital apparently in perfect health within three weeks after the date of the opening into the pleura. The same facts, though varying in their details, have been observed in several other instances.

As soon as perforation has occurred, the air, acting on the pleural membrane, excites irritation, and ordinarily gives rise to the outpouring of fluid; so that the signs of hydropneumothorax may be discovered within a very few hours. Orthopnœa generally exists at first, but it often ceases after a time, and the patient lies, as he does in simple pleurisy, on his back, with his head somewhat raised, and his body slightly inclined towards the

[1] See Hospital Case-book.

affected side. Distress and anxiety are depicted on the countenance; the complexion is pale and dusky, and the lips are more or less livid; the voice is weak; the skin moist, and often covered with a cold clammy perspiration; the pulse is quick and feeble, and the respiration extremely hurried.

The physical signs of simple pneumothorax are sufficiently characteristic. There is evident convexity of the affected side, with obliteration, widening, and even bulging of the intercostal spaces, immobility or diminished movement of the chest-walls, and inaction or diminished movement of the intercostal muscles, contrasting forcibly with the increased play of the opposite side of the chest and the energy of its intercostal action.

Palpation informs us that the vocal fremitus is diminished or altogether annihilated, that the intercostal spaces are more than usually elastic and resilient, and that the heart is more or less displaced.

Mensuration confirms our impression respecting the enlargement of the affected side, and the increased width of the intercostal spaces.

Percussion elicits a clear tympanitic resonance, which sometimes changes its character and becomes amphoric and of a metallic quality, over the trachea and larger bronchi. When the dilatation of the chest and the tension of the chest-walls are excessive, the true tympanitic character of the sound is diminished or lost. As the mediastinum, the heart, and the diaphragm are more or less displaced, the area of clear resonance on percussion may extend considerably beyond its normal limits, and thus may transgress the middle line of the sternum; when pneumothorax occurs on the left side, the pericardial dulness may be wholly replaced by tympanitic resonance.

Auscultation furnishes different results according as the amount of effused air is larger or smaller. If the quantity of air be small, the respiratory sounds are weak and distant, and the vocal resonance is weak; if it be great, so that the lung is thoroughly compressed, the respiratory sounds and the vocal resonance are almost or altogether absent, except in the interscapular region at the root of the larger bronchi, where diffused blowing respiration and diffused but loud vocal resonance may still be audible. In some instances, the voice, the cough, and even the sound of inspiration may be accompanied by a metallic or amphoric echo, especially when the communication with the bronchi is free; and in others, even where little or no respiratory murmur is present, there may be diffused vocal resonance over the entire chest. The heart's sounds are usually transmitted feebly through the distended pleura, but sometimes, as with the voice and cough, they may give rise to an echo of a metallic quality.

The signs of hydropneumothorax, as the title implies, are in part those of pleurisy and in part those of pneumothorax. The former, such as dulness on percussion, and the other signs referable to the presence of fluid, are met with in the lower or more dependent portions of the chest; the latter including tympanitic resonance, at the upper part, to which air necessarily rises. Their precise character will be manifest by reference to what has been already stated respecting the signs of pleurisy and pneumothorax respectively; and it need only be added, that their relative position in the chest will be found to vary according to the posture of the patient.

But there are certain signs resulting from the coexistence of air and fluid in the same cavity, which are not met with in connection either with pleurisy or pneumothorax, but which frequently occur with and are very characteristic of hydropneumothorax. I allude—1st, to fluctuation, which is felt by the patient as well as by the observer when the body is abruptly jerked or shaken; 2dly, to the ringing, splashing sound—the succussion-sound of Hippocrates[1]—which is heard under the same circumstances; 3dly, to the remarkable metallic tinkling which sometimes accompanies succussion of the patient, but is also apt to accompany cough or inspiration, or a sudden change in the patient's posture. The character and mechanism of these sounds will be seen by reference to the former part of this work;[2] and it need only be stated further that the two former may be heard as well when the communication between the bronchi and the pleura has closed as when it remains open; but that metallic tinkling is not usually found to accompany inspiration unless the opening be free, so that bubbles of air can pass through the fluid during inspiration.

The quantity of air and fluid, and their proportion relatively to one another, vary greatly, and thus of course the distension of the side and the urgency of the symptoms arising therefrom are also found to vary. This difference, as far as relates to the quantity of air, is due, I believe, to a peculiarity in the form of the opening into the pleura. In some instances it is valvular, and permits the free ingress of air, but closes instantly against its egress; whilst in other cases it is fistulous, and admits of air being expelled as freely during expiration as it entered during inspiration. In the former, air necessarily accumulates in the

[1] "Hippocrates de Morbis," lib. ii, 45.
[2] See chapter x, pp. 156–165.

pleura and causes enormous distension; in the latter, it exercises little influence in producing distension. Thus it happens that in certain cases, but not invariably, enlargement of the affected side and displacement of the organs reach their utmost limits in hydropneumothorax, and so does amphoric resonance of the voice, the cough and the respiration—the amphoric character being usually most marked in proportion as the ear approaches the seat of perforation.

The general symptoms of hydropneumothorax do not differ notably from those of pneumothorax, and, therefore, need not be recapitulated.

The prognosis of pneumothorax and hydropneumothorax is always uncertain; and if an opinion is asked as to the issue of any particular case, it should be given with due caution and reserve: *cæteris paribus*, it is most favorable in traumatic cases in which the chest-walls are punctured, and a communication is thus established between the air and the pleural sac, but in which, nevertheless, the lung is sound and is not wounded; and it is most unfavorable in perforative cases resulting from disease in the tissue of the lung. But the greatest uncertainty attaches to the question, irrespective of that arising from the cause of the disease. The most favorable cases will sometimes terminate fatally in the course of a few days, whilst cases apparently the most unfavorable go on steadily to recovery. I have seen five instances of complete recovery from pneumothorax in phthisical patients, and several in which the patients recovered from the immediate effects of the attack sufficiently to return to their homes, though air and fluid still coexisted in the pleural cavity. Only last year, a man under my care in the York Ward of St. George's Hospital, in whom an empyema emptied itself through the lung, recovered rapidly without a single drawback,[1] and I have known several other instances[2] of the same kind.

There are no statistics which throw any light upon the prognosis of pneumothorax, nor have I met with any feasible explanation of the differences observed in its issue in different cases. Certain it is, that although the habits of life and constitutional peculiarities exercise their influence in this as in other disorders, they do not afford an adequate explanation of the vast differences observed in the course of the disease. My own experience in-

[1] Hospital Case-book, xxxviii, p. 178.

[2] One case has been already referred to (page 194). In this case, which was under my care in the Crayle Ward of St. George's Hospital, an empyema emptied itself through the lung, and yet the patient was enabled to leave the hospital without any trace of air in the pleura, and apparently in good health, within three weeks from the date of perforation. See Hospital Case-book for April 22d, 1865.

clines me to believe that, *cæteris paribus*, those cases are most apt to run an untoward course, which are accompanied by great accumulation of air, by great displacement of the thoracic organs, and great consequent interference with the functions of life; and that, therefore, an unfavorable issue is most probable when the opening into the pleural sac is of a valvular form, or becomes so in consequence of the superimposition of false membrane. Indeed, I question whether in some such cases it may not be desirable to relieve the tension of the side by puncturing the chest-walls, and allowing the air to escape.

When pneumothorax does not prove rapidly fatal, the opening into the pleura may either close or remain pervious. The former condition occurs when the case terminates favorably. But closure of the opening is not necessarily a sign of a favorable issue. In some fatal cases the opening may close; in others it may remain patulous. In certain instances, again, the opening may remain patulous for months or years, and may ultimately close, and the patient may recover; in others it may remain pervious for an equally long time without any material declension of the patient's health, though recovery does not eventually take place. Laennec mentions the case of a consumptive patient in whom all the signs of hydropneumothorax, with a pervious opening into the pleura, existed at the expiration of six years; and I have noted three in one of which these signs were present at the end of eleven months; in another, at the expiration of nineteen months, and, in the third, the case referred to below,[1] after the lapse of twenty-seven months. Such cases, however, are quite exceptional, and can be regarded as little else than medical curiosities.

The treatment of pneumothorax is similar to that recommended for pleurisy, due allowance being made for the nature of the malady from which the affection has originated. Opium, in full doses, with sulphuric ether, and other diffusible stimulants and antispasmodics, may be necessary to tranquillize the nervous system, overcome the collapse, relieve the dyspnœa, and subdue the agonizing pain by which the accession of the disease is often marked; but as soon as the first shock of the attack is overpast,

[1] A man named Lacey was admitted into St. George's Hospital, on February 18th, 1846, suffering from empyema of the left side, in whom this fact was strikingly exemplified. Fourteen months after the chest was punctured the side was discharging freely, and two years and three months afterwards there was still a discharge daily. Five years after the operation I saw him again. At that time the opening in the side was quite closed, and the side, which had been much collapsed, had expanded considerably. How long the discharge had continued it was impossible to ascertain with accuracy, but he stated that it had persisted nearly three years.

and reaction has set in, our principal reliance, in traumatic cases, must be on opium internally, and on turpentine and poppy fomentations and blisters externally. Venesection has been recommended, and, in some instances, has afforded relief; but it must be remembered that, in the majority of cases, pneumothorax arises in persons whose lungs are organically diseased, and who can ill bear loss of blood. In such cases, therefore, if bloodletting be practised, a small quantity only of blood should be drawn. For the same reason, the constitutional effects of mercury should be avoided; and if mercury is given at all, it should be administered cautiously, and in combination with medicine calculated to sustain the patient's strength. In short, the treatment which I have found most efficacious in tuberculous cases, consists of the exhibition of full doses of opium or other sedatives, followed, in the course of a few days, by quinine and cinchona, the mineral acids, cod-liver oil, and a light but nutritious diet, aided from the first by turpentine and poppy fomentations, linseed-meal poultices, blisters, and the application of mercurial ointment on the side, after the manner recommended in pleurisy occurring in weakly subjects.

In non-tuberculous cases, tapping of the chest appears to be desirable whenever the accumulation, whether of air or fluid, is such as to produce great oppression of the breathing, and interfere with the functions of life. But in tuberculous cases, the question of tapping can seldom arise, inasmuch as it can only be regarded as a palliative, and, in many instances, would not be justifiable, even when viewed in that light. Nevertheless, if the mischief in the lungs is not very extensive, nor far advanced, there cannot be any reason to decline performing the operation, if the urgency of the dyspnœa or the displacement of the thoracic organs is such as to demand operative interference. It does not render matters materially worse, and it sometimes affords great temporary relief.

Hydrothorax.

Hydrothorax is a term applied to dropsy of the pleura, or, in other words, to a non-inflammatory accumulation of serous fluid in the pleural sacs. It commonly results from passive congestion of the pleural vessels, consequent on disease of the heart and obstruction to the pulmonary circulation, or on Bright's disease of the kidneys, or other disorders which impoverish the blood and tend to congestion of the capillary system. But sometimes the congestion assumes an active form, and trenches closely on true inflammation. In these cases, there is more or less uneasiness in the chest, and often some accession of febrile action; but these soon pass off; and when death occurs, no plastic exudation or other inflammatory product is found in the chest. The effusion, as in the former cases, consists of nothing more than a thin, clear, yellowish-green, or straw-colored, transparent serum, in which some masses of amber-colored, gelatiniform lymph are sometimes found floating.

Hydrothorax generally commences, and often continues for some time, without any symptom to attract attention beyond gradually increasing shortness of breath and a sense of suffocation when the patient attempts to lie down. There is little or no cough, no stitch in the side, no pain or tenderness on pressure as in pleurisy, no febrile heat of skin, no marked acceleration of the pulse. The patient's posture is characteristic. When the effusion is very scanty, so that he is able to lie down, he invariably reclines on his back, rather than on either side; but after a short time there is always more or less orthopnœa, and as effusion increases, the difficulty and distress of breathing become excessive, all the accessory muscles of respiration are brought into play, the lips and face become extremely livid, and the extremities also cold and livid; the countenance is anxious, and the skin is bedewed with a clammy perspiration. Indeed, these symptoms are manifested to a far greater degree and occur more rapidly than in pleurisy. This arises from the fact that pleuritic effusion is usually confined to one side of the chest; whereas, hydrothorax, being commonly referable to causes which operate through the general circulation, is apt to occur simultaneously on both sides, so that the functions of respiration and circulation are more seriously disturbed in the latter than in the former affection.

The physical signs are almost identical with those which have been described as accompanying pleurisy, with the exception that in hydrothorax, as the pleura is not roughened, no friction-fremitus is felt, and no friction-sound heard; that percussion dulness occurs on both sides of the chest, instead of being limited to one side; that the area of dulness shifts its position with the varying posture of the patient more rapidly and more markedly than in pleurisy; and that as effusion takes place on both sides, there is little or no lateral displacement of the heart and mediastinum as in cases of abundant pleuritic effusion on one side only of the chest.

The treatment of hydrothorax must be regulated by the nature of the mischief from which it arises. It is merely a symptom of other mischief, and unless that mischief can be remedied, it is vain to expect to get rid of the fluid. If it be due to valvular or other disease of the heart, or to disease of the kidney, or to some diathetic disorder, our aim must be to alleviate the complaint by which it is occasioned, relieve the circulation, and so put a stop to further effusion. If we are unable to do this, our utmost efforts will not avail to counteract the continued outpouring of fluid; whereas, if we can mitigate the primary disease, and arrest the effusion, we have a reasonable prospect of being able to promote absorption of the fluid already effused. Dry cupping, repeated blistering, and the use of iodurctted lotions and ointments externally, and internally the administration of diuretics, absorbents, and purgatives, are the means to be employed on these occasions. Tapping is seldom needed, and ought never to be adopted unless effusion on both sides of the chest continues in spite of remedies, and threatens immediate suffocation. Its performance, under such circumstances, might serve to afford temporary relief and prolong life, and, therefore, would be justifiable, even though it be incapable of effecting a cure.

Hæmothorax.

Hæmothorax, or hemorrhage into the pleural cavity, never occurs as an idiopathic affection, but is occasioned sometimes by a fracture of the ribs, or by a wound of the chest-walls, and sometimes by the outpouring of blood consequent on disease of the thoracic walls or the viscera which they inclose. Thus it may result from an aneurism bursting into the pleural sac, from the giving way of the visceral pleura under the pressure of pulmonary apoplexy, and from the rupture of vessels accompanying ulceration in various forms of disease of the lungs, the pleura, or the chest-walls.[1] These latter causes, however, are of such rare occurrence that, in a large majority of cases, the complaint is of traumatic origin.

Practically this affection calls for very few remarks, inasmuch as it is not productive of any characteristic physical signs or general symptoms, and admits of very little relief by treatment. The physical signs resemble those observed in pleurisy, except that pleuritic friction-sound is not heard in hæmothorax; again they are similar to those which accompany hydrothorax, except that they are confined to one side of the chest, and that the dul-

[1] Sir Thomas Watson (loc. cit., vol. ii, p. 3, ed. 1) records a remarkable case in which an enormous hæmothorax which had caused enlargement of the left side of the chest, effaced the intercostal spaces, and pushed the heart over to the right of the sternum, resulted from scrofulous ulceration, which had destroyed two of the ribs and laid open one of the intercostal arteries.

ness on percussion does not shift its position with the varying posture of the patient so constantly and so completely as in that disease. Further, in the majority of cases, the accession and full development of the physical signs are more sudden than in either pleurisy or hydrothorax. The general symptoms consist of little more than dyspnœa and labored breathing, the necessary results of compression of the lung and interference with its action. But when extreme distress of breathing occurs suddenly, during the progress of visceral disease, or after injury of the chest, and, though accompanied by the physical signs of effusion on the one side of the chest, is not attended by any evidence of inflammatory action, but rather by pallor, faintness, and failure of the pulse, there is seldom much difficulty in arriving at a correct conclusion respecting the nature of the malady.

The treatment of hæmothorax may be summed up by the statement that everything must be done to obviate the tendency to death. The collapse is sometimes so great when the escape of blood first takes place, that diffusible stimulants are absolutely necessary to sustain life; afterwards, when reaction has occurred, pleurisy may be set up, and treatment calculated to relieve that disorder may be required accordingly. The only question which can arise, is as to the propriety of puncturing the chest at the very outset of the hemorrhage. My own opinion is adverse to the practice, inasmuch as it is obvious that the admission of air into the pleura may be prejudicial, and that the removal of the pressure which the extravasated blood creates must favor the continuance of hemorrhage, and so endanger the patient's safety. I am bound, nevertheless, to add that my objections to the operation are purely theoretical, and that M. Roux[1] and others have practised it with success. The question of performing the operation at a later period, when pleurisy has occurred and serous effusion has taken place, must be decided on the principles already discussed, when the subject of tapping the chest in pleurisy was under consideration.

[1] Reported by Sédillot in p. 107 of his work, "De l'Empyème."

CHAPTER II.

PNEUMONIA, OR INFLAMMATION OF THE LUNGS.

We have hitherto confined our observations to pleurisy, or inflammation of the investing membrane of the lungs. We must now pass on to the consideration of pneumonia, or inflammation of the substance of the lungs, the true pulmonary tissue.

The attack is generally preceded by restlessness, followed by shivering, prostration of strength, feverish heat of skin, increased frequency of pulse, hurried respiration, and short, dry cough, with a stitch or catching pain in the side, about on a level with the nipple. The restlessness and uneasiness increase; the breathing becomes more hurried and oppressed, varying in frequency from 30 to 60 in a minute; and even if the cough were dry in the first instance, it is after a time accompanied by the expectoration of stringy, adhesive, and often rusty-colored mucus. The patient lies on his back, or slightly inclined to one side or the other, supported by pillows; the face is commonly flushed, the countenance anxious; the nostrils are dilated, and in full action; the lips are more or less livid, the tongue is coated with a white or yellowish-white fur, the urine is high-colored and loaded with lithates, and the pulse full and frequent, varying from 100 to 120.

Such are the usual symptoms of pneumonia. They are subject, however, to every possible variation, and possess so little of a distinctive character, that a physical examination of the chest is needed, in order that they may be interpreted correctly. With that aid, fortunately, there is little difficulty in recognizing their true character, and tracing every step in the progress of the disease.

But before discussing the physical signs of pneumonia, it may be well to look a little more in detail into certain circumstances connected with its invasion, and into the variations observed in its general symptoms.

And first, as to the circumstances connected with its invasion. Inflammation of the lungs generally occurs as a primary affection whilst the patient is apparently in good health. In some instances, however, it arises in connection with bronchitis, the inflammation spreading gradually from the larger to the smaller bronchi, and so by degrees to the air-vesicles and tissue of the lungs. In other instances it appears to be occasioned by the congestion and irritation to which the lung is subjected during the progress of febrile disorders. Its accession therefore may be sudden, or it may be gradual and insidious.

Next, as to the variations observed in the general symptoms. Shivering is the most constant forerunner of pneumonia; but it may be altogether absent, even in severe cases. Therefore, although it is generally observed, the mere fact of its non-occurrence will scarcely justify an opinion as to the non-existence of the disease.

Pain in the side is another symptom of variable occurrence. It is usually present to a greater or less degree, so that M. Grisolle, who examined 301 cases, with a special view to this inquiry, reports it in no less than 272 cases; and in 94 cases which occurred in the physicians' wards of St. George's Hospital, during the period of my registrarship, it was noted in 73 cases. Nevertheless, the most formidable pneumonia may exist without it; or it may be severe and catching at one period of the disease, and slight or altogether absent at another.

It has been stated—I know not on what authority—that pain is not met with in cases of pure, uncomplicated pneumonia, but arises only when the pleura becomes involved by an extension of inflammation from the contiguous lung-structure. Observation has led me to doubt the accuracy of the statement; for although it is true that pain is more constantly present in pleurisy than in pneumonia, yet instances are not wanting in which wide-spread pleuritic inflammation has been unattended by pleuritic pain. On the other hand, cases of pneumonia are met with in which no evidence of pleurisy can be obtained by auscultation, and in which, nevertheless, the stitch in the side is very severe; nay more, I have traced cases to the dead-house of St. George's Hospital, in which pain in the side had existed during life, and in which, after death, the closest inspection has only served to confirm the impression previously entertained as to the absence of pleuritic inflammation. Therefore, whilst admitting that acute, catching pain in the side is suggestive of pleuritic complication, I am bound to maintain that it does not necessarily indicate its existence. It may be, and generally is, attributable to pleurisy; but in some instances it is referable to intercostal neuralgia, and it ought not to be regarded as of pleuritic origin, unless a careful physical examination of the chest unmistakably attests the correctness of such an opinion. The mere presence of a metallic or quasi-ægophonic resonance of the voice is not sufficient to decide the question.[1] Pleuritic friction-sound should

[1] See pp. 120–124 of this treatise.

be heard, or some other evidence of effusion into the pleural cavity obtained, before such a conclusion can be justified.

In corroboration of the view that the pain which accompanies inflammation of the lungs is not always occasioned by pleurisy, is the fact that its seat does not usually coincide with that of the pulmonary inflammation. It is but fair to conclude that pleurisy, arising in connection with pneumonia, would occur in that portion of the membrane which is in apposition with the inflamed lung-tissue; and, undoubtedly, when well-marked pleuropneumonia occurs, the pain is commonly referred to that portion of the chest at which evidence of pneumonia and pleurisy is to be obtained. But, in the cases now under consideration, in which proof of pleuritic inflammation is confined to the existence of slight ægophonic resonance of the voice, the pain seldom corresponds with the seat of pneumonia. It is usually felt on a level with, or a little below the nipple; but it may occur in any other portion of the thoracic walls, and it is aggravated by causes which influence the pain of intercostal neuralgia just as much as that of pleurisy. It is increased by cough and by a full inspiration; by sudden change of posture, and by suddenly and forcibly raising the arm on the affected side; by pressure upon or percussion of the intercostal spaces over the painful part, and, in short, by whatever brings the intercostal muscles into play, or exerts any pressure upon them. Thus, it generally happens, that the patient cannot lie on the affected side, and prefers lying on his back. In certain cases, however, I have known a patient lie most comfortably on the painful side, being relieved by the uniform pressure thus exerted, though suffering, of course, some increased oppression of the breathing in consequence of his position. It is obvious, then, that in the existence of pain there is nothing absolutely characteristic of the disease under consideration, and little even to aid us in our diagnosis.

Difficulty of breathing is another symptom which varies greatly in severity. Sometimes it is extreme, and is usually most urgent between the sixth and tenth days of the attack, sometimes it is so slight that the patient is hardly conscious of it; and between these two extremes every degree of embarrassment of the breathing may be observed. It always takes place, however, to a greater or less degree, and begins to manifest itself within a few hours after the commencement of the attack. The respirations may amount to 24 or 26 in a minute, almost from the very outset of the attack, and may even number 35 or 40, without any apparent difficulty of breathing; but when they attain, as they sometimes do, to 60 or 70 in a minute, it is obvious that suffocation is imminent: the face becomes livid or pale, the nostrils are dilated and in full action, the greatest anxiety is depicted on the countenance, and the patient is almost unable to speak.

Strangely enough, the dyspnœa is not proportioned to the extent of lung affected by inflammation, and does not appear to be necessarily connected with mischief in any particular portion of the lung; so that, in many cases, there may be less frequency of breathing when an entire lung is inflamed, than in others when a comparatively small portion of the lung-tissue is implicated. This holds good even when the inflammatory mischief is unaccompanied by pain. The fact appears to be, that some persons possess a relatively large surplus quantity of lung-tissue, which is not brought into play during tranquil breathing; and they are, therefore, less embarrassed by the inflammation and consequent inaction of a certain portion of their lung than are those who, possessing relatively small lungs, ordinarily employ the whole, or the greater part, of their respiratory apparatus. Hence, it happens that, although extreme dyspnœa is always a symptom of the gravest import, it is not, necessarily, of fatal augury; and that, viewed alone, without reference to other symptoms, it is a very uncertain guide to the extent of the pulmonary inflammation.

There is yet another point relative to the respiration, which is deserving of special notice, viz., that the increase of its frequency is out of all proportion greater than that of the pulse. Thus, at the very outset of the attack, the respirations will be often 30 in the minute, whilst as yet the pulse does not exceed 80 or 84; and when the pulse rises to 100 or 110, the respirations will be found to be 45, 50, or even 60. In like manner, with the decrease of the disease, the respirations fall in frequency more rapidly than the pulse, and soon resume their natural relative proportions. Indeed, the variations in the ratio existing between the pulse and the breathing is in no disease more marked than in pneumonia; and in none is more information to be derived from it in regard to the progress and probable issue of the disorder.

The cough of pneumonia is not in any way characteristic. Commencing almost with the first invasion of the disease, it is not always proportioned either to the extent or the intensity of the pulmonary inflammation; it is generally frequent, but seldom occurs in paroxysms; it is short and dry in the earlier part of the attack; looser and accompanied by expectoration at a somewhat later period; slight or altogether absent towards the close of fatal cases. The expectoration at first consists of nothing more than ordinary bronchial mucus, but at the expiration of a day or two, and sometimes within a few hours, it assumes a truly pneumonic character, and becomes viscid, semi-transparent, often rusty-colored, and tenacious, adhering so strongly to the vessel containing it, as to admit of the vessel being turned upside down, without becoming detached from its sides. The rusty-colored expectoration consists of mucus intimately mixed with blood—not streaked with it as in bronchitis, but thoroughly mixed and amalgamated with it—so that it acquires a yellowish or reddish-yellow, or even a red color, according to the quantity of the blood. If the disease be not very intense, the expectoration never attains the degree of viscidity or the depth of color above referred to, but, though still tenacious, and adherent to the sides of the vessel, moves from one part to another, as the vessel is tilted. If the disease, however severe, progresses to a favorable termination, the sputa become more abundant, less adhesive, and less highly colored, passing through the various shades of orange, until at length they become greenish or whitish, and resemble the expectoration of ordinary catarrh. If, on the other hand, the disease be hastening to a fatal termination, the expectoration becomes scanty, less tenacious, and of a darker or dullish brown hue, resembling the juice of prunes. Sometimes, if the type of inflammation be low, or if it be connected with tubercular deposit, the mucus may be tinged or streaked with blood; and in another class of cases, which of late years have been more frequent than formerly, the expectoration is neither blood-tinged nor rusty-colored, but consists throughout of nearly colorless, stringy, and more or less frothy mucus. In fact, the character of the expectoration varies greatly in different cases, and even at corresponding stages of the complaint in cases which otherwise resemble each other closely; so that it is impossible from mere inspection to draw any inference as to the stage at which the disease has arrived, or as to the precise condition of the lung. Under the microscope the rusty-colored sputa of pneumonia are seen to consist of mucus-corpuscles, epithelium, exudation-cells, granular matter, blood-discs, oil-globules, and casts more or less perfect of the ultimate subdivisions of the bronchi. Sugar has been detected in it during the height of the inflammation; and Dr. Beale has shown that an excess of chloride of sodium is constantly present.[1] Bamberger reports that, during the inflammatory period, the sulphuric acid in the sputa is remarkably increased, while the alkalies in combination with phosphoric acid, which enter largely into catarrhal sputa, are almost wholly absent.[2] During the advanced stages of inflammation of the lungs,

[1] "Med.-Chir. Trans.," vol. xxxv.

[2] See "Year-book of Med. and Surgery for 1862," New Sydenham Society.

when the respiration is very frequent, the expired air is colder than natural, and contains less than the ordinary proportion of carbonic acid. This has been clearly proved by the experiments of Nysten, and is doubtless attributable to the impermeability of the lung, and to the consequent diminished activity of the chemical changes taking place in the pulmonary apparatus.

Heat of skin is one of the most constant symptoms of inflammation of the lungs. The temperature of the surface is almost always raised, and the skin is often pungently hot and burning. Sometimes the thermometer, placed in the axilla, or under the tongue, will rise on the first day to 100 or 102, and on the third or fourth day to 105°; but after the sixth or seventh day the temperature begins to fall, and when the breathing is much oppressed, and the disease is tending to a fatal termination, the temperature of the body sinks, the surface becomes cold, a clammy perspiration breaks out, and a vesicular eruption of sudamina sometimes makes its appearance. A decrease of temperature always precedes the falling of the pulse and the commencement of resolution.

The pulse varies greatly in different cases and at different periods of the disease. About the third or fourth day of the attack it generally beats 120 or 130 per minute, and in cases tending to a fatal issue it may attain to 160 beats per minute; but in some persons the disease may run its course with a pulse not exceeding 60 or 70. In most of such cases there is a remarkable absence of feverish excitement, and the symptoms have always appeared to me to denote passive congestion rather than truly active inflammation. Certain it is that in the more formidable types of the disease the pulse is always frequent, and in the first instance generally full and resistant; and even when the breathing has become oppressed and the pulmonary circulation embarassed, it is still rapid, though small and feeble, and sometimes intermittent. In one instance only have I chanced to meet with a quiet pulse coincidently with serious and extensive pulmonary inflammation; and in the case alluded to, the pulse increased in frequency after the first violence of the attack had abated. I am unable even to offer a conjecture as to the cause of the slowness of the pulse in these cases, but the suggestion offered by Dr. Walshe as to the pulse in such patients being still slower in health, is not in accordance with the results of general observation. The same slowness of the pulse is sometimes observed in pericarditis, peritonitis, and other inflammatory disorders; and in those cases, just as in the case of pneumonia above alluded to, I have known the frequency of the pulse increase with the subsidence of the inflammatory mischief.

The blood in pneumonia is highly charged with fibrine, so much so that the fibrinous element amounts sometimes to 13 parts per 1000, and on the average to 7.3 parts.[1] Consequently, when venesection is practised, the clot is highly buffed and cupped; and when the circulation through the lung is much impeded, there is a great disposition to the formation of fibrinous coagula in the right cavities of the heart. In some cases the presence of these coagula forms a serious aggravation of the danger, and renders recovery impossible.

The brain sympathizes with the general disturbance, and "cerebral symptoms" not unfrequently result. These consist of headache, which is amongst the earlier and more common accompaniments of pneumonia; of delirium, which, though less common, is by no means of unfrequent occurrence;[2] and of convulsions and coma, which are rarely, if ever, met with, except towards the close of fatal cases. These head symptoms when severe, are always alarming; for they denote that the nutrition of the brain is seriously interfered with; and inasmuch as this can take place only through the medium of the blood, it is obvious that the vital fluid must be either highly charged with a *materies morbi*, or else must be very imperfectly arterialized. In either case it is unfit for the proper maintenance of the vital functions, and in its effects affords a measure of the severity of the disease, of the extent and intensity of the mischief in the chest, and of the irritability of the nervous centres. My opinion, however, is opposed to the belief that the delirium occurs merely as a consequence of the impeded state of the pulmonary circulation. I am rather inclined to regard it as invariably dependent on a poisoned blood, the morbid action of which is rendered more surely operative by its imperfect arteralization, and by some unusual excitability of the brain.

Herpes labialis is apt to appear about the third or fourth day of the attack, especially in young and middle-aged patients, and is usually regarded as of good augury. But I am inclined to doubt the correctness of this opinion. The mortality is certainly less in cases in which an eruption of herpes takes place; but this is explicable by the fact that the eruption commonly occurs at an age when the rate of mortality from pneumonia is always low. In the few instances in which I have observed it

[1] See Simon's "Chemistry," Sydenham Society's publication, vol. i, p. 260.

[2] A remarkable instance of violent maniacal delirium in connection with inflammation of the upper lobe of the right lung occurred in a man (Job Warren, æt. 24), admitted under my care into the Cambridge Ward of St. George's Hospital, February 7th, 1862. It came on suddenly without previous headache, and subsided under the use of full doses of opium and ether.

in early or advanced life, the issue of the cases has not been favorable.[1]

The digestive organs are usually deranged to a greater or less extent, but not in any constant or characteristic manner. Thirst is sometimes a prominent symptom, more especially among children.[2] The appetite is impaired or altogether lost; vomiting occurs sometimes, but not generally; the tongue is usually furred, and the bowels are often but not invariably costive.

The urine is generally scanty during the height of the disease, deep-colored, loaded with lithates, and of specific gravity varying from 1015 to 1030, or even higher. The amount of urea is always increased at the commencement of the disease, and especially from about the fourth to the sixth days, after which it gradually decreases, and may fall below the normal amount. Sometimes the urine contains a few crystals of lithic acid and of oxalate of lime, and occasionally a small quantity of albumen; the inorganic salts are deficient, the chloride of sodium especially so—a fact observed in other acute inflammatory disorders, and dependent in great measure on the nature of the food consumed. It has been found, however, that as the chlorides decrease in the urine they increase in quantity in the sputa,[3] and *vice versâ*, and it is therefore probable that their deficiency in the urine is partly attributable to their determination to the inflamed lung.

It will be admitted, then, that there is nothing in the general symptoms of pneumonia to guide us with certainty to a correct diagnosis. If the sputa happen to be extremely adhesive, semi-transparent, and rusty-colored, and present under the microscope the characters already described as appertaining to pneumonic expectoration, it is fair to conclude that inflammation of the pulmonary tissue is present. But inflammation of the lungs may be very severe, and yet the sputa may not be characteristic; so that it is only by reference to the physical signs that a correct conclusion can be drawn as to the nature of the disease and the stage at which it has arrived.

It has been already stated that the physical signs are dependent on the morbid changes to which inflammation gives rise in the lung, and it may be well, therefore, to collocate a descrip-

[1] Herpes occurred in 182 out of 421 cases of pneumonia observed by Geissler; and of the herpetic cases 17 only died, whereas 70 of the non-herpetic patients succumbed to the disease (see Froriep's "Notizia," vol. iii, No. 7, 1861, quoted in "Med. Times and Gazette," February 1st, 1861).

[2] Rilliet et Barthez, p. 99.

[3] See Dr. Beale's paper in "Med.-Chir. Trans.," vol. xxxv.

tion of the morbid changes and physical signs which are usually observed at different stages of the disease.

By common consent, the course of pneumonia has been divided into three stages, corresponding to three well-marked and distinctive conditions of the lung. The first is that of *engorgement or splenization;* the second, that of *red hepatization;* the third, that of *gray hepatization, or diffused suppuration of the pulmonary tissue.* The peculiarities of each condition and the physical signs by which they are each accompanied will be seen on reference to the subjoined description.

Morbid Anatomy.

First Stage—Engorgement, or Splenization.

The pulmonary capillaries are gorged with blood, and the air-cells are loaded with bloody serum, which has exuded into them, and has been rendered frothy by the admixture of air.

The inflamed lung is of a livid red or violet color externally, and its surface is less smooth and glistening than natural; it does not collapse like a healthy lung, and is heavier and less crepitant under the finger; moreover, it is less elastic than natural, so that it pits on pressure like an œdematous lung.

When cut into, a quantity of reddish, frothy fluid escapes, and the cut surface is seen to be spongy and of a brick-red color. A gentle stream of water directed on it will wash away the greater part of the exudation, and render this condition more apparent. The lining membrane of the bronchial ramifications is of a deep red color.

The permeability of the lung is lessened, not destroyed; so that, although the specific gravity of the lung-tissue is increased, the most engorged portions will float when placed in water. Its consistence is found to be diminished, and its tissue is softer than natural, and easily torn, like the spleen. Hence the term splenization.

If resolution takes place at this stage of the complaint, the sero-sanguinolent fluid which has been poured out is absorbed, the congestion gradually passes off, and the lung returns to its normal condition.

In certain instances, which during life are unaccompanied by crepitation, there is little or no frothy fluid in the air-cells, but the interlobular cellular tissue is the seat of inflammation, and is distended with a reddish-colored fibrinous exudation, which presses upon and, for the time, obliterates the air-vesicles.

Second Stage—Red Hepatization.

Not only are the pulmonary capillaries gorged, but the air-cells are distended with fibrinous exudation, which produces consolidation of the pulmonary tissue, so that the lung does not collapse when the chest is opened. The surface of the lung is still of a brownish-red or livid-red color, but the hepatized portions no longer contain air, no longer crepitate under pressure, and sink when placed in water.

When an incision is made into the lung, the fluid which escapes is usually thicker and less in quantity than in the state of engorgement; it is of a reddish-brown or claret color, and almost free from admixture of air. The lung-tissue is no longer spongy in appearance; it is evidently more compact and solid than natural, and is sometimes firm and resistant; but more commonly it is soft and friable, or easily broken down; so much so, indeed, that it may be quite rotten, and may be reduced to a state of pulp by the slightest pressure.

Its cut surface is frequently smooth, and resembles the cut surface of liver; more commonly it is granular, and occasionally it is pulpy and almost diffluent. Sometimes it is of a uniform claret-red color, but at others mottled or variegated, partly from the presence of black pigment, and partly from the contrast afforded by the interlobular cellular tissue, and by the coats of the bronchial tubes and vessels, which are less red than the surrounding tissue.

Its torn surface is always granular, and when the granulations are carefully examined with a lens of low power, they are seen to be swollen air-cells, filled with blood-tinged coagulated fibrine, which sometimes may be detached from the cell-walls to which it is adherent. In some rare instances the interlobular cellular tissue appears to be the principal seat of mischief, and is seen to be infiltrated with fibrinous exudation, in which case the air-vesicles are compressed and almost obliterated, instead of being distended with exudation, and the torn surface of the lung does not present the usual granular appearance.

As the second passes into the third stage of the disease, the hepatized lung becomes less dark colored, and traces of a gray or yellowish matter may be perceived here and there mixed with the reddish fluid which escapes when the cut surface is gently scraped with a scalpel.

When resolution takes place at this stage of the complaint, a serous fluid is poured out, which leads gradually to the separation and liquefaction of the fibrinous matter previously effused, and as the liquefied matters are expectorated the lung becomes more pervious to air.

Ultimately the exuded matter may be thoroughly got rid of by expectoration, the engorgement of the pulmonary capillaries may pass off, the color and sponginess of the lung return, and its tissue may reassume its normal appearance.

Third Stage—Gray Hepatization.

This consists in diffuse suppuration of the pulmonary tissue, and is sometimes termed purulent infiltration. In this stage the coloring matter of the blood disappears, the solid fibrinous exudation is broken up, and is gradually replaced by liquefied exudation-matter and true pus. The affected portion of the lung is of a gray or dirty gray color, or mottled, from admixture of black pigment or red coloring matter; it is dense and impermeable to air, sinking instantly in water, and is soft and rotten, breaking down under pressure into a yellowish-gray pulpy mass. When an incision is made into it, its whole texture is seen to be infiltrated with pus; puriform matter, unmixed with air, oozes from every divided bronchus, and from almost every air-cell, and emits a faint, disagreeable odor; and when the liquid matter has been removed by washing, the granular texture so characteristic of the second stage of the disease, is seen to have almost disappeared; here and there, however, it still partially exists, the granulations being of a gray color.

Sometimes, though rarely, a circumscribed abscess results from pneumonic inflammation, so rarely, however, that when Laennec published the second edition of his work, he had met with only six examples of it; and, amongst the cases which proved fatal in St. George's Hospital during the ten years ending December 31st, 1850, twenty-two instances only of pulmonary abscess were met with, including cases of secondary inflammation. When it does occur,[1] it forms an irregular-shaped cavity, with ragged walls, containing puriform matter, and surrounded by rotten lung-tissue, infiltrated with pus. When it has discharged its contents it may collapse and cicatrize, and thus, if the portion of lung involved be large, it may lead to retraction of the chest-walls over the collapsed parts.

Another occasional result of pneumonic inflammation is gan-

[1] M. Grisolle is of opinion that the formation of pneumonic abscess is connected with a debilitated constitution. He says (loc. cit., p. 332)—"Sur seize individus qui ont succombé avec des abcès pulmonaires primitifs, et pour lesquels on a tenu compte de l'état constitutionnel, je n'en trouve que deux ayant les attributs d'une bonne constitution."

grene, either circumscribed or diffused—most commonly circumscribed and very limited in extent. Its occurrence, however, is even more rare than ordinary abscess. I have only met with it on three occasions, and it was not once noted amongst the 305 cases of pneumonia referred to by M. Grisolle.[1] Indeed it is doubtful whether its occurrence is not attributable to some constitutional taint, some peculiarly distempered condition of the blood leading to embolism of the nutrient vessels, and not to the violence of the pneumonic inflammation. Certain it is that the arteries leading to the gangrenous parts are often filled with coagula. The gangrenous portion is of a greenish-brown color, and is sometimes soft and wet, more often diffluent, and emits an intensely fetid odor. Even during life, the abominable stench is characteristic of mortification. It results in the gradual liquefaction and elimination of the dead matter, and the consequent formation of a cavity which may collapse and cicatrize like an ordinary pneumonic abscess.

In some instances pneumonia terminates in induration of the lung: the exudation-matter, instead of undergoing the process of softening, hardens, and subsequently contracts; rendering the lung-tissue tough and impervious to air, and the bulk of the lung smaller than natural. On two occasions I have met with cases in point. The physical signs of pulmonary consolidation were very persistent, and a post-mortem examination showed that simple consolidation and contraction of a lung had occurred without pleuritic inflammation. Such cases, I believe, are not examples of ordinary pneumonia, but are connected with an exudation more than usually fibrinous into the interlobular cellular tissue; undoubtedly they are extremely rare; and, in some of the instances referred to by authors as lungs contracted by pneumonia, I believe the lungs to have been collapsed, and in others to have been compressed by pleuritic effusion.

Physical Signs.

First Stage.

Inspection.—The costal movements are not materially diminished except when catching pain in the side leads to their being restrained.

Palpation.—No material alteration can be perceived in the vocal fremitus.

Mensuration.—The size of the affected side is not increased.

[1] Grisolle, loc. cit., p. 345.

Percussion.—A peculiar amphoric quality of resonance is often emitted by that portion of the chest at which inflammation is just beginning;[1] but this ceases soon after the commencement of exudation, and is replaced by dulness more or less marked, according to the amount of exudation and the extent of lung involved.

Auscultation.—During the early period of engorgement, and before exudation takes place, the breathing is weak in the affected parts; exaggerated in their immediate vicinity. As soon as fluid is poured out, the respiratory sounds are obscured or replaced by the characteristic small crepitation of pneumonia—a sound as of a multitude of minute crackles, which occur in a volley towards the close of inspiration, and are not affected by coughing or expectoration. Neither bubbling râles nor rhonchi are audible, unless bronchitis be also present.

When the air-cells become completely filled and the minute bronchial ramifications obstructed by exudation, the crepitation ceases; and when, as often happens, inflammation extends to the smaller bronchi and capillary bronchitis is set up, small bubbling râles are heard accompanying expiration as well as inspiration. In some few instances, especially when pneumonia arises in connection with acute rheumatism, tubular breathing is rapidly induced, but crepitation never occurs—a fact which I have verified on several occasions, and believe to be attributable to the occurrence of exudation into the interlobular cellular tissue, and consequent immediate occlusion of the air-cells. The mere non-occurrence of crepitation, therefore, is not a certain proof of the non-existence of inflammation of the lung-tissue.

The vocal resonance is usually somewhat intensified.

If the disease is checked at this stage, and resolution takes place, crepitation gradually ceases, and is replaced by the respiratory murmur, whilst at the same time the chest recovers its normal resonance on percussion.

Second Stage.

Inspection.—The costal movements are diminished on the affected side; that of elevation, however, less than that of expansion. On the unaffected side they are somewhat increased.

Palpation.—The vocal fremitus varies. It is usually above the average standard, but, in some instances, it falls below it, and occasionally, when the bronchi are plugged with exudation it is altogether abrogated. Of this I have no doubt; for, on

[1] See pp. 68–75 of this treatise.

two occasions in which during life I noted entire absence of vocal thrill, I had the opportunity of verifying by post-mortem examination the existence of mere pneumonic consolidation.

Mensuration.—I have never been able to satisfy myself as to the existence of any increase in the size of the affected side, but M. Grisolle asserts that he has detected slight enlargement in some rare instances.[1]

Percussion.—The sound is usually dull, and the sense of resistance to the finger considerable. In some instances, however, the diminution of clearness is not great, and in others, percussion may elicit a clear but shallow resonance. This occurs only when a superficial layer of the lung contains air, whilst solidified lung lies behind it.[2]

Auscultation.—The respiration is of a hollow character, sometimes diffused throughout the hepatized portion, but more commonly tubular in its centre, and harsh, diffused, and blowing towards its periphery. Over those portions where tubular breathing exists there is entire absence of râles and rhonchi, and intense vocal resonance is heard, which is usually of a metallic, ringing character, and often *quasi* ægophonic. Towards the confines of the hepatized portion the vocal resonance is less intense, and loses its ringing and bleating character. Just beyond the limits of the hepatization, fine crepitation is often audible.

Sometimes, when the fibrinous exudation is very great and the bronchi become obstructed, neither tubular breathing, nor rales nor rhonchi, are audible over the hepatized lung; there is little or no vocal resonance, and there is absolute dulness on percussion. Such examples, however, are somewhat uncommon. The heart's sounds are frequently transmitted through the consolidated lung with unnatural intensity; but the facility of their transmission appears to vary with the precise species of exudation, and the consequent condition of the consolidated lung. Through some lungs which are extensively hepatized the heart's sounds are not transmitted at all; through others, also extensively hepatized but still vibratile, not only are the cardiac sounds transmitted, but a distinct fremitus may be conveyed to the hand.

[1] "Traité pratique de la Pneumonie," par M. Grisolle, pp. 226–7.

[2] See pp. 68–75 of this treatise. The cases referred to by various authors in which this resonance is attributed to the proximity of an air-distended stomach, or to the presence of a large bronchus behind a thin layer of solidified lung, are all, I believe, explicable by the condition of lung above described. If the proximity of an air-distended stomach would occasion this resonance, it ought to be present in almost every case of pneumonic consolidation of the lower lobe on the left side.

As soon as resolution has commenced, and liquefaction of the solid exudation-matter has taken place, other signs begin to manifest themselves. The respiration is then accompanied by returning crepitation, or more often by moderate-sized rales, resulting from air passing to and fro through the fluid which occupies the air-cells and smaller bronchial ramifications. If any of the larger bronchi contain fluid, the rales may not only become large and coarse, but, if the surrounding lung-tissue remains condensed, may even give rise to the impression of gurgling.

As resolution proceeds, the respiration ceases to be tubular, and becomes at first more diffused, then simply coarse and blowing, and ultimately natural; the vocal resonance becomes less intense, and loses its *quasi* ægophonic character; and the percussion-note, which at first is dull, assumes by degrees its normal character. Indeed, the clearness and resonance of the percussion-note returns more quickly than it does after pleurisy.

Third Stage.

The physical signs in the early part of this stage of the disease resemble those already described as belonging to the preceding stage; in short, they are simply the signs of consolidation. It has been questioned whether resolution can take place when the lung has reached the stage of gray hepatization, and it is manifestly impossible, from the nature of the case, to adduce conclusive evidence of the fact. Be this as it may, it is certain, that in the event of its occurrence, it would be accompanied by signs very closely resembling those observed during resolution in the second stage of the disease, and differing from them only in the larger size of the rales.

When a cavity is in process of formation, whether as the result of simple abscess or of circumscribed gangrene, there is little or no physical indication of the fact. The percussion-sound, of course, is dull if the abscess is seated superficially, and respiration is almost, if not quite inaudible over the part; but there is no positive evidence of the mischief which is going on. It is otherwise, however, as soon as the abscess has formed a communication with the bronchi, and part of its contents have been evacuated. There are then, superadded to the ordinary signs of consolidation, the signs resulting from the presence of a cavity, and the admixture in it of air and fluid. The percussion-sound may then be dull or of a shallow amphoric character, according to the size and position of the cavity, the condition of its walls, and the amount of air which it contains; respiration of a hollow character may or may not be audible, and may or may not be accompanied by large-sized bubbling or gurgling (when this occurs it is usually of a metallic quality), and the vocal resonance, though varying with the freedom of communication with the bronchi, is generally much augmented, and may even attain a pectoriloquous character.

If the cavity be large and seated superficially, retraction of the chest-walls over the affected part will be perceptible after a time; but, under no other condition does retraction of the thoracic parietes occur during the acute stage of pneumonia. The experience of Messrs. Grisolle, Woollez, and others, coincides with mine in this particular. Therefore, as retraction has never been observed, except under the circumstances referred to, in any instance in which a post-mortem examination has proved the absence of pleuritic effusion, it is fair to conclude that, in the cases observed by Drs. Stokes and Walshe during the progress of supposed uncomplicated pneumonia, pleuritic complications must have coexisted.

Gangrene gives rise to the physical signs which accompany abscess of the lung; and, physically, is not distinguishable from ordinary softening and excavation of the lung-tissue.

Induration of the lung is attended by diminution of the costal movements, by dulness on percussion, by the absence of vesicular, and the presence of hollow-sounding tubular breathing, and by increase of vocal resonance. Ordinarily there is absence of râles and rhonchi, but if bronchitis or œdema of the lung occurs, these sounds, of course, exist; and if the air-tubes become clogged by secretion, there may be absence rather than increase of vocal resonance. After a time, as contraction of the indurated lung takes place, retraction of the chest-walls occurs over the affected parts.

Drs. Stokes, Walshe, and some other authorities maintain that intense arterial injection of the lung exists in pneumonia prior to the stage of engorgement—a condition characterized, after death, by "dryness and bright vermilion color," and distinguishable during life, by the occurrence of respiration which is harsher, rougher, and of higher pitch than natural. My opinion, however, accords with that of Skoda, who is opposed to this supposition. Theoretically, I am unable to conceive how mere arterial injection and dryness of the pulmonary membrane can produce the signs enumerated as belonging to this condition; and practically, my observations incline me to assert with M. Grisolle, that "weakness of the respiratory murmur, attended by loss of purity and softness," usually characterizes the commencement of the disease. Indeed, this is only what might have been expected from the dryness of the pulmonary membrane and the diminution of the size of the ultimate bronchial ramifications, consequent on its tumescence. Further, it is inconsistent with the result of my observations to suppose that the bright vermilion color of the lung sometimes observed after death in cases of pneumonia, is attributable specially to pneumonic injection of its tissue; for it is not always met with in pneumonia, and sometimes it is observed in fatal cases of anæmia, and in other diseases which, during life, had not been accompanied by symptoms of pulmonary mischief. Rokitansky and others attribute it to anæmia,, and it certainly does occur most frequently in cases which have been characterized during life by anæmic symptoms. Assuredly it is not the result of inflammation of the lungs, and cannot be adduced as evidence of its existence; neither is it productive of any physical signs or accompanied by any special general symptoms on which to base a line of treatment.

The existence of fine crepitation in the lung is often spoken of as pathognomonic of acute inflammation of the lungs; but it ought to be understood that this sign, if viewed alone, does not afford a safe criterion as to the presence of the disease in question. In some instances it is absent throughout the attack; and in many more a crepitant rale is heard in the earlier stages of the disorder, which is undistinguishable from that which accompanies capillary bronchitis. True, the crepitation which is met with in pneumonia is generally more abundant, more rapidly evolved, and is finer and of a drier character than that of capillary bronchitis; true, also, that it is commonly confined to inspiration, and is heard chiefly towards its close. But it is equally true that these characters are by no means constant, and post-mortem investigations enable me to testify that whilst, on the one hand, the fine bubbling rale of capillary bronchitis, which accompanies expiration as well as inspiration, is met with in cases which speedily result in pneumonic consolidation of the lung-tissue; so, on the other, a fine dry, crepitating rale, which is confined, or nearly so, to inspiration, and is undistinguishable by its mere character from the crepitation of pneumonia, may long persist without being accompanied by pulmonary consolidation. In short, it is only by the concurrence of different signs that it is possible to arrive at a trustworthy conclusion. Fine crepitation, occurring coincidently with intense heat of skin and rusty-colored expectoration, warrants the strongest suspicion of inflammation of the lungs, and justifies the adoption of active treatment; and if it be accompanied by marked alteration in the ratio of the pulse and respiration, and is speedily followed by dulness on percussion, and tubular breathing, the existence of acute inflammation of the lung-tissue cannot be doubted. But crepitation, however fine, if not attended by an alteration in the ratio naturally subsisting between the pulse and the respiration, and not speedily followed by tubular breathing, cannot be relied upon as indicative of pneumonia. If it occurs without these symptoms, it is commonly indicative of capillary bronchitis with scanty secretion; whereas, if under the same conditions it is accompanied by dulness on percussion, it is probably due to rapid œdema, or else to congestion of the lungs connected with some febrile hæmic disorder, cardiac disease, or the deposit of tubercle, all of which may produce an outpouring of non-plastic fluid into the air-vesicles and terminal bronchi.

The seat of the disease will sometimes assist in determining the question; for whereas both lungs are prone to be affected in capillary bronchitis, and in passive congestion resulting from heart disease, or hæmic disturbance, inflammation is very apt to be confined to the right lung, or, at all events, to one lung. Thus of 139 cases which I have noted with a view to this inquiry, 77 were instances in which the right lung only was affected; 51 in which the left lung only suffered; and 11 in which both lungs were more or less implicated. An analysis of 1710 cases recorded by different authors, affords strong support to this observation; for in 908 cases the right lung only was inflamed; in 523 cases the left lung only; in 279 both lungs were attacked.[1] And when it is stated further, that of the entire number of cases of double pneumonia a large proportion commenced as cases of single pneumonia, the symptoms having existed for many days in one lung before they made their appearance in the other, it will be evident that the mere topography of the disease will often serve to decide any doubt which may exist as to the real import of a crepitation.

Some authors have asserted that, at the outset of the attack, inflammation is usually confined to the base of one lung; and have attempted from the precise position of the crepitation, to draw an inference as to the real nature of the disease in doubtful causes. But careful and extended observation has shown that no one part of the lung is specially prone to suffer; that the proportion of cases in which the base is inflamed is only about as four to three in relation to those in which the upper part is affected, and that in a considerable number of cases the

[1] See Grisolle, loc. cit., p. 28.

middle part of the lung is the first seat of the invasion.[1] Further, it has been ascertained that in certain epidemics the upper lobe is more apt to be attacked than the lower;[2] as if the precise seat of the disease is determined, not so much by the predisposition of particular portions of the lung to this special form of inflammation, as by the tendency of the materies morbi, under certain atmospheric or epidemic influences, to excite irritation in particular portions of the respiratory apparatus. However this may be, it is certain that the precise position which the mischief occupies in the lung is not a trustworthy criterion as to the nature of the mischief; and perhaps the only fact which can be definitely announced in reference to this subject is, that when the upper portion of the lung is the primary seat of inflammation, the disease will be found, in the great majority of cases, on the right side.[3]

Inflammation of the lungs is more prevalent in winter than in summer,[4] but, relatively to the number of persons attacked, it proves little more fatal at one season than at another.[5] Indeed, the danger of its attack varies with the intensity of the prevalent epidemic influence more than with the season or state of the atmosphere. At one time a large proportion of attacks prove fatal, whilst at another, even under the same treatment, and though an equally large extent of lung be implicated, death is the exception rather than the rule. *Cæteris paribus*, the danger is proportioned to the age of the patient, the extent of lung

[1] M. Grisolle gives the following statement relative to 264 cases of pneumonia which fell under his own observation, viz., that in 133 cases the lower lobes were affected, the upper in 101 cases, and the middle third of the lung in 30 cases (loc. cit., p. 34). M. Andral states that in 88 cases which he examined, the lower lobe was inflamed in 47 instances, the upper lobe in 30 instances, and the entire lung in 11 cases ("Clin. Med.," vol. iii, p. 470).

[2] See Stokes on "Diseases of the Chest," p. 319.

[3] This has been the case in 13 out of 16 cases which have occurred in my own practice, and it coincides with the experience of others. Thus M. Barth reports that in 19 cases in which he met with pneumonia of the upper lobe, the inflammation was confined to the right side in no less than 18 instances; and in like manner M. Briquet states that in 18 observations which he had made in similar cases inflammation was found affecting the right side only in 14 cases ("Archives de Médecine," 3me série, t. vii, p. 494).

[4] See reports of Registrar-General; also "Dict. de Médecine," art. "Pneumonie," par M. Chomel. M. Grisolle (loc. cit., p. 137) gives the following analysis of 296 cases, viz.:

20	occurred in	January.	13	occurred in	July.
40	"	February.	3	"	August.
47	"	March.	5	"	September.
62	"	April.	2	"	October.
40	"	May.	22	"	November.
8	"	June.	34	"	December.

[5] Grisolle, loc. cit., pp. 529–30.

affected, and the stage of the disease at which the treatment is commenced; and in this as in other serious disorders the danger varies with the habits of life and constitutional powers of the individual. Habitual drunkards, and those who have indulged in excesses of any kind, are least able to resist the depressing influence of the disease, and less capable of bearing the remedies which are necessary for its subjugation. In early infancy and extreme old age the attacks generally prove fatal; between the ages of six and twelve, death very rarely occurs; between the ages of sixteen and thirty, the mortality is from 6 to 9 per cent., whilst at other periods of life the mortality ranges between one-fifth and one-seventh of those attacked.[1] Again, those patients who submit themselves to medical treatment within the first three days of the attack recover in the proportion of twelve to one; whereas amongst those who are not treated until the fourth day, the mortality rises to one in eight; and amongst those who are not seen until the seventh day or afterwards, to one in three, or even to one-half.[2]

In my own practice the mortality has not ranged so high as the statistics just quoted would have led one to anticipate; for although patients are admitted into the hospital, of all ages, and in all stages of the complaint, I have lost only 4 out of 78 cases, and of these 4, 1 died the day after admission, and the other 3 were in an advanced stage of the disease before the treatment was commenced. Whether this favorable result is attributable

[1] M. Valleix reports, as the result of his observation at the "Hospice des Enfans trouvés" at Paris, that pneumonia is almost certainly fatal to new-born children; and Messrs. Rufz and Gerhard state that of 27 cases of pneumonia occurring in children between the ages of two and five years, no less than 25 proved fatal. ("Journal des Conn. Méd.-Chir.," t. iii, p. 105.) Between the ages of six and fifteen, the same observers met with only 1 death in 80 cases; and of 116 cases occurring in persons between the ages of sixteen and thirty, 8 only had a fatal termination. Between the ages of thirty and sixty the mortality rises with each decade from one-seventh to one-fifth, and beyond that age the mortality is much greater; of 129 cases of pneumonia occurring in persons between the ages of sixty and ninety, no less than 77, or about three-fifths, proved fatal. (See Grisolle, loc. cit., pp. 518–20.) M. J. J. Leroux in his "Cours sur les Généralités de la Médecine," tom. vi, gives the following account of 364 cases of pneumonia:

Age.	Patients attacked.	Deaths.	Per cent.
13 to 30	182	17	9.3
30 to 40	58	15	25.9
40 to 50	47	16	34.0
50 to 60	55	23	41.8
60 to 70	16	9	56.2
70 to 75	6	5	83.3
	364	85	

[2] See Grisolle, loc. cit., p. 551.

to my supporting the patient and avoiding depressing treatment, it is difficult at present to decide; but, inasmuch as the statistics from which the more unfavorable results are deduced relate to cases in which copious venesection and other actively depressing treatment was employed, it seems not improbable that such may be the case.[1]

It is commonly supposed that the extent of the lung implicated has an important bearing on the issue of the case, the result being most favorable in those instances in which a small portion only is affected. It is needless to say that *cæteris paribus* this proposition holds good, but I question whether under ordinary circumstances the gravity of an attack is directly proportioned to the extent of lung inflamed. In some instances it is difficult to rescue from death a patient who is suffering only from pneumonia of one lung, whilst other persons whose lungs are both inflamed, and apparently to an equal extent, pass through the attack without any symptoms to cause serious alarm. The contrast is often so strongly marked, that I have been led to refer it to some difference in the intensity of the exciting cause of the disease, or to some peculiar susceptibility to its influence. And I am the more inclined to believe this to be the true explanation of the phenomenon from the fact that in the worst cases the depression is very great from the very outset of the attack, and before any large extent of the lung-tissue is seriously involved.

The duration of the disease varies in favorable cases from about eight to twenty-one days, the patient in the average of cases being convalescent about the fourteenth or sixteenth day. In fatal cases, death commonly occurs between the sixth and twenty-first day of the attack, the cases being quite exceptional in which the patient does not recover if he outlive the end of the third week. Further, true relapse is rare, so much so, indeed, that amongst the ninety-four cases which were noted in St. George's Hospital during the period of my registrarship, there was only one well-marked example of it; and if the term be restricted to cases characterized by a recurrence of rigors, and of rusty-colored expectoration, I question whether the proportion is ever larger than 3.57 per cent., as recorded by M. Grisolle.[2]

[1] This opinion is countenanced by a paper in the "British Med. Journ." for August 23d, 1862, by Dr. J. Hughes Bennett, and by another by the same author in the "Lancet" for 1865, in which he reports having treated 129 cases of pneumonia after much the same plan as I usually adopt, but with even greater success. He lost only 4 out of his 129 patients.

[2] Grisolle, loc. cit., p. 456.

Pneumonia is rarely followed by tuberculization of the lung, and even when set up in a lung already partially infiltrated with tubercle, it does not usually lead to softening of the tuberculous matter, or to any material extension of the mischief. In this respect it contrasts remarkably with bronchitis, and establishes the fact of its being physiologically as well as pathologically a totally distinct disorder.

The treatment of inflammation of the lungs has varied greatly at different epochs, and even now is a subject on which most discordant opinions are expressed. Venesection was formerly a very favorite remedy; and up to the time when Rasori and Laennec had recourse to the aid of tartarized antimony, it was the physician's sheet-anchor in the treatment of this disease. The common practice, within the memory of living men, was to take from twenty-four to thirty-six ounces of blood at the very outset of the attack, and to repeat the bloodletting to sixteen or twenty ounces, once, twice, or even three times; and, strange as it may seem to those who have the management of cases in the present day, marked benefit appears, in some instances, to have resulted from this treatment.[1] Laennec, however, found that even in his day patients could not bear the loss of so much blood; and, at the present time, there are those who maintain not only that venesection is inadmissible, but that brandy and stimulants are needed throughout the attack. Certain it is that the "medical constitution" must have undergone some change since the days of excessive bloodletting; for the type of disease for some years past has been of a low character, and patients cannot bear a great loss of blood, and do not require it for the subjugation of inflammation. But careful observation in the wards of St. George's Hospital has led me to believe that, in some cases at least, moderate venesection is of essential service in relieving the pain and modifying the severity of the attack. In cases of sthenic pneumonia characterized by intense heat and dryness of the skin, a full, resistant pulse, rusty-colored expectoration, and great oppression of the breathing, bloodletting, had recourse to at the beginning of the attack, not only affords immediate relief to the breathing, but appears to remove the extreme tension of the system, and to promote secretion. After ten or twelve, or sixteen ounces of blood have been taken, the pain in many instances ceases, expectoration takes place more easily, and alters in character; the skin becomes moister, and evidence is afforded of the action of remedies which before had proved inoperative. Moreover, the effect is so well marked and

[1] Grisolle, loc. cit., pp. 574-596.

immediate, that no reasonable doubt can exist as to how it is brought about. But, I am bound to say that the cases in which I have judged it expedient to employ phlebotomy are very few in number. The patient's constitution is often shattered by anxiety, excesses, or other causes; the epidemic influence is frequently of a low type, and from one cause or another, cases of true sthenic pneumonia are rarely met with. In other cases, venesection from the arm is inadmissible, and bloodletting, if practised at all, must be confined to leeching and cupping.

It has been attempted by means of numerical returns to establish the necessity for bloodletting in all cases of inflammation of the lungs; and, even in the present day, not a few persons maintain that venesection always shortens the duration of the disease, mitigates its severity, and diminishes the mortality resulting from it. There are others who speak with equal confidence as to its being unnecessary in any case, and who adduce statistics in support of their view. But in this, as in most other cases where statistics are appealed to in proof of the efficacy of any particular plan of treatment, sufficient care is seldom taken to guard against the various sources of error. Thus it happens that the conclusions drawn from this source usually prove fallacious. And so it is in the present instance. Putting aside considerations connected with each patient's constitution, and with the great variety in the character of the epidemic influence in different years, and at different seasons and in different localities, the facts already mentioned (pp. 244, 245), relative to the variations in the mortality according as the disease occurs at different ages, and is or is not early combated by treatment, nullify the result of all statistical observations from which a due regard to their influence is excluded. And when it is remembered that physicians living at the same time, and observing the disease under similar atmospheric and other conditions, have arrived at precisely opposite conclusions from the same numerical method of calculation, it needs not an elaborate argument to prove how little reliance is to be placed on the results. Common sense points out that what is serviceable in one case will prove mischievous in another, under different conditions of age, sex, constitution, and the like; and experience not only indorses this view, but proves that under certain circumstances bloodletting is a palliative of extreme value. It also shows that it is unnecessary to let blood in most cases, and that excessive bloodletting, under whatever circumstances practised, impairs the strength, leads to great impoverishment of the blood, arrests the actions on which the absorption of exudation-matter depends, and not only exposes the patient to risk, but even, under the most favorable conditions, induces a tardy convalescence. It is needless, therefore, to add that phlebotomy should not be resorted to, except when its employment is most clearly indicated, and that even then it should be employed very cautiously. In my own practice at St. George's Hospital and elsewhere, I have had recourse to it in pneumonia three times only within the last seven years; but I am satisfied, nevertheless, that in each of those cases it was the means of affording great relief, if not of saving life, and that cases do occur from time to time in which its employment is absolutely necessary to the well-being of the patient.

It is impossible to lay down a general rule as to the amount of blood which may be safely taken, or, indeed, to dogmatize in the slightest degree respecting the use of venesection in pneumonia. All that can be confidently stated is, that the indications for bloodletting should be strongly marked before it is undertaken at all; that the amount of blood to be lost must vary in each case according to the type of the disorder and the constitution of the patient, but that at the present day it is seldom if ever necessary to draw more than from ten to sixteen ounces in order to relieve the pain, and that eight or ten ounces will usually suffice; that bleeding, if resorted to at all, should be practised early, inasmuch as it may relieve the local congestion by which the early stage of the disease is accompanied, but cannot remove, and may even cause deterioration of the matters which are effused in the subsequent stage of the disorder;[1] that it is seldom necessary to draw blood from the arm oftener than once, and lastly, that advanced age or the existence of pregnancy or menstruation need form no bar to its employment when the general symptoms demand its aid.

It is right to add, that whenever local pain exists, and bleeding appears necessary, cupping or leeching may be employed in addition to general venesection, inasmuch as the local abstraction of blood relieves pain rapidly, and may even cause its immediate cessation. In mild attacks, or in low types of the disease, in children of tender age, and in weakened constitutions, it should supersede all other forms of bloodletting. Dry cupping between the shoulders, aided by the application of leeches to the immediate seat of pain, and subsequently of a linseed poultice covered with oiled silk, forms in many instances the most efficient means of affording relief when the local pain is severe.

[1] It has been clearly shown by Louis and others, that recovery takes place sooner when bloodletting is employed within the first four days than when it is resorted to later in the attack.

To Rasori and Laennec we are indebted for the introduction of the plan of treatment which, next to that of venesection, has obtained the greatest reputation in pneumonia—I mean the administration of large and repeated doses of tartarized antimony. Rasori, commencing with grain doses, gave as much as twenty-four or twenty-six grains in the twenty-four hours, and trusted solely to its action for effecting a cure; whilst Laennec, though following Rasori's example in giving the remedy in large and repeated doses, conceived it to be advisable to resort to moderate venesection in aid of the antimonial. He, therefore, commenced by a bleeding from the arm of from eight to sixteen ounces. Others have followed the same plan of treatment; and although nausea, vomiting, diarrhœa, and griping pain in the abdomen have sometimes resulted from the exhibition of the antimony, and in some few instances death has appeared to be occasioned by it, yet, ordinarily, it has ceased to produce vomiting or purging after the first day, and has been tolerated as perfectly as when given in smaller doses. But the amount of success resulting from the treatment has not been uniformly great. At first, Laennec reported of it most favorably, but subsequently, under different atmospheric or other conditions, it failed to exert any remarkably curative action, and the proportion of death rose as high as that attained under other methods of treatment. Thus, of 46 patients treated by Laennec, 17 died; and of 140 cases treated by Chomel, 40 proved fatal. Further, Andral, who carefully watched this plan of treatment, and tested it largely, expresses his doubts as to whether the disease is ever benefited by these large doses of tartarized antimony.

My own opinion is quite in accordance with the doubts expressed by Andral. In our own country, and within my time, tartar emetic has been seldom given in the doses introduced by Rasori; but I have seen quite enough of its action in the doses recommended by Dr. Walshe and others, to induce me to hesitate in recommending their administration.[1] More than once in hospital practice I have seen extremely dangerous depression, with profuse cold clammy sweats, produced by its exhibition in smaller doses than those just alluded to; and even when such doses have failed to occasion unpleasant symptoms, they have

[1] Dr. Walshe (loc. cit., p. 426) says—"The salt should at first be given in doses of half a grain combined with dilute hydrocyanic acid, paregoric, and tincture of orange-peel every hour for the first three or four hours, and the dose then increased at intervals of two hours to one grain; in the course of twelve hours the quantity may be raised to two grains, its repetition made less frequent, say every fourth hour."

not appeared to exercise a more decidedly curative influence than smaller doses exhibited at longer intervals. Therefore, in my own practice I have seldom given more than a quarter of a grain at intervals of three or four hours, according to the circumstances of the case, and in most instances have found an eighth or a sixth of a grain sufficient. In such doses the remedy does not nauseate after the first time of its administration, but operates very manifestly in mitigating the severity of the symptoms. Indeed, an instance rarely occurs in which its administration is not called for at the outset of the attack, and it may be given in combination with salines and moderate doses of digitalis until relief is obtained. Exceptional cases are sometimes met with in which the character of the pulse, or the occurrence of profuse perspiration, speedily indicates serious depression of the vital powers; and in these, of course, its administration must be discontinued. It should be clearly understood, however, that the exhibition of antimony is no bar to the use of stimulants, and in like manner that the necessity for stimulants does not preclude the use of antimony. The two remedies may be often combined advantageously, and some of the most striking recoveries from pneumonia which it has been my lot to witness have taken place under their conjoint influence. In most instances the vital depression, which is observed in these cases, is referable to the force of the disorder rather than to the influence of the remedy; but if a few doses of brandy do not serve to counteract it, the prudent course is to omit the antimony. Speaking generally, however, it may be fairly stated that no single remedy is so useful as tartar emetic, and that, administered judiciously, with due regard to the exigencies of the case, it modifies and represses the morbid action in the lungs, and rarely produces unpleasant symptoms.

In some instances, however, tartar emetic either fails in exerting a curative action, or is not well borne by the system. Cases in which hepatization of the lung proceeds with extreme rapidity, in which crepitation either does not exist at all, or is of very short duration, giving place after a few hours to intense tubular breathing, and those again which are marked by extreme depression almost from the first, are rarely benefited by tartar emetic. The former often accompany the rheumatic diathesis, and are seen most strongly marked in connection with acute rheumatism. The latter are met with in cachectic persons, and especially under peculiar atmospheric conditions. The most efficient treatment in the one class of cases consists of calomel and opium in repeated doses, together with digitalis and full doses of alkalies or neutral salts. The latter is to be combated by salines and stimulants, with digitalis or the veratrum viride, and calomel should be given only under special circumstances. In either case the external application of turpentine fomentations is useful throughout, and blisters assist in relieving the local pain, and in promoting the absorption of the plastic exudation as soon as the first fury of the attack is overpast; but if employed at the very outset of the disease they appear to increase the feverish excitement, and are productive of evil rather than of good. Throughout the attack, fomentation of the chest is of the greatest benefit; and when turpentine stupes can be no longer borne, the whole of the affected side should be encased in a jacket poultice covered with oiled silk.

The action of mercury in pneumonia is a question on which opposite opinions have been entertained. There are persons who deny that it ever exerts a beneficial influence; and there are those, on the other hand, who laud it as the most potent remedy we possess when the disease has reached the stage of hepatization. My own opinion does not accord with either of these dogmas. Observation at the bedside has fully convinced me of the efficacy of tartarized antimony in many cases of inflammation of the lungs long after hepatization has commenced; but it has equally satisfied me of the superiority of mercury in certain forms of the disease. Indeed, if my views are correct, the efficacy of mercury is especially conspicuous in those cases of pneumonia in which tartar emetic is of least avail; in other words, in those instances of the disease which are accompanied by a more than ordinarily plastic exudation—cases in which, as already stated, crepitation either does not exist at all, or is replaced in a few hours by intense tubular breathing. In these, calomel and opium, in combination with salines and small doses of tartar emetic, will often produce results which are not attainable by any other means.

Of late years it has been proposed to treat inflammation of the lungs by the exhibition of repeated doses of alcoholic stimulants internally, whilst turpentine stupes are being applied externally; and reports have been made from time to time in the journals, of cases so treated by Dr. Todd and others. But I suspect that these reports are calculated to convey an erroneous impression as to the result of such treatment in ordinary cases of pneumonia. They speak of the cases as being examples of the ordinary sthenic form of the disease, and of the treatment—which consists essentially of half-ounce doses of brandy at intervals of an hour—as being eminently successful. My own observation of this method of treatment has been too limited to justify my expressing a positive opinion as to its efficacy; but I may state that in four out of the only five cases in which I have seen it tried, it failed very signally in sustaining its reputation. Not only did it not afford relief, but in each instance it produced aggravation of the symptoms, and was discontinued in consequence. And when it is remembered, in connection with this fact, that a large proportion of the cases in which the alcoholic treatment has been successfully employed have ranged between the ages of fifteen and thirty—a period of life during which as already shown, the mortality from pneumonia, under whatever treatment, is at its *minimum*—the inference, I think, may be fairly drawn, that the success which has been attributed to the action of the so-called remedy has been really due to other causes. In these, as in other disorders, stimulants are often needed in aid of other remedies; but from what I have seen and what I have learned from the testimony of others, I am satisfied, that now as heretofore, their action, either for good or evil, in each particular case, depends on whether they are needed by the system, and are administered judiciously or injudiciously. It should be added, however, that the necessity for stimulants in certain cases, and at certain stages of the disease, as it has prevailed of late years in London and other large towns, is more apparent, and arises more frequently than the experience of former years would have led us to expect. Few cases are met with which are not benefited by moderate doses of stimulants at some period of their course, and if the slightest symptom of exhaustion is observed, their assistance should be had recourse to without delay. When their action proves serviceable the heat of the skin diminishes and the frequency of the pulse falls from the commencement of their administration.

There are yet other remedies which it behooves me to mention, although I cannot from personal observation adduce any facts in proof or disproof of their efficacy. I allude to the administration of sulphate of copper in combination with small doses of opium internally, and to the application of ice or freezing mixtures to the surface of the chest. Sprinkhardt and Sauer, of Pesth, are both warm in their praise of the former remedy when administered in the proportion of half a grain every hour, and assert that its efficacy is vastly superior to tartarized antimony; and Niemeyer and Smoler are equally decided in lauding the virtues of ice externally.[1] The employment of these remedies is opposed to preconceived opinions; but, nevertheless, if experience confirms the statements of their advocates, they will prove most valuable allies. At present the

[1] See Schmidt's "Jahrbucher," vol. cxiii, p. 337.

facts on which their reputation rests are too few to justify more than a carefully conducted experimental inquiry.

Another remedy which has been highly praised is the veratrum viride. Ritter, Roth, and other continental authorities have spoken highly of the efficacy of veratrum album; but American writers maintain the superiority of the veratrum viride. Dr. Rogers[1] reports a case where it "acted like a charm;" and Dr. Cutler, who is perhaps its warmest advocate, has reported[2] another in which equally beneficial results were observed. The preparation usually employed is the tincture, of which from 5 to 8 minims are given every two or three hours, until nausea is produced, or the pulse falls below the natural standard, when the dose should be diminished or repeated less frequently. There cannot be a doubt that when administered in this manner, the drug exercises a remarkable control over the circulation; and in the few instances in which I have employed it to combat pneumonia, it has appeared to exert a beneficial influence. At present, however, my experience of its action is too limited to justify a positive opinion as to its efficacy; but I have seen enough to convince me that it deserves a further trial.

In pneumonia, as in other acute diseases, the secretions must be carefully regulated, and a moderately restricted regimen enforced. Solid food must be avoided, but good beef-tea may be given from the first, and the patient should be encouraged to drink freely of milk and whey, in which an egg may be beaten up. Excessive purgation is worse than useless, as tending needlessly to exhaust the strength; but gentle laxatives should be employed when necessary, and diuretics, such as nitre and acetate of potash, prove useful as eliminants.

There are some varieties of inflammation of the lungs which demand special notice, not only as differing from ordinary cases of the disease in their symptoms and progress, but also as requiring different treatment, and leading to different pathological results. Amongst these may be mentioned—

1st. The cases already alluded to as characterized by inflammation of the interlobular cellular tissue—cases which are especially prone to occur in rheumatic persons. These are always difficult of diagnosis, and are apt to be confounded with pleurisy. They are to be distinguished from ordinary pneumonia by the comparative absence of crepitation, and by the fact that the crepitation which exists is heard during expiration as well as

[1] "Amer. Med. Monthly," June, 1858, p. 410.
[2] "Boston Med. and Surg. Journ.," vol. lvi, p. 511.

during inspiration; and from pleurisy, by the rapidity with which strongly marked tubular breathing ensues, and continues, day after day, unimpaired in intensity; by the increase of the vocal fremitus, and by the increased resonance of the voice. Sometimes the inflammation is of a suppurative character, but more commonly the inflammatory exudation is fibrinous; and as the disease becomes chronic, the fibrine gradually solidifies and contracts, and produces the condition of lung described as "cirrhosis," by Dr. Corrigan. As the disease progresses, the bronchi undergo dilatation, the pulmonary tissue is compressed and becomes impermeable, the lung contracts, and the side of the chest falls in. As a result of this condition, dyspnœa occurs, the costal movements are diminished, the percussion-note becomes dull or else somewhat amphoric, the vocal fremitus is increased, the vocal resonance is abnormally great, the respiration is of a hollow, blowing character, and if bronchitis or œdema exist, is accompanied by sonorous rhonchi and large bubbling râles.

In the acute stage of the disease, I believe nothing is of more service externally than leeches and blisters, followed by poultices kept applied over the entire side of the chest; and, as internal remedies, calomel and opium, with full doses of alkalies and the neutral salts, iodide of potassium, and occasional doses of colchicum. In the more advanced stages, when the exudation-matter has solidified, and has already produced compression of the lung-tissue, I do not believe that any medicine will serve to remedy the mischief which has occurred; and our chief aim should be to sustain the health, and promote free and easy expectoration, if bronchitis happens to coexist.

2dly. Another class of cases deserving of notice is that in which inflammation, instead of spreading through an entire lobe of the lung, is confined to certain lobules, the intervening lobules remaining sound. This form of disease, which has been termed "lobular pneumonia," was formerly considered to be of frequent occurrence in infancy, and even now is so spoken of by some writers. But careful research has shown, that in many cases the solidification of the pulmonary lobules, which was formerly attributed to pneumonia, is due, when occurring in infancy, to simple collapse of the lung. This is proved by the fact that after death, in such cases, the solidified lobules may be inflated almost to their natural size; whereas hepatized lobules do not admit of inflation. Therefore, although, as already stated, this form of pneumonia is sometimes met with in infancy, it does not occur so often as was formerly supposed. In adult life it is observed occasionally, but it rarely, if ever, occurs as an idiopathic affection; most commonly it arises in connection with embolism or pyæmia, and the formation of secondary abscesses in the lungs.

The physical diagnosis of this form of disease is extremely difficult. Inspection, palpation, and mensuration afford only negative information, and percussion fails to give any assistance, so long as mischief is confined to two or three lobules. Indeed, the only physical indication of the disease is to be found in the respiration, which is exaggerated in the parts adjacent to the consolidated lobules, and sometimes absent, sometimes harsh and blowing, immediately over the seat of consolidation.

3dly. Cases in which the disease is of secondary origin, or, in other words, arises during the progress of other forms of disease. This, which is often termed intercurrent pneumonia, is a most common and dangerous form of the malady, and is apt to supervene at all ages, from earliest infancy to extreme old age. It is the principal cause of death in many cases of measles, hooping-cough, and other infantile disorders; it is apt to complicate typhus fever, rheumatic fever, variola, and other acute blood disorders, as also meningitis, and other inflammatory diseases in the adult; and it is a frequent attendant upon renal anasarca, and upon phthisis pulmonalis, cancer, and other chronic disorders connected with local organic mischief and a distempered condition of the blood.

In all these cases the physical signs are much the same as are met with in ordinary idiopathic pneumonia, except. perhaps, that in the more adynamic forms of disease the exudation is more serous, and less plastic in its nature; and that as it is more profuse, and not so readily got rid of by expectoration as in ordinary cases of pneumonia, there is a tendency to obstruction of the air-passages, with entire absence of breathing, rather than to tubular breathing, over the affected portions of the lung. In a practical point of view these cases differ altogether from those of ordinary pneumonia. They partake, in many instances, more largely of congestion with consequent œdema than of active inflammation; and the mischief is dependent on and kept up by the irritation of the *materies morbi* out of which the primary disease arises. Hence, *inter se*, they differ greatly in character, and are more purely congestive, or more strictly inflammatory, according to the precise nature of disorder from which they take their origin. In all cases, however, "sublatâ causâ tollitur effectus," and the only rational treatment is that which endeavors to subdue the primary disease, whilst at the same time it aims at counteracting the secondary local mischief, by dry cupping, blisters, turpentine fomentations, jacket poultices, and other topical applications. In the pneumonia of fever, brandy and quinine may be the appropriate internal remedies; alkalies in that of rheumatism; soda, colchicum, and magnesia in that of gout; compound jalap powder, bitartrate of potash, gin, and other diuretics in that of Bright's disease. I do not wish to imply by this statement that calomel, antimony, and digitalis are never needed in these cases; but I do mean that they are often useless, or worse than useless; and even when required for the subjugation of local inflammation, should be kept subservient to the general treatment which is directed against the original disorder. I have seen too much of the mischief produced by an opposite course of treatment to doubt as to the correctness of this general rule. Indeed, it may be stated generally, that the only safe and efficient treatment is that which is calculated to mitigate or remove the blood disorder, which is the source of all the mischief, aided by such external applications as serve to alleviate local irritation or to counteract internal inflammation without unduly depressing the patient.

4thly. Cases in which the disease, though running on rapidly to consolidation, is unattended by the ordinary, general symptoms of pneumonia, and is, therefore, said to be "*latent*." Their peculiarity is simply that which their name implies, and which renders the mischief liable to be overlooked. They are met with chiefly in infants,[1] or in elderly persons, or in persons suffering from adynamic disorders, and the mischief partakes of the character of congestion resulting in sero-plastic effusion rather than of active inflammation. These cases constitute a large proportion of those described by Piorry under the title of "hypostatic pneumonia." Great perversion of the ratio naturally subsisting between the pulse and the respiration ought always to excite suspicion of the disease; and when once this is aroused, a physical examination of the chest will serve to reveal the nature and extent of the mischief; for although the general symptoms are absent, the physical signs of pulmonary consolidation are fairly marked, and differ little in their nature from those which accompany ordinary cases of pneumonia.

The treatment of this form of disease consists chiefly of measures calculated to relieve the local congestion and maintain the force of the circulation; and in many instances, from the nature of the case, these take the form of diffusible stimulants

[1] M. Guersent, Physician to the Hôpital des Enfans, says that "three-fifths of the children who die in the hospitals between birth and conclusion of the first dentition die of pneumonia in a latent form." "Dict. de Médecine," vol. viii, p. 92.

internally, and counter-irritants, derivatives, and fomentations externally. Dry cupping is always useful, and turpentine fomentations are of great service; whereas venesection, leeches, and tartarized antimony are seldom admissible, and are generally productive of serious mischief.

5thly. Cases in which pneumonia occurs in a chronic form. This variety of disease, when unaccompanied by the deposit of tubercle or other adventitious product in the lung, is extremely rare. It is met with in two distinct forms, the one being commonly the sequel of acute pneumonia, the other occurring as a chronic affection without having been preceded by symptoms of the acute disease. The first is characterized by infiltration of the lung-tissue, with a tough, grayish, solid matter which does not usually exhibit a tendency to soften, but leads gradually but slowly to contraction and shrinking of the lung and retraction of the chest-walls. The lung in this case is drier and harder than in a corresponding state of ordinary hepatization, is smooth when cut, and often creaks under the scalpel. In most instances probably this form of so-called chronic pneumonia is the second stage of the first-described variety of the disease, in which the interlobular cellular tissue is principally affected, and its peculiarity is referable to the exudation being more than ordinarily fibrinous in character. The second variety ought, perhaps, to be described as fibroid degeneration rather than as chronic pneumonia, and should be classed under the head of Phthisis, as being more nearly allied to that disease than to simple inflammation of the lungs. The lung is infiltrated by a reddish, solid material, which after a time becomes distinctly fibroid in character, loses its deep-red color, and in many places assuming a granite-gray tint, causes the lung to become strikingly mottled.[1] The lung in these cases, as in those last mentioned, is firmer and drier than in hepatization, it often creaks under the knife, and scarcely yields a trace of fluid when scraped. Sometimes its cut surface is smooth, at others markedly granular, in which case the granulations are more strikingly manifest when the lung-tissue is torn. These cases, I believe, form a connecting link between chronic inflammatory consolidation and tuberculous infiltration.[2]

The general symptoms are those of failing health and strength, with slight but habitual shortness of breath, cough, and scanty catarrhal expectoration unaccompanied by spitting of blood—

[1] See a paper "On Fibroid Degeneration of the Lung," by Dr. Sutton, in vol. xlviii of the "Med.-Chir. Trans."

[2] See Andral, "Clin. Méd.," t. ii, p. 310.

and oppression—not pain—at the chest. If the disease supervenes on an attack of acute pneumonia, the patient does not regain his strength, but, on the contrary, becomes weaker and loses flesh; his pulse remains quick, the appetite fails, and he is troubled with feverishness, especially towards evening.

The physical signs are somewhat deceptive. The expansile movement of the chest is always impaired, and there is dulness on percussion; but the other signs are by no means constant. In the first variety of the disease the chest-walls may fall in, in consequence of the shrinking of the affected lung; and this condition may be accompanied by harsh, blowing, and more or less hollow-sounding respiration in some parts of the chest, whilst in others the breathing is weak, or altogether absent, or accompanied by occasional rales; the vocal resonance is increased, and the vocal fremitus intensified. Sometimes, however, if the earlier stage of the complaint has been overlooked, these signs may not be present; the lung, as yet, has not contracted, and, consequently, there is no falling in or retraction of the chest-walls; the bronchi remain obstructed, so that there is entire absence of breathing-sounds, whether healthy or morbid; the voice is not transmitted through the chest-walls to the ear, and vocal fremitus cannot be felt. In short, the physical signs more nearly resemble those met with in pleurisy than those which are ordinarily attendant on pneumonia; and it is only by noting the absence of the physical signs of pleurisy, and by carefully watching the progress of the disease, that it is possible to arrive at a correct diagnosis. In the second variety of the disease the difficulty of arriving at a correct opinion is even greater, for the upper lobes of the lung are specially prone to be attacked, and then the case may simulate tubercular phthisis. A case in which an entire lung was implicated fell under my observation eleven years ago. The patient, a young lady, aged twenty, had long been short-breathed, but otherwise had enjoyed fair health, until within three days of her death. She had never had an acute attack of pulmonary inflammation, and she died of constriction of the larynx and trachea, consequent on the pressure of an enlarged thyroid. The condition of the chest was ascertained by dissection, Mr. Prescott Hewett very kindly performing the post-mortem examination. No trace of fluid or of pre-existing inflammation was found in the pleura, but one lung was perfectly solid. It was as large as the other lung when partially inflated, and did not collapse when the chest was opened, but filled the whole of one side of the thorax. It sank at once when placed in water. Its color was grayish-red; its consistence so firm that it creaked under the scalpel, and its cut surface was smooth, and emitted scarcely a trace of fluid even when scraped with the knife. It did not contain a particle of tubercle; but the bronchial and mesenteric glands were enlarged, and contained cretaceous matter. The bronchi were filled with exudation-matter. An instance in which the lower lobe of the right lung was similarly affected, and gave rise to precisely the same physical signs, was observed by M. Requin.[1] Such cases are apt to be mistaken during life for cases of old-standing pleuritic effusion; and, in whatever stage they are examined, whether before or after contraction of the lung has taken place, the diagnosis is always difficult. When the upper lobe of the lung is thus affected, the case is apt to simulate consumption; for not unfrequently some softening occurs and vomicæ are formed at the apices of the lungs, in which case there is flattening of the infra-clavicular region of the affected side, with imperfect expansion, dulness on percussion, harsh, hollow breathing-sound, prolonged expiration, and increased vocal and tussive resonance. The precise character and order of progression of the general symptoms, coupled with the general aspect and family history of the patient, and with the stationary condition of the parts, as determined by several examinations at considerable intervals of time, form the only clue to the real nature of the case; but they will often suffice to excite suspicion in the mind of the careful and experienced physician, and will induce him to express a cautious opinion. Fortunately, the diagnosis is not of much practical importance, inasmuch as these cases require treatment similar to that which proves useful in tubercular phthisis. Cod-liver oil, tonics and alkalies internally, daily oleaginous inunction, the use of ioduretted and counter-irritant applications to the chest, a generous diet, proper exercise, and fresh air, are the remedies most likely to prove serviceable. In one case which I saw in consultation with Dr. C. J. B. Williams, and which I had the opportunity of watching for three years, this plan of treatment, aided by a twelve months' residence in Egypt, entirely removed all trace of the disease.

Gangrene of the Lung.

In my observations on pneumonia, the occurrence of gangrene of the lung was briefly alluded to, and reference was made to the morbid appearances and physical signs to which it gives rise. Mortification, however, must be regarded as an accidental complication or occasional sequence of pneumonia, rather than as a

[1] See Grisolle, loc. cit., pp. 350–1.

natural consequence of the disease, for it is rarely met with in sthenic cases, in which pulmonary inflammation is most intense, and sometimes occurs in instances in which pneumonia does not exist.

There are two distinct varieties of gangrene of the lung, viz., first, the diffused, or uncircumscribed variety; and secondly, the circumscribed. These differ greatly in their anatomical characters, but they both appear to be connected with a depraved condition of the blood analogous to that which produces the various forms of idiopathic gangrene. This, doubtless, in most cases, is their proximate cause, the local inflammation being only the accidental or exciting cause. It must be admitted, however, that the exciting or determining causes of the disease appear to be of very different force and character in the two varieties. In the diffused form of the disorder, the slightest impediment to the pulmonary circulation suffices to excite its occurrence—nay, in some instances, uncircumscribed gangrene occurs without any obvious pre-existent local obstruction; whereas the circumscribed form is rarely met with, unless intense pulmonary inflammation of a low type has existed, or secondary abscesses have formed in the lungs, or obstruction of the nutrient vessels has been brought about by the pressure of an aneurismal tumor,[1] or of tuberculous, cancerous, or other adventitious deposits, or still more rarely, by the presence of minute fibrinous concretions detached from the endocardium on the right side of the heart, and carried onwards with the blood to the lungs. It seems fair to conclude, therefore, that in the former variety of the disease the blood is more deteriorated than in the latter—a view which is borne out by the class of cases in which the two varieties of gangrene occur respectively.

Mortification of the lung is extremely rare; so much so indeed, that in the post-mortem records in St. George's Hospital for the ten years ending 1850, it was noted in no more than nineteen instances; and I have met with only seven other cases in which its general characters and physical signs were thoroughly well marked during life. It may occur at any age, and appears from M. Boudet's inquiries to be more common in children than in adults, and more so in adults than in persons much advanced in years.[2] When it does occur, the gangrenous process may be

[1] A remarkable case of pulmonary gangrene occurring as the result of aneurismal pressure occurred under my care at St. George's Hospital, in January, 1860. For full details see "Hospital Post-mortem Register and Case-book" for February 3d, 1860, under the head of "Henry Barnes;" also "Trans. Pathol. Soc.," vol. xi, p. 62.

[2] The subjoined table, which throws important light on the subject of

confined to the lung, or may be associated with mortification in other parts of the body.

It is not an unusual accompaniment of cancrum oris; accordingly, M. Boudet's table shows that gangrene of the lung coexists with gangrene of other parts of the body more frequently in infancy than in after years.

The symptoms of gangrene vary considerably, according to the form which the disease assumes. In the diffused, or uncircumscribed form, the progress of the disease is often extremely rapid. The mischief is ushered in by utter prostration of strength; a small, weak, and frequent pulse, pallor and anxiety of countenance, and all the symptoms of rapidly failing power; coupled with these, there is frequent but feeble cough, and profuse frothy, diffluent expectoration, of a peculiar greenish color, and intensely fetid gangrenous odor. Ere long the vital powers seem utterly oppressed or exhausted, the pulse fails, the features become collapsed, the expectoration ceases from want of power to expectorate, and the patient rapidly sinks.

The progress of circumscribed gangrene is somewhat different, and is marked by a train of symptoms which vary greatly at different periods of the disease. At first, the general symptoms are simply those of slight pneumonia, or pulmonary congestion; but they are attended by an amount of prostration which is quite disproportioned to the extent and apparent intensity of the local mischief. Sometimes, indeed, the invasion of the disease is even more insidious; there is nothing to draw attention to the condition of the chest, and depression of the vital power is the only prominent symptom. After a time, however, expectoration commences; and the sputa, which at first are of a muco-purulent character, sometimes tinged with blood, emit a disagreeable odor; but as soon as a free communication is established between the air-passages and the sloughing tissue of the lung, they not only acquire an intensely fetid gangrenous odor, but assume an ap-

gangrene, was published, as the result of his own observations, by M. Boudet, in the "Archives de Médecine," September, 1843:

Age.	Number of post mortem examinations.	Cases of gangrene of the lung.	Ratio of cases of gangrene of lung to post-mortem examinations.	Cases of various spontaneous gangrene.	Total cases of gangrene.	Ratio of cases of gangrene to post-mortem examinations.
Children, .	135	5	1 to 27	9	14	1 to 9
Adults, .	156	2	1 to 78	4	6	1 to 27
Aged persons, . .	220	2	1 to 110	7	9	1 to 24

pearance more or less characteristic of the disease. They lose their muco-purulent character, and become extremely liquid, or sero-purulent, and of a dirty greenish or ash-gray color. At the same time, the breath acquires an offensive, putrid odor, the pulse becomes feeble and rapid, and there is every evidence of great and increasing prostration. Not unfrequently, the patient passes rapidly into a state of collapse, and sinks in a few hours, or it may be days, without the occurrence of any other symptom. Sometimes, though rarely, profuse hemorrhage occurs,[1] and terminates in death. Sometimes the sloughing implicates the large bronchi. and the œsophagus; sometimes it occasions perforation of the pleura, and gives rise to fatal pneumothorax; and it has been even known to cause perforation of the diaphragm, and to produce subcutaneous emphysema, by making its way through the two agglutinated layers of the pleura. At other times the disease is apt to pass into a somewhat chronic condition; the symptoms show that the patient's power of resistance is considerable; there is no longer rapid failure of the vital power, but the expectoration, which continues profuse and fetid, loses its extremely liquid character, and becomes more distinctly purulent; there is hectic fever, with night-sweats and emaciation, and, after weeks, or it may be months of suffering, the patient dies utterly exhausted. In other cases the disease progresses more favorably; the sputa decrease in quantity, become decidedly purulent, and lose their fetor; the heat of skin and hectic fever subside, the appetite returns, the patient gains flesh, and complete recovery ensues. These, however, are exceptional cases, and probably do not constitute above eight or ten per cent. of the whole. In all cases which do not speedily terminate in death, the fetor of the breath is apt to vary greatly. At one time it is insufferable, at another, nay, sometimes within a few minutes afterwards, it may be simply disagreeable; and then after the lapse of a longer or shorter period it may again resume its poisonous quality. This alteration in the fetor and sweetness of the breath was very remarkable in a case admitted under my care into the York Ward of St. George's Hospital in October, 1858.[2]

The only rational explanation appears to be, that temporary obstruction of the bronchi leading to the sloughing tissue, shuts off the source of the gangrenous odor, which reappears as soon as the obstruction is removed, whether by the act of respiration or of coughing.

[1] See St. George's Hospital "Post-mortem Records" for 1847, p. 67. Hemorrhage took place into a gangrenous cavity in the right lung.

[2] See "Hospital Case-book," xxxvii, p. 325.

The physical signs of the mischief which is going on are often obscure, and never distinctive; indeed, they are simply those of local pulmonary consolidation, followed by evidence of the breaking up of the lung-tissue, and subsequently of pulmonary excavation; they are not marked by any character calculated to throw light on the nature or cause of the disorganization. Sometimes, but not invariably, the mischief is preceded by the ordinary signs of pneumonia, and it is generally accompanied by the signs of bronchitis as soon as the breaking up of the lung-tissue has commenced.

The morbid appearances observable after death vary with the form which the disease has assumed. In the diffused form the lung-tissue is extremely congested, and in a state of serous and sanguineous engorgement; its condition, indeed, resembles that of the first stage of pneumonia, except that it is more humid, and breaks down more easily. The affected portions are of a livid red, or else of a greenish, or brownish-black, or ash-gray color, more or less diffluent, and often converted into a sanious, putrid fluid, which escapes when the lung is cut, and emits an intolerably fetid, gangrenous odor. Not unfrequently, some portion of the sphacelated tissue has been broken down and removed by expectoration during life, in which case an irregular excavation is found, the walls of which are composed of ragged, pulpy, mortifying tissue. In the uncircumscribed form of disease the parts in process of mortification blend gradually and insensibly with the sounder portions of the lung; whereas in the circumscribed variety the gangrenous portion of the lung is surrounded by inflamed tissue, infiltrated by exudation-matter, which is more or less plastic, and maps out the limits of the affected portions. In this latter form of the disease the mortified tissue is usually found soft and diffluent, as in the uncircumscribed variety; but sometimes a sphacelated spot may be met with, harder and more compact than the tissue of sound lung, and forming a sort of core in the midst of the softer diffluent matter. Sometimes the cavities which are formed in this variety of disease, instead of being pulpy and ragged, become lined with a false membrane, which keeps up a purulent or muco-purulent secretion, and occasionally the density of this membrane appears to prevent their collapse and cicatrization. Be this as it may, these gangrenous excavations exhibit very little tendency to cicatrize, and free secretion from their internal surface has been known to go on for many years.[1]

[1] I have seen a smooth false membrane lining old tuberculous cavities and keeping up profuse secretion for years—in one case for no less than nine years. See also Walshe, loc. cit., p. 451.

In the diffused form of the disease the gangrene usually attacks a much larger portion of the lung than in the circumscribed variety; indeed, an entire lobe is not unfrequently affected, and sometimes the greater part of one lung is implicated.[1] In the circumscribed, as also in the uncircumscribed form of the disease, the mischief is usually limited to one lung, except in cases in which it ensues on secondary abscesses. In these it is scattered throughout both lungs, and is found most commonly towards their periphery; whereas in other cases the posterior portions of the lungs, and especially of the right lung, are those which are most prone to suffer.

The diagnosis of pulmonary gangrene is not in all cases an easy matter. It rests upon the peculiar fetor of the breath and expectoration, coupled with the sero-purulent character of the sputa and the physical signs of softening and excavation of the lung-substance; and as fetor not to be distinguished from gangrenous fetor may arise in cases in which no evidence of pulmonary softening is to be obtained, and in which there is reason to believe that mortification of the lung-tissue does not exist, as it may also occur from sloughing of the internal surface of tuberculous and other cavities not primarily gangrenous, and as the character of the sputa will not alone serve to establish the true nature of the mischief, it is obvious that in most cases no great confidence can be felt in a diagnosis as to the existence of this disease, unless made by a very competent observer, and verified by a post-mortem examination. Nevertheless, when the signs already enumerated as indicative of gangrene supervene *suddenly* in a person suffering from adynamic congestion or inflammation of the lungs, and who was previously free from bronchitis and tuberculous disease, there can be little doubt as to their true character, especially when gangrenous fetor of the breath has *preceded*, and is subsequently accompanied by the signs of pulmonary excavation.

Another source of difficulty, however, is the occurrence of gangrene of the lung without any perceptible fetor. I have never met with such a case; but Cruveilhier has described this variety of the disease; and Dr. Walshe, though he does not refer to instances in point, speaks of it as if it were a matter within his own cognizance. At all events, its occurrence must be extremely rare, and when it does occur it is not likely to be recognized during life.

Fortunately, the difficulties incidental to the diagnosis of gangrene of the lung are not of material importance in practice;

[1] Laennec, loc. cit., p. 222.

for whether the disease exists in an idiopathic form, or whether fetor be caused by sloughing of small portions of the bronchial mucous membrane, or by mortification of portions of tubercular or other cavities in the lung, the aim of the physician in all cases must be to support the patient and obviate the tendency to death. There is nothing specific in the treatment to be employed. Whenever, and in whatever form of disease, gangrenous fetor arises, the vital powers are low and require support, and the amount and nature of that support must be determined by the precise condition of the patient, and not by any theoretical consideration as to the extent and nature of the mischief.[1] In the acute stage, quinine and bark in full doses, with ten or twelve minim doses of nitro-muriatic acid, opium, brandy, and diffusible stimulants are the appropriate remedies; and when the affection passes into a more chronic state, cod-liver oil, the balsam of Peru in drachm doses, turpentine, and full doses of the compound tincture of benzoin, may be advantageously added. Ammonia has been recommended, but in my experience has not proved so useful as the mineral acids. Chlorate of potash has also failed in the only instance in which I have seen it employed; but its efficacy is so remarkable in cancrum oris, with which gangrene of the lung is often associated, that it deserves a further trial in full and repeated doses. It is not incompatible with the other remedies recommended, and there is no reason why it should not be administered in combination with them, or better still, as a drink dissolved in barley-water during the period of their administration. Bloodletting is inadmissible, and dry-cupping and blistering, though not excluded from our list of agents by the nature of the disease, are of questionable utility.

[1] In evidence of this fact I may cite a very interesting case which occurred under my care at St. George's Hospital, in the year 1860. Stephen Deacon, æt 42, was admitted into Hope Ward on January 11th, having been seized with a paralysis of the left side. The attack came on suddenly, accompanied by headache, three days before admission, whilst he was at work as a laborer. It had not been preceded by any apparent illness, and before his admission he had partially regained power over his leg, though his arm still remained motionless. His heart was healthy, the kidneys were sound, and he had no cough. On the 13th he was seized with dyspnœa and great vital depression, and on the 14th began to cough and expectorate an enormous quantity of extremely fetid matter mixed with blood, which emitted a gangrenous odor. After the lapse of a few hours the sputa lost their puriform character, and became sero-purulent and of a dirty ash-gray color. Notwithstanding the paralysis, full doses of brandy, with bark, and ten-minim doses of nitric acid were administered; and in the course of ten days the cough had almost ceased, the expectoration was devoid of offensive odor, and the paralysis had in great measure passed off. On the 25th he left the hospital, having quite recovered, except in regard to the arm, which still remained weak.

The fetor of the breath may be corrected by rinsing the mouth with chlorinated washes or a weak solution of permanganate of potash; and the inhalation of chlorine will tend to modify the septic changes going on in the lung, and so to lessen the depression of the vital power. The vapor of turpentine is another agent which is reported to exercise a distinctly remedial power, and experience has convinced me that air impregnated with carbolic acid is decidedly serviceable.[1] It seems probable that ozonized water, or small quantities of dilute carbolic acid pulverized, and so inhaled, would prove beneficial; and although as yet I have had no experience of its action in this complaint when so employed, I should not hesitate to have recourse to its aid in severe or obstinate cases. The diet should be nutritious, and as generous as the digestive powers of the patient will permit. Wine; soup or strong beef-tea, or warm milk thickened with isinglass; arrowroot, made with beef-tea instead of water; eggs beaten up with brandy, blancmange, and essence of meat, are among the articles of diet which are most suitable on these occasions. Meat, with pale ale or bottled stout, may be given as soon as the patient's stomach can bear it.

Œdema of the Lungs.

Œdema of the lungs is not a condition of frequent occurrence. Arising as it does from the same class of causes which occasion œdema in other parts of the body, it should be regarded as a secondary affection, symptomatic of mischief elsewhere, rather than as a primary or idiopathic disease of the pulmonary tissue. It is a common accompaniment of Bright's disease of the kidneys, and of the dropsy which occurs in sequel of scarlatina; it is apt to accompany typhus fever, scurvy, and other diseases in which there exists a depraved and fluid condition of the blood; and it is sometimes the result of extreme pulmonary congestion from whatever cause arising, and thus is found associated with acute bronchitis and pneumonia, with obstructive or regurgitant disease of the mitral valve, with obstructive disease of the aortic valves, and with pressure on the pulmonary veins, whether caused by aneurism, or by cancerous or other morbid deposits in the chest. It is speedily induced by section of or pressure on the par vagum, probably as a result of defective innervation, and is usually attendant on the pulmonary congestion which precedes slow death.

By whatever cause produced, œdema of the lungs is characterized by infiltration of the air-cells and interstitial areolar

[1] See "Med. Times and Gazette," April 15th, 1845, and "Wiener Zeitschrift," 1853.

tissue with a thin serous fluid. When the œdema is passive and connected with chronic mischief, the fluid effused is usually colorless; whereas when the œdema is acute, and arises from active congestion of the lung, it is more commonly tinged with blood. It generally affects both lungs, and is diffused throughout their structure, or limited to the inferior or posterior portions, in which the laws of gravitation would naturally cause the fluid to accumulate. Oppression at the chest, with increased frequency of respiration, amounting sometimes to excessively hurried breathing, slight cough, and a thin, but sometimes rather tenacious frothy expectoration, are the general symptoms by which its accession and course are marked. The morbid anatomy of the parts, and the physical signs to which the disorder gives rise, will be evident from an inspection of the subjoined table:

Chronic and Passive Œdema.

Morbid Anatomy.	*Physical Signs.*
Chronic œdema—The lung is of a *pale grayish* color; collapses slowly and imperfectly; is inelastic and doughy; pits readily on pressure; and scarcely crepitates on being handled. On section, a large quantity of almost colorless serum, unmixed, or nearly so, with air, oozes from the surface; the lung-tissue is tough and resistant, and portions of it sink instantly in water without being previously subjected to pressure. The lining membrane of the bronchi is often of a dark, livid color, consequent on chronic vascular congestion.	*Inspection* affords no information. *Palpation* is said sometimes to disclose increased vocal fremitus. This has not been the result of my experience. *Percussion.*—The sound is duller than natural, and the parietal resistance increased. Skoda speaks of a tympanitic sound on percussion, but I have never met with it. *Auscultation.*—The respiration is weak or almost absent, or else harsh and coarse, and accompanied by bubbling râles, according to the precise degree of infiltration of the portion of the lung auscultated. If, as often happens, bronchitis coexists, sonorous and sibilant rhonchi will be heard.
Acute œdema.—The lung is usually *red;* collapses slowly and imperfectly; is inelastic, but less so than in chronic œdema, and is more crepitant on pressure; on section, a large quantity of a pale red, frothy serum escapes from the cut surface; the lung-tissue is red, soft, and easily broken down, and the lining membrane of the bronchi is of a red color, such as is seen in cases of active congestion or inflammation.	*Acute œdema.*—Inspection, palpation, and percussion yield the same results in acute as in chronic œdema; but on auscultation the vocal resonance is generally louder than in the chronic variety, and the râle which attends it is much finer. Indeed, it ordinarily resembles the fine bubbling râle of capillary bronchitis, and is to be distinguished from it only by the lesser viscidity of the sputa, and the absence of the general signs of bronchitis. In some instances it so closely resembles the crepitation of pneumonia, that the absence of tubular breathing and other symptoms which usually attend the progress of that disease, alone enable us to discriminate between them.

The treatment of pulmonary œdema must be in strict relation with the cause from which it originates. When acute, and dependent on Bright's disease or scarlatinal dropsy, diuretics and saline purgatives are especially indicated, and the administration of squills and digitalis should not be neglected; an emetic in some instances proves extremely serviceable, and so do vapor and hot-air baths, and cupping or dry cupping on the loins. Its occurrence betokens the necessity for a tonic and stimulant treatment as soon as the chest is somewhat relieved; and nothing answers better than the tincture of the sesquichloride of iron, in combination with digitalis, stimulating diuretics, and a light vegetable bitter. If it occurs during the course of typhus fever, or other disorders of an adynamic type, diffusible stimulants internally, and blisters and mustard poultices to the surface of the chest, are commonly found to be the most efficient agents; whereas, if it is associated with much pulmonary congestion, repeated dry cupping, aided by digitalis, expectorants, and saline purgatives, prove our most valuable and trustworthy allies.

Pulmonary Hemorrhage, Hæmoptysis, and Pulmonary Apoplexy.

Pulmonary hemorrhage and spitting of blood[1] is a subject of fearful interest to the physician, and of serious, nay vital importance to the patient. Sometimes, indeed, it is the result of idiopathic congestion of the lung, and may be regarded as a primary affection of not a very serious character. But such an event is exceedingly rare. Much more commonly it is attributable to engorgement of the pulmonary vessels, induced by organic disease of the lungs or heart, and is altogether a secondary affection—a mere symptom of structural mischief in the chest. Unfortunately, too, the maladies of which it is an index are of a serious nature, and usually tend to an untimely death. It is a frequent harbinger of consumption in its varied forms, and is often one of its early symptoms; so that too much attention cannot be bestowed on the investigation of every point relative to its occurrence.

Considerable difference of opinion exists as to the source and

[1] Throughout this section I apply the term pulmonary hemorrhage indifferently to all outpourings of blood which take place from any portion of the respiratory apparatus, from the epiglottis and larynx to the pulmonary parenchyma. During life it is often impossible to decide from which portion of the tract the bleeding proceeds, and even after death it is sometimes difficult to decide the question. It is therefore useless to attempt to separate the different varieties for the sake of description.

significance of the bleeding which leads to spitting of blood. Some persons imagine that when the larynx, trachea, or bronchi are congested, blood may exude from the mucous surface after the manner in which it oozes from the posterior fauces and the mucous membrane of the mouth; nay more, they affirm that exudation from the surface of the air-passages is a very frequent cause of bleeding, even in cases in which no local structural mischief exists. Others, without denying the possibility of hemorrhage from the bronchial mucous membrane, assert, not only that it is of rare occurrence, but that it seldom, if ever, takes place except when the blood is altered in character, as in purpura, scurvy, and similar disorders, or where the mucous membrane of the air-passages is ulcerated. These writers refer the bleeding to the structure of the lung itself, and they support their opinion by reference to the minute dark spots which are often found in the lungs after death in cases in which spitting of blood has existed during life—spots which are evidently caused by the outpouring of blood consequent on the giving way of the pulmonary tissue.

My own opinion occupies a middle place between these two extremes. Careful observation has led me to believe that, except in cases of diseased heart, or of purpura, and other hemorrhagic disorders, bleeding seldom arises from the bronchial mucous membrane, unless that membrane is the seat of ulceration, or is acted on by aneurismal or other tumors, or by tubercular or other adventitious deposits, which cause pressure upon the bronchial vessels, and thus mechanically lead to distension and rupture of the capillaries. But inasmuch as I have frequently seen blood ooze from the congested mucous lining of the mouth and throat, and on several occasions have traced cases to the dead-house of St. George's Hospital in which spitting of blood had occurred during life, and in which, nevertheless, no organic disease of the lung existed, and no evidence of any disruption of the pulmonary tissue could be discovered after death, I see no sufficient reason to doubt that, in some instances at all events, hemorrhage may take place from the congested bronchial membrane. On the other hand, the large amount of blood which is sometimes poured out, the rapidity with which the outpouring occurs, and the evidence which, in certain instances, may be obtained after death, of the giving way of the pulmonary tissue, are facts which lead irresistibly to the conclusion that the tissue of the lung is a frequent source of bleeding. And when, further, it is considered that simple ulceration of the bronchial mucous membrane is of extreme rarity, that in many instances of pulmonary hemorrhage there is entire absence even of bronchial congestion, and that the delicacy of the lung-tissue is such as to render it very liable to give way under conditions productive of extreme congestion and mechanical obstruction to the pulmonary circulation, the conclusion is inevitable, that the proper tissue of the lung is a much more frequent source of hemorrhage than the bronchial mucous membrane.

But whatever its source, pulmonary hemorrhage is always a symptom of grave importance, as indicating interference with the circulation of vital organs. In some instances it may result from wounds or other mechanical injuries of the lung; in some, the heart may be the cause of the mischief, giving rise, by its diseased and irregular action, to undue congestion of the lung; in others, the pressure of aneurismal tumors, or of tubercle, cancer, or other adventitious matter, may produce the same effect; in others, again, pneumonic congestion, or abscess, or gangrene, may be the cause of the bleeding; in others, ulceration of the larynx, trachea, or bronchi; whilst in certain instances, extreme congestion of the bronchial mucous membrane, more especially, when connected with a spanæmic condition of the vital fluid, appears to lead to the outpouring of blood, irrespective of any organic or permanent lesion. But mischief of a serious nature must have proceeded to some extent before hemorrhage can occur. Even in those instances in which spitting of blood is commonly supposed to be independent of local mischief, and to be of little importance, as when it occurs after violent straining efforts, or in connection with diminished atmospheric pressure during the ascent of lofty mountains, my own observation has led me to take a more serious view than that usually entertained, and to believe that in these, as in other cases, there is usually some latent mischief in the chest—some local cause of pulmonary congestion—some mechanical interference with the capillary circulation through the lungs.[1] In

[1] In this observation I am borne out by Dr. Walshe. In a sensible note (loc. cit., p. 470), he remarks—"Boussingault, D'Orbigny, and Roullin make no reference to hæmoptysis as having occurred in their ascents of the Andes; De Saussure observed nothing of the sort in the ascent of Mont Blanc; and Mr. Albert Smith, whose medical education gives value to his testimony, insisting in his narrative on the difficulty of breathing experienced, is silent concerning expectoration of blood, and yet his party and guides numbered twenty-four." A large additional mass of similar evidence might now be adduced from the experience of members of the Alpine Club, as also from that of aeronauts who have rapidly attained to very high altitudes in balloons. Messrs. Glaisher and Coxwell, who in this way rose to the unexampled altitude of six miles, and nearly lost their lives through the rarefaction of the atmosphere, make no mention of spitting of blood in their detail of the symptoms which they experienced. For full details of this ascent see "British Med. Journ." for Dec. 13, 1862, p. 625.

several instances which have come before me, in which persons apparently in good health have spat blood, under one of the conditions above specified, I have known the symptoms of consumption set in, and run on rapidly to a fatal termination. And although it is true that individuals are sometimes met with who have had hæmoptysis more or less profuse, and who, nevertheless, have attained to a good age, without the occurrence of phthisical symptoms, this fact does not invalidate the conclusion that their lungs are more or less organically diseased. The same fact is sometimes observed in persons who are descended from a consumptive stock, and who eventually die of phthisis. Indeed, nothing is more common than for spitting of blood to precede by many years the fatal development of tubercular disease; and its occurrence only proves what most persons are prepared to admit, that tubercular disease of the lungs, if not extensive, may exist for an indefinite period in a quiescent state, without giving rise to symptoms of consumption, or, in other words, that persons who have tubercles in their lungs, if placed under circumstances favorable to their health, may live on for years in the enjoyment of fair average health, and, as post-mortem records clearly show, may die at an advanced age of some other disorder, without having manifested any symptom of consumption, with the one exception of spitting of blood. In short, if those cases are excluded in which the sputa are slightly specked or streaked with blood—as in certain instances, of severe congestive bronchitis—or are rusty-colored, from admixture with it, as is often seen in acute pneumonia, the only exceptions I am disposed to admit to the fearful significance of hæmoptysis, are those in which it results from blows on the chest, or from mechanical injury to the lung, or in which it occurs in women, vicariously to the menstrual discharge.[1] In these cases, it *seems* occasionally to take place independently of structural pulmonary disease; but post-mortem examination would of course be needed to verify the fact, and render it a matter of certainty. In default of this proof, it is only reasonable to suppose that although the pulmonary congestion may be referable to disorder of the menstrual function and vicarious action, yet that the bleeding is usually attributable to the existence of latent tubercles, causing local pressure and obstruction of the pulmonary circulation.

It follows, from what has been already stated, that hæmoptysis,

[1] Most practitioners must have met with cases in point, but I would refer those who have not to a remarkable instance of vicarious menstruation from the lungs recorded by Pinel, and quoted in Sir Thomas Watson's "Lectures," ed. i, vol. ii, p. 140.

or the discharge of blood by expectoration, must always be regarded, as suggestive of organic disease of the chest. It does not necessarily indicate the fact, but it is sufficient to excite the gravest suspicion, and to render imperative a close and careful examination into the condition of the thoracic organs. It matters not whether the quantity of blood be small,—mere specks or streaks,—or whether it amounts to drachms or ounces; in either case its presence affords just cause for alarm. For although, as already stated, the researches of pathology oblige us to admit that hemorrhage may arise from a variety of causes, they have also served to establish the fact that it is most frequently connected with tuberculization of the lung; and to such an extent does this hold good, that spitting of blood is justly regarded as one of the earliest and most important signs of pulmonary consumption.[1]

The quantity of blood expectorated in different cases of pulmonary hemorrhage varies extremely. Sometimes it is so small as to escape observation; and at times, more especially in disease of the heart, is undistinguishable by the naked eye, although the microscope at once reveals blood-corpuscles in the sputa; sometimes, though rarely, it is so profuse, and is poured out so rapidly, that the patient is suffocated by it, or else dies from the effect of syncope; more commonly, it is ejected slowly, a mouthful at a time, and in quantity varying from a few streaks to a teaspoonful or several ounces.

It is commonly supposed that the quantity of blood expectorated affords a trustworthy measure of the significance of pulmonary hemorrhage, and that a few slight streaks of blood are of little or no importance. They may arise, it is often said, from the throat, or may be the result of simple bronchitis. On the other hand, some physicians attach a fatal significance to every streak of blood in the sputa, and are disposed to doubt the correctness of those who would attribute such hemorrhage to simple congestion of the bronchial mucous membrane. Unfortunately too, the matter, though of some practical importance, does not admit of very positive decision. There cannot be a doubt that, in the majority of cases in which spitting of blood occurs, however small the quantity of blood may be, tubercles in an active or latent state are present in the lungs. The revelations of the stethoscope, corroborated as they are by inspection in the dead-house, abundantly attest this naked fact. But this is not the whole question at issue. The point to be determined is not simply whether spitting of blood is ever met with

[1] For facts corroborative of this statement see chap. iv of this treatise.

when there is no organic disease of the chest, but in what proportion of cases it occurs independently of structural mischief, and whether any clue to the condition of the chest can be obtained from the quantity of blood expectorated. And here we find ourselves completely at fault. Experience undoubtedly overrules the dicta of those who would deny the occurrence of hemorrhage as a result of simple pulmonary congestion; for on several occasions I have traced cases to the dead-house of St. George's Hospital, in which there has not existed any organic disease in the chest, and in which, nevertheless, hæmoptysis, to a greater or less extent, has been observed during life. At the same time it justifies the most positive statement that such an occurrence is comparatively rare. From the nature of the case, it is impossible to obtain trustworthy statistics to throw any light on the proportion of cases in which spitting of blood occurs irrespectively of organic mischief, or to show the relative significance of slight streaks of blood and of drachms or ounces of blood in the expectoration. Nevertheless, as on the one hand there is abundant evidence to prove that slight streaks of blood are often indicative of the presence of tubercles, and on the other cases have been met with in which a considerable quantity of blood has been ejected by coughing, as the result of simple pulmonary congestion, it is fair to conclude that no reliance can be placed upon the mere quantity of blood as a test of the condition of the lungs and air-passages. In short, spitting of blood is itself a fact of grave clinical significance; but in most instances the quantity of the blood ejected adds little or nothing to the information it affords. The only exception to this general rule is in the case of very profuse hæmoptysis. This, if unconnected with suspended menstruation, can only arise from excessive congestion of the lungs, or from the giving way of a large vessel. In the former case, it may be occasioned by mechanical causes, as by the pressure of tubercle or other matter in the lung, or by the regurgitation of blood through a diseased mitral orifice; in the latter it may result from ulceration of one of the pulmonary vessels produced by the presence of tubercle or other foreign matter in the lung, or from the oozing or bursting of an aneurismal tumor; it cannot be attributable to mere idiopathic congestion.

Again, the color of the blood in hæmoptysis is not to be depended on as a proof either of the source of the hemorrhage, or of the condition of the parts from which it originates. The blood is usually florid, but it may be dark-colored, or almost black; or it may be partly of an arterial, and partly of a venous tint; it is generally frothy, from admixture of air, and is manifestly ejected by the act of coughing; but if it be more profuse, it may escape in gulps from the mouth in a non-aerated condition, or may even give rise to reflex actions, which may induce the patient to declare that it was ejected by vomiting, and not by coughing.

Indeed, the whole question of pulmonary hemorrhage is involved in considerable difficulty. Not only does doubt sometimes arise as to whether spitting of blood is or is not connected with structural changes within the chest, but much uncertainty may be felt as to whether blood ejected from the mouth is derived from any portion of the respiratory apparatus. Even when the blood observed in the sputa results from pulmonary hemorrhage, patients first become conscious of its presence when it reaches the posterior fauces. They are apt, therefore, to assert, with the greatest confidence, that it is derived from the throat, and not from the chest. In like manner, they will sometimes refer it to the gums, the mouth, or the posterior nares. The mere appearance of the blood will not serve to elucidate the subject; but a close investigation into the state of the mouth and the pharynx, will generally show whether the hemorrhage is referable to either of the sources indicated by the patient. And as the determination of the question is a matter of great importance, the student should make a careful examination, so as not to be misled by the earnestness of the patient's assertions.

Another source of difficulty in the diagnosis of hæmoptysis is the occurrence of hæmatemesis, or vomiting of blood from the stomach. It has been already stated, that in severe cases of hemorrhage from the lungs, blood is ejected rapidly by mouthfuls, and that its ejection is sometimes accompanied by efforts which the patient is unable to distinguish from vomiting. Hence, persons who have had a severe attack of hæmoptysis will often assert that they have suffered from "vomiting of blood;" and even when closely questioned on the subject, will deny that they brought up the blood by coughing. But if care be exercised in conducting the inquiry, facts may generally be elicited calculated to throw light on the real nature of the malady. In cases of spitting of blood, the family and personal history of the patient will generally indicate the probability of disease of the chest; and in corroboration of the information thus obtained, the face will usually be found flushed, and the pulse excited, full, and bounding; the seizure will have been preceded by more or less shortness of breath, a slight hacking cough, a sense of weight or tightness in the chest, and a tickling at the top of the larynx; the blood will be frothy and usually of a florid color; and although the greater part of it may have been brought up in the course of a few minutes, yet blood-streaked or blood-stained sputa will continue to be expectorated for a considerable time after "the bursting of the bloodvessel." On the other hand, hæmatemesis is usually preceded by a sense of weight and uneasiness at the pit of the stomach, and is sometimes followed by tenderness on pressure; the patient's face and lips are blanched, and his pulse is seldom so excited as in hæmoptysis, and is weak, rather than full and bounding; the blood ejected is non-aerated, and, unless the hemorrhage occurs rapidly from ulceration of an artery in the stomach, is of a dark venous hue; it is brought up suddenly, and the vomiting is not followed, as in hæmoptysis, by the expectoration of blood-streaked or blood-stained sputa, and the stools are dark-colored and pitch-like in appearance, from the quantity of blood they contain—a circumstance which never occurs in hæmoptysis, unless blood has been accidentally or wilfully swallowed. Thus, the general symptoms alone will suffice, in most instances, to stamp the case as one of hæmoptysis; but no positive conclusion ought to be drawn, and no opinion expressed, until the chest has been carefully examined. For although, on the one hand, spitting of blood may occur before any distinct evidence can be obtained of structural mischief in the lungs, and on the other, hæmatemesis may take place in persons whose lungs are structurally diseased, yet these coincidences are so rarely met with, that the finding of liquid bubbling râles in some portion of one or both lungs would go far towards solving any doubt as to the blood being derived from the chest.

When pulmonary hemorrhage occurs, it may or may not give rise to consolidation, or to rupture of the tissue of the lungs. Ordinarily, when the quantity of blood poured out is small, it is ejected by cough before coagulation takes place, and no local ill results ensue; but when the bleeding is more profuse, and especially in regurgitant disease of the mitral valve, when great impediment exists to the onward flow of blood through the lungs, the air-cells and terminal bronchi may become distended with coagulated blood, forming firm nodulated masses, of from half an inch to four inches in diameter, blackish red, or even black in appearance, impervious to air, inelastic, and excessively firm to the touch. The masses thus formed constitute what has been termed "pulmonary apoplexy"—a singularly bad and inappropriate term, but one which was introduced by Laennec, and has obtained unusual currency in the profession. This form of local mischief is not produced by hemorrhage from any particular source; it may occur whenever blood is present in the air-passages, and finds its way into the air-vesicles. Sir Thomas Watson gives a case in which it resulted from blood which trickled down, through the windpipe, into the lungs, during an attack of bleeding which resulted from ulceration of the lingual artery;[1] and most persons who have had frequent opportunities of witnessing post-mortem investigations, must have met with cases in which it has arisen in connection with hemorrhage in cases of consumption. Nevertheless, it may be stated, that disease of the left side of the heart is the principal, though not the invariable cause of its occurrence. If the outpouring of blood into the air-cells does not produce rupture of the pulmonary tissue, the apoplectic masses, though irregularly shaped, are clearly circumscribed and sharply defined, consisting of one or more lobules, firmly blocked up with fibrine; whereas if, as sometimes happens, the texture of the lungs gives way, and admits of blood infiltrating the interlobular areolar tissue, the pulmonary apoplexy is diffused, and the characteristic sharp outline of the lobular form of the mischief is wanting. The former variety of mischief is by no means uncommon, and usually accompanies enlargement of the left cavities of the heart, and inefficiency of the mitral valve. The latter is rare, and is seldom met with except as the result of mechanical injury to the chest, or the bursting of an aneurism, or the giving way of a large vessel, in consequence of ulceration. The one is commonly found affecting both lungs, the other is almost always confined to one lung; the first generally implicates several portions of the lungs, the last is usually confined to one portion of the lung. Neither the one form nor the other is necessarily accompanied by hæmoptysis, though spitting of blood is usually an attendant symptom of both, and affords the principal evidence we can obtain of their occurrence.

When the bleeding is very profuse, the blood may not only break up the tissue of the lung, but may burst through its serous envelope and escape into the pleural cavity. Many cases are on record of death produced in this manner.[2]

The local effects produced by outpourings of blood in the lungs will be seen by reference to the subjoined description; the physical signs to which they give rise are subsequently given in detail:

Morbid Changes.

The tissue of the lungs and the bronchial mucous membrane

[1] See "Watson's Lectures," ed. i, vol. ii, pp. 146–7.

[2] See Dr. Patterson's "Observations on Pulmonary Apoplexy proving Fatal by Rupture of the Periphery of the Lung by Effusion of Blood into the Pleura," in the "Edin. Med. and Surg. Journ." for January, 1846.

are much congested, and sometimes, though rarely, there is ulceration of the mucous surface. Tuberculous or other deposit in the lung, or aneurismal or other tumors, pressing on the lungs and producing distension and rupture of the capillaries, or, in some rare instances, ulceration of a large vessel, or insufficiency of the valves on the left side of the heart, are seen to be the cause of its occurrence.

When pulmonary apoplexy exists, it is usually found in the lower and posterior portion of the lungs, though it is sometimes near their anterior surface. It causes a distinct projection of the pleura if it is situated near the surface of the lung, and if, as sometimes happens, the pleura gives way, blood may be found in the pleural cavity. The air-cells implicated in the apoplectic mass are blocked up with coagulated blood, and there may or may not be laceration of the lung-tissue. If there is no giving way of the lung-structure, there are seen on the surface of the lung one or more sharply defined nodulated masses of solidified tissue, varying from half an inch to four inches in diameter. These masses are of a blackish-red color, inelastic, and excessively firm to the touch, do not crepitate on pressure, and sink instantly when placed in water. On section, they present a dry and nearly homogeneous surface, sometimes slightly granular, from which a small quantity of dark, grumous blood can be obtained by scraping. In some few instances, a small clot of loosely coagulated blood exists in the centre of the apoplectic clot. The adjacent lung-tissue is sometimes inflamed, in which case it is loaded with a sero-sanguineous, but frothy fluid. Under these circumstances washing with water will remove the inflammatory exudation, and expose the outline of the apoplectic mass.

If laceration of the lung-tissue has occurred, the defined outline of the lobular form of the disease will be absent, in consequence of the blood having passed in all directions into the interlobular cellular tissue; and the apoplectic mass will consist of coagulated blood, of various degrees of firmness, intermixed with portions of broken-down and disorganized pulmonary tissue. In some rare instances suppuration and gangrene will have occurred in the apoplectic mass.

As resolution takes place, the tint of the coagulum becomes lighter, and passes into a brownish and then into a yellowish red. The effused matter gradually softens, and is converted into a yellowish or rusty-colored fluid, which is removed by expectoration and absorption. The bronchi once permeable, air is readmitted, the proper lung-tissue reappears, and the abnormal firmness and density of the affected portion pass off.

Physical Signs.

When the hemorrhage from the lungs is slight, and pulmonary apoplexy does not occur, neither *inspection*, *palpation*, nor *percussion* affords any information as to the seat or amount of the bleeding. Auscultation, however, will usually reveal a thin bubbling râle at the spot whence the blood is derived.

When the bleeding is more profuse, and accompanied by pulmonary apoplexy, an abundant bubbling râle will be heard at the seat of effusion, and will continue until coagulation takes place. Then of course all bubbling will cease, and the existence or non-existence of physical signs of the mischief which has occurred will depend upon the seat and extent of that mischief.

If the bleeding be slight and deeply seated in the lung, all signs of its existence will cease with the coagulation of the blood. The amount of blood effused will not seriously impair the motion of the chest, and neither inspection, palpation, percussion, nor auscultation will avail to discover its seat.

If the nodulated masses of solidified lung-tissue lie superficially, percussion and auscultation may possibly yield some evidence of the mischief; but positive information can be rarely thus obtained if the nodules are few and small.

When the hemorrhage is more profuse, and the patches of pulmonary apoplexy are larger and lie superficially, palpation, percussion, and auscultation may all furnish evidence of the seat of mischief.

In this case percussion will elicit a dull tone over a space corresponding to the extent of lung implicated.

Auscultation will detect a diminution or entire absence of the respiratory murmur over the affected portions, and coarse breathing, possibly accompanied by bubbling, in their immediate vicinity.

When the apoplectic mass is large, and is traversed by a good-sized permeable bronchus, hollow breathing and increased resonance of the voice may be heard, and palpation will make us aware of increased vocal fremitus.

With the progress of resolution, the hollowness of the breathing and the increase of the vocal resonance cease, râles of all kinds begin to be heard over the affected parts, and ultimately are replaced by natural vesicular breathing.

If, instead of undergoing resolution, pulmonary apoplexy is followed by pneumonic abscess or gangrene, the signs of these several affections will be met with instead of the signs of resolution.

Thus, then, it will be seen that the most careful physical examination of the chest will often fail in throwing much light upon the source of pulmonary hemorrhage, and that were it not for the occurrence of hæmoptysis, we should often remain in ignorance of its existence. But it is otherwise in regard to the diseases from which the spitting of blood originates. When once the appearance of blood in the sputa has directed attention to the condition of the chest, the physical signs will usually enable us to determine the precise character of the mischief present; and in the great majority of cases they will teach us that, whatever the seat of the bleeding, its occurrence is attributable to tubercular disease of the lung, or to disease of the left side of the heart.

The treatment of spitting of blood must be varied according to the extent and probable cause of the hemorrhage. If the heart be the organ at fault, digitalis, aconite, or the veratrum viride, which exercise a controlling influence over its action, should be administered internally, together with a few doses of calomel to unload the liver, and saline purges to draw off watery evacuations from the bowels, and so to relieve the circulation; at the same time venesection may be had recourse to, and cupping or dry cupping between the shoulders should be employed to mitigate the local congestion. Perfect rest must be strictly enjoined, all mental excitement avoided, and the diet restricted to milk, barley-water, iced lemonade, and other cooling beverages. In the most frequent form of spitting of blood, viz., that which is attendant upon tubercular disease of the lung, when the quantity of blood ejected is small—not exceeding a teaspoonful—and where the hemorrhage is unattended by febrile excitement and symptoms of general engorgement of the lungs, little heed need be paid to its occurrence; the bleeding is nature's mode of relieving temporary local congestion, and will prove beneficial rather than hurtful. The diet need not be altered, nor tonic medicines discontinued; and the utmost that is necessary or likely to prove useful in the way of special treatment, is dry cupping between the shoulders, a dose of calomel or blue pill if the liver be inactive, a saline purge to produce a free, watery evacuation, and the administration of the mineral acids in combination with digitalis, tonics, and cod-liver oil.

If the hemorrhage, though scanty, be attended with feverish heat of skin and excessive vascular action, it will not be safe to continue the use of tonic medicines or a stimulating diet. The patient must be kept low for some time, and active measures must be taken to subdue the feverish excitement of the system. He should remain in bed, with his head and shoulders elevated; a free circulation of cool air should be kept up in the room in which he lies, his mind should not be disturbed, and silence should be enjoined. Iced water should be his drink, and his diet restricted to barley-water, cold milk, whey, and cold beef-tea. A free action of the liver must be sustained; and cooling saline purgatives should be given, with a view not only to relieve the vessels, but prevent the effort of straining at stool. The excitement of the nervous and vascular systems should be tranquillized by salines, with tartar emetic, digitalis, the veratrum viride, and opium; and if local congestion be excessive, the aid of dry cupping or cupping, or even venesection, must be had recourse to. In the latter case, small and repeated bloodlettings, from six to ten ounces, according to the constitution of the individual, appear to answer better than a single full bleeding, and venesection from the arm has proved more serviceable in my hands than the loss of blood by cupping; the blood is drawn more quickly, and the effect on the pulse is more decided. But whether venesection from the arm or cupping be employed, dry cupping between the shoulders ought never to be neglected. In most instances, it will supersede the necessity for bloodletting, and it is at all times a valuable adjunct to venesection in removing pulmonary congestion. Leeches in these cases prove comparatively useless.

If the hemorrhage be profuse, the question cannot arise as to what measures are best suited to the patient's constitution, or the malady under which he is laboring. The sole point to be decided is how to stay the bleeding, and obviate the tendency to death. A variety of remedies have been employed for this purpose, amongst which I may mention, as having received the greatest amount of testimony in their favor, venesection, cupping, dry cupping, the application of ice down the spine, tartar emetic, acetate of lead and opium, gallic acid, sulphuric acid, alum, matico, ergot of rye, and turpentine. My own experience leads me to testify most strongly in favor of repeated dry cupping, aided by the internal administration of full doses of digitalis, either alone or in combination with turpentine.[1] If gallic acid is had recourse to, it should be given every hour, in eight- or ten-grain doses, until the hemorrhage is subdued, or until a dark-green color in the sputa indicates its action on the system; or, if lead be employed, it should be given in doses of two or three grains, combined with dilute acetic acid and laudanum, after the method recommended by the late Dr. Anthony T. Thomson, and the dose should be repeated every hour or every two or three hours, according to the urgency of the symptoms.

[1] ʒjss to ʒij of the tincture, or gr. vj to gr. viij of the powder daily.

If the aid of turpentine is invoked, it should be administered at brief intervals, in half-drachm doses.

If any symptoms of sinking arise, they must be met by diffusible stimulants, and must be treated on general principles, without reference to the nature of the mischief in the chest, Meanwhile, the administration of those remedies which are calculated to control the bleeding should be perseveringly continued.

A few words may be added by way of caution. During convalescence from an attack of hæmoptysis, and for some time afterwards, the patient should be advised to keep perfectly quiet and free from bodily or mental excitement. Especially is this the case when the hemorrhage has resulted from disease of the heart. Nothing is more likely to determine its recurrence than public speaking, or whatever puts a strain upon the organs of respiration; or than violent bodily exertion, which accelerates the circulation and increases vascular action; or than mental excitement, which also hurries the heart's action, produces pulmonary congestion, and thus proves equally prejudicial. These are dangers which may be avoided if the patient is duly warned against them; and it behooves us therefore to place them clearly before him ere he is permitted to return to the ordinary avocations of life.

Acephalocysts in the Lungs.

Hydatids sometimes exist in the lungs, having been developed in the pulmonary tissue, or having found their way there from the liver. When developed in the lung, they may remain there for some time without giving rise to notable disturbance; but as they gradually enlarge, they exert pressure on the surrounding tissues, and may occasion hæmoptysis, bronchitis, pneumonia, or even gangrene. In some instances they have been known to cause perforation of the pleura and give rise to pneumothorax; and they may also make their way through the diaphragm.

If the chest be examined whilst as yet they are in a quiescent state, and are small, and have not occasioned local irritation, the breathing may be coarse and harsh in their immediate vicinity, and there will probably be prolongation of the expiratory sound; but if they chance to be deeply seated in the lung, these symptoms will be masked by the normal sounds emitted by the healthy lung-tissue which lies between them and the chest-walls.

As they enlarge and excite bronchitis, pneumonia, or pleurisy, the general symptoms and physical signs of those diseases will be present; and unless some portion of an acephalocyst be expectorated, or the hooklets of the echinococcus be discovered by the microscope, there will be nothing to point to the true nature of the disease.

Not unfrequently, the general symptoms are those of rapidly progressive phthisis—cough with muco-purulent expectoration, spitting of blood, night-sweats and emaciation—while the physical signs may at first consist of dulness on percussion, absence of breathing over the seat of the cyst, and râles and rhonchi with prolongation of the expiratory sound in its vicinity. When, after a time, the cyst bursts, there may be hollow breathing, with gurgling and pectoriloquy, just as in any other cavity which has established a free communication with the upper air-passages. Here, again, the discovery of a portion of a cyst, or of the hooklets of the echinococcus in the sputa, will alone enable us to determine the nature of the disease.

When the cyst makes its way into the lungs from the liver, the pulmonary mischief will have been preceded by hepatic symptoms, and almost certainly by a sharp attack of pleurisy. At length the patient is somewhat suddenly attacked with extreme distress and difficulty of breathing; his countenance becomes anxious, his features collapsed, the skin clammy and more or less livid, and the extremities cold; incessant paroxysmal cough supervenes, vomiting occurs, and by degrees he becomes more or less deeply jaundiced. Then come symptoms of acute pneumonia and pulmonary consolidation, followed by the physical signs of excavation, the expectoration at the same time passing through every variety of tint and consistence, from that of slightly rusty-colored mucus to that of deeply bile-tinged, muco-purulent matter, or of a dark brown colored fetid fluid, containing shreds of lung-tissue and entire hydatids, or portions of hydatid cysts.

In this case, as in those already referred to, the discovery of the hydatid cysts, or of the hooklets of the echinococcus, constitutes the only distinguishing mark of the disease, as the general symptoms and physical signs might be attributable to simple hepatic abscess making its way out though the lungs.

In all instances in which acephalocysts are met with in the lungs, the issue of the case is extremely doubtful. When they are developed in the pulmonary organs, statistics seem to show that although the symptoms which attend their rejection are very severe, and may continue for many months, yet that recovery may be brought about in nearly half the cases; whereas, when they make their way into the lungs from the liver, the constitutional disturbance is so great as to leave little hope of recovery. Nevertheless recovery has been noted in a few exceptional cases.

Treatment is of little avail in these cases of intrathoracic hydatids. We know of no means of destroying the acephalocyst or inducing its expulsion. Possibly, however, an exception may be made in reference to turpentine. Knowing as we do how powerful are the anthelmintic properties of this drug, and the readiness with which it finds its way into the circulation, it seems fair to conclude that if given in repeated doses for a considerable length of time it might lead to the death of the acephalocyst, and thus promote the recovery of the patient. An opportunity has not occurred to me to put this treatment to the test, but on general grounds it appears to deserve a trial. In the event of its failure, the only rational means to be adopted for our patient's relief are those which are calculated to mitigate pain, assist expectoration, and sustain the failing power of the system. Opium, ether, the various expectorants and alcoholic stimulants are of the greatest service, and counter-irritants are useful in relieving the local inflammation. As soon as the first severity of the symptoms has subsided, tonics must be given freely, and the strength sustained by a nutritious diet.

CHAPTER III.

BRONCHITIS.

THIS disease is essentially an affection of the bronchial mucous membrane, and may arise from any cause, whether mechanical, constitutional, or epidemic, which excites irritation, congestion, and inflammation of that membrane. Thus it may follow the inhalation of the fine metallic particles which result from needle-grinding and other similar occupations; it is a prominent symptom of the so-called "hay-asthma," a disorder attributable to the inhalation of the pollen of the *Anthoxanthum odoratum*, or sweet-scented spring grass,[1] and it is a frequent accompaniment of the local congestion induced by heart disease, aneurismal pressure, and other similar causes; it occurs as a consequence of the disordered condition of blood which accompanies tuberculosis, albuminuria, gout, and other constitutional disorders; and

[1] See a paper by Mr. Gordon in "Med. Gazette," vol. iv.

it is a common attendant on continued fever, the various exanthemata, and those epidemic and endemic influences which occasion influenza and common catarrh.

The disease may occur either in an acute or in a chronic form. The acute disease is generally ushered in by symptoms of a so-called "cold"—chilliness, followed by heat of skin, the temperature ranging from 99.5° to 102.5°, general lassitude and aching of the limbs, uneasiness about the frontal sinuses, sneezing, running at the nose, sore throat, and hoarseness. These symptoms are soon followed by a sensation of roughness and tickling in the windpipe, with frequent dry cough, more or less tightness or uneasiness behind the sternum, and soreness diffused over the front of the chest. The cough aggravates the pain in the chest, and when severe and dry, as it usually is at first, produces a distressing sense of tearing; but in the course of a few hours, or after the lapse of a few days, it commonly becomes looser, and is then accompanied by the expectoration of a thin, saltish, frothy mucus, sometimes streaked with blood. This gradually increases in quantity, changes its character, and for some days becomes glairy, semi-transparent, and of a faintly yellowish color. Subsequently it assumes a grayish or greenish yellow tint, and is characterized by more or less opacity and viscidity. If the attack has been severe, or if the patient be weak, the secretion after a time becomes distinctly muco-purulent, and, in some instances, may even lose its glairiness altogether, and present the character of thoroughly opaque nummulated sputa. This is the course of events when the attack is passing off, which it usually does in favorable cases between the fourth and tenth days of the disease. But it often happens that the expectoration loses its opaque and puriform character, and again becomes frothy, glairy, and tenacious. This indicates a return or an extension of the inflammation; so that the precise character of the bronchial secretion forms a valuable guide to treatment. The respiration and the pulse are both increased in frequency; the former comparatively more so than the latter, but the precise frequency of the respiration, and the existence or non-existence of actual dyspnœa, varies with the extent of the bronchial affection, and the freedom with which the secretion is expectorated. When the larger and medium-sized air-passages are alone affected, and expectoration is free, and secretion not excessive, the oppression of the breathing seldom amounts to actual dyspnœa; but if the air-passages are widely implicated, and the mischief extends to the capillary air-tubes, the symptoms are much more urgent. There is then a dreadful sense of tightness and oppression at the chest, with dyspnœa and excessive restlessness; the patient sits erect in bed, or with his body bent forwards; the countenance is anxious, the pulse quick, the face flushed, and the skin hot, and sometimes moist, while distension of the jugular veins, and more or less lividity of the lips, cheeks, and general surface, extending even to the fingers' ends, betoken impediment to the circulation through the right side of the heart, as a consequence of the widespread mischief in the lungs. If the disease terminates favorably, the symptoms of oppressed breathing gradually pass off; the cough becomes less constant; the expectoration continues free, the pulse and the respiration decrease in frequency, and the body resumes its normal temperature. But in cases tending to a fatal termination a different train of symptoms is observed: exhaustion becomes a prominent feature of the case; the patient unable any longer to support himself erect, sinks gradually in bed, until his head is scarcely raised above the shoulders; the lividity of the countenance increases, drowsiness ensues, the pulse rises in frequency, but, in consequence of the increasing impediment to the circulation and the formation of clots in the right cavities of the heart, it becomes irregular or intermittent, and decreases in force; the temperature of the body falls below the normal standard, and the surface becomes covered with a cold, clammy perspiration. Meanwhile the cough decreases; the breathing, though more rapid, is carried on more tranquilly; the sputa almost cease, in consequence of the want of power to expectorate, and the air-passages become gradually more and more loaded. After lying for some time in an unconscious or semi-conscious state, varied only by wandering or occasional delirium, and sometimes by slight convulsions, the patient dies ultimately from suffocation, or from apnœa resulting from the arrest of the circulation through the lungs in consequence of the coagulation of the blood in the pulmonary arteries and in the right cavities of the heart.

The prognosis of acute bronchitis is always doubtful, especially in elderly persons. When the capillary air-tubes are involved in the mischief the disease is of extreme gravity, at whatever age it may occur; but in boyhood and in the middle period of life recovery usually takes place if the medium-sized bronchi only are implicated, and the patient applies early for advice. In infancy and in old age it is of more serious import, and a very guarded opinion should be given as to its issue. In childhood and early manhood recovery is the rule rather than the exception, and the issue is determined mainly by the time at which the treatment is commenced, and the care and judgment displayed in the management of the case. But in advanced years the condition of the heart and the existence or non-existence of emphysema have an all-important bearing on the progress and issue of the case. However mild the attack may appear at its onset, it must not be regarded too lightly. When the lungs are emphysematous and the heart is feeble and dilated, comparatively little disturbance of the respiration is needed to throw a grave impediment in the way of the heart's action. Distress of breathing, with lividity of the surface, readily ensues, and a tendency is soon manifested to the formation of clots in the right cavities of the heart. The prognosis, therefore, under the circumstances referred to, should always be guarded, even from the first, and still more so as soon as the slightest lividity of the lips and congestion of the capillary vessels of the face ensues. If under these conditions the heart's action becomes irregular and intermittent, the result in the majority of cases will be fatal.

The morbid changes induced in the bronchial tubes, and the physical signs resulting therefrom, will be more readily understood in their mutual relation, when viewed together at the different stages of the disease.

Morbid Changes.

1st. *Dry Stage.*

This stage is marked by congestion of the vessels of the bronchial mucous membrane, with arrest of its natural exhalation or secretion, and consequent redness, dryness, and roughness of its surface; there is fulness or thickening of the substance of the membrane, dependent in part on turgescence of the vessels, in part on infiltration of the submucous tissue, which give rise to diminution more or less irregular in the calibre of the affected air-tubes. These changes are seldom visible to the naked eye beyond the fourth or fifth division of the bronchi; but they may extend to the smallest ramifications of the air-tubes, as in so-called capillary bronchitis. They are not confined to any portion of the lungs, but where the disease is not associated with tubercles, the lower and posterior parts are their most frequent seat. There is reason to believe that throughout the complaint, but especially in its early stages, there may be, and often is, spasmodic contraction of the circular muscular fibres of the bronchi.

2d. *Moist Stage.*

The congestion and dryness of the bronchial lining membrane, which mark the first stage of the attack, is followed by the secretion of mucus more or less frothy, viscid, and tenacious, as already described. The air-tubes are generally somewhat dilated; their lining membrane is swollen, of a dark venous color, and often softened, and ulceration may be detected occasionally on its surface; the mucus or muco-purulent secretion is more or less closely adherent to the membrane. Microscopically, the thin watery portion of the secretion consists of a serous fluid, loaded with young abortive, epithelial cells; the thicker semi-transparent portion consists of a tenacious fluid, resembling white of egg, containing mucus corpuscles, and patches of small-sized epithelium; the opaque variety contains epithelium, pus, and mucus globules, some few blood-discs, exudation-cells, and granular matter. In the rare form of the disease, which has been styled "plastic bronchitis," lymph is poured out, forming a false membrane, which is ejected as a cast of the bronchial tubes. In fatal cases, the trachea and bronchi are found blocked up by secretion resembling that expectorated during life, the lungs are more or less distended with air, in consequence of its inability to escape through the obstructed tubes, and the right cavities of the heart are filled with clots. Emphysema frequently complicates the morbid changes which result from this disease, and so occasionally do pulmonary collapse and pneumonia.

Physical Signs.

1st. *Dry Stage.*

Inspection shows nothing more than hurried respiration.

Palpation seldom affords much information, but rhonchial vibration is sometimes perceptible.

Percussion.—The sounds are normal.

Auscultation.—The respiratory murmur is simply exaggerated over the non-affected portions of the lungs; but over the affected portions the sound of respiration is coarse, dry, and harsh, accompanied and ultimately replaced by rhonchi, or sounds of vibration which, *cæteris paribus*, are of a higher or lower pitch according to the size of the tubes in which they take their origin. The expiration is somewhat prolonged; the vocal resonance is not materially altered.

2d. *Moist Stage.*

Inspection reveals hurried breathing, with increase in the abdominal, and decrease in the thoracic expansion. If the dyspnœa be excessive, the lower portion of the sternum may even fall in during inspiration. The expiratory movements are prolonged and manifestly inefficient.

Palpation detects the existence of rhonchial vibration. The vocal fremitus varies; sometimes it exceeds, at others falls short of the normal standard.

Percussion.—The sounds are almost normal, except in certain instances, in which congestion or œdema of the lung, with excessive collection of secretion in the air-tubes leads to dulness at the posterior and inferior portions of the chest; in others, which occur more especially in children, in which obstruction of a large bronchus by a plug of thickened mucus causes pulmonary collapse, with corresponding local diminution in the clearness of the percussion-note; and in others, again, in which emphysema exists, and the clearness and fulness of the percussion-sound is thereby abnormally raised.

Auscultation.—The pulmonary breathing is impaired or suppressed, and replaced by rhonchi, which are grave or high-pitched, and by rales or sounds of bubbling, which are larger or smaller according to the size of the tube in which they originate. In capillary bronchitis, a fine or minute bubbling is heard during inspiration and expiration, most commonly at the bases of both lungs, posteriorly. The vocal resonance is not materially altered.

Acute idiopathic bronchitis commonly affects both lungs simultaneously and attacks the middle, lower, and posterior parts; whereas bronchitis connected with the deposit of tubercles ordinarily selects the upper lobes. Its prognosis is extremely serious at all ages; for although, when it is confined to the larger tubes, it does not usually prove fatal except to infants and to aged debilitated persons, yet when it is extensively diffused throughout the lungs, and involves the smaller air-passages, constituting what is termed "capillary bronchitis," it is attended in all cases by excessive dyspnœa, and is fraught with danger even to the middle-aged and vigorous adult. The latter, if seen early in the attack and treated judiciously, may be rescued in most instances from the effect of its fury; but so serious is the disease when occurring in infants and in aged, debilitated persons, whose hearts are fatty and feeble, that it proves fatal to above one-half of those attacked. The inflammation is most severe and most widely diffused when the expectoration is profuse; and the danger is great in proportion to the frequency of the respiration, the lividity of the surface, the feebleness of the heart, and the frequency and irregularity of the pulse.

Capillary bronchitis has been mistaken for inflammation of the lungs in consequence of the sound of bubbling heard in the former disease being sometimes as fine and crepitant as in cases of pneumonia. But the absence of severe rigors at the outset of the attack, the comparative coolness and occasional moistness of the skin, the clearness or resonance of the percussion-note over the affected portions of the lungs, the absence of tubular breathing, the non-existence of rusty-colored expectoration, and the occurrence of rales during expiration as well as during inspiration, together serve as distinguishing marks of the former malady.

The treatment of acute bronchitis must be varied, not only with the type and severity of the attack, but with the age, strength, and constitutional power of the patient. If the attack declares itself as a "common cold," there is seldom much necessity in the first instance for any active treatment. A full dose of opium, either alone or combined with a tumbler of white wine whey, or some other hot alcoholic stimulant, will sometimes suffice to check the disease; or, if there be so much febrile excitement as to render such treatment inexpedient, a hot air or hot water bath or a hot leg bath, followed by an ammoniated saline draught, combined with a full dose of sweet spirits of nitre, a few drops of the liquor morphiæ, and from twenty to forty minims of the vinum antimonii, will often induce a copious action of the skin and kidneys, and thus carry off and put an end to the attack. In some persons a few full doses of quinine will have a similar effect. But whether the disease commences as a "cold," or declares itself in any other form, serious symptoms oftentimes arise which require more active means for their subjugation. In the sthenic type of the disease occurring in the adult, venesection may be practised if the febrile action runs high, and the violence of the other symptoms is great; but in town practice it is rarely needed, and in most cases would prove dangerous. Even when bloodletting seems desirable, observation inclines me to recommend cupping between the shoulders in preference to general venesection. The local abstraction of eight or ten ounces of blood in this manner, repeated if necessary, and followed by dry cupping, serves in most cases to relieve the congestion of the vessels and to mitigate the severity of the symptoms. This is shown by the diminished frequency of the respiration. Relief is often obtained even whilst the cupping-glasses are filling; and all that can be accomplished by bloodletting in this disease may be effected in this manner more safely and with less exhaustion to the patient than by the use of general venesection. But bloodletting, however practised, must be employed with extreme caution, and must be confined to young and robust persons of a healthy constitution, for in the aged, the unhealthy or asthenic, it soon leads to exhaustion, and thus aggravates the danger by favoring the tendency to coagulation of the blood in the right cavities of the heart and by increasing the difficulty of expectoration. Even in the strong and vigorous, care must be taken not to withdraw too much of the vital stimulant, for the principal danger to be apprehended in the advanced stages of the disease is, that the patient may not have the requisite strength to give a hearty cough and rid his lungs of the phlegm which oppresses them. In children I believe it is never necessary to open a vein, for leeches on the chest form a convenient and effective substitute for phlebotomy.

Next to bloodletting, the most powerful and efficient remedy we possess is tartarized antimony. Administered in solution, in doses of a sixth, or a quarter of a grain, every three or four hours, it appears to conduce to free secretion and generally to mitigate the symptoms of the disease; and although it is difficult, if not impossible, to obtain statistical confirmation of the fact, I believe that, employed in the manner recommended, it is quite as efficacious, and much safer as a remedy, than when given in the heroic doses (four to twelve grains daily) recommended by some authors. Even forty-minim doses of the vinum antimonii will sometimes produce so much effect on the pulse as to render it necessary to reduce the quantity of the remedy, or to omit it altogether; and on more than one occasion I have known larger doses produce alarming depression without a corresponding beneficial effect. In acute cases, however, and at the outset of the attack, when the pulse is forcible, and more especially when the capillary air-tubes are implicated, tartarized antimony should form an ingredient of every prescription; and if carefully watched, its administration is never attended with danger except to infants and young children. Even to children it may usually be given without fear, but it sometimes proves wonderfully depressing to infants; and from two cases which I have myself witnessed, and from others communicated to me by medical friends, I am inclined to think that, even in very moderate doses, it may sometimes prove almost poisonous. Therefore when it is given to infants its action should be watched with more than ordinary care.

With these provisos, it may be stated that, at the outset of the attack, the administration of tartarized antimony in a saline diaphoretic draught is advisable; and of all salines perhaps the liquor ammoniæ acetatis is the most efficacious. Digitalis is often useful as an adjunct to the mixture, and in some instances produces a magically good effect. Indeed, if given in fifteen or twenty-minim doses, its influence is generally so beneficial that its administration ought always to be resorted to in severe cases. While the cough remains dry, and before free secretion has commenced, the air-passages should be fomented by the inhalation of air charged with the steam of water, the bowels should be freely but not violently acted on, the secretions stimulated if necessary by mercury, and mustard poultices or turpentine stupes applied to the chest. As soon as secretion from the air-passages is fully established, aqueous inhalations cease to be grateful to the patient, and are apt to prove oppressive. At this stage a large blister should be applied to the sternum, or to either side in front, and when it has risen the blistered surface should be covered with a bread-and-linseed poultice. Some persons recommend the application of a blister between the scapulæ, on the ground that it draws off more fluid in that situation than when it is applied in front; but the distress it causes by preventing the patient from lying on his back in bed much more than counterbalances any extra benefit which would thus accrue. Blisters, therefore, should be confined to the anterior surface of the chest, whilst its posterior surface is stimulated by means of mustard poultices and turpentine fomentations.

As soon as the first severity of the attack is subdued and the expectoration becomes thicker and less copious, the quantity of tartarized antimony and digitalis may be decreased, and then, if expectoration continues free, squills, or ipecacuanha, with paregoric, or a few minims of the liquor morphiæ, will be found useful adjuncts to the mixture. At a still later stage of the complaint, when the febrile symptoms have subsided, the tartarized antimony and digitalis should be omitted altogether, and in some instances it answers well to support the patient's strength by full doses of quinine or bark combined with squill and chloric ether, while the cough is quieted by morphia and ipecacuanha. If there is any difficulty in expectorating, the compound squill pill may be given, or a draught containing full doses of cinchona with five grains of carbonate of ammonia and thirty or forty minims of the compound tincture of benzoin or of the balsam of Peru. This will generally facilitate expectoration, and so relieve the dyspnœa; but if any difficulty is experienced in unloading the chest, recourse should be had to the action of an emetic, followed by the administration of the decoction of senega, or the mistura ammoniaci, both of which are powerfully stimulating expectorants, and may be given in aid of other remedies. In this stage of the attack I have often afforded relief by recommending the inhalation of steam charged with carbolic acid, and if there is any lividity of the face or lips the inhalation of oxygen is of the greatest service. Under no circumstances should opium be given if the secretion is copious and expectoration difficult, as it tends to paralyze the action of the air-tubes, and so interferes with the ejection of the muco-purulent matter. Nitric or sulphuric ether, the spirit of chloroform, and the ethereal tincture of lobelia, are useful adjuvants in cases which are attended by bronchial spasm, and ipecacuanha proves specially useful in children. If the patient be of a gouty habit, a few drops of colchicum wine, combined with fifteen or twenty grains of the bicarbonate of soda, prove a capital expectorant; and if of a strumous or consumptive disposition, the most efficient remedy in checking the bronchial secretion and restoring the patient's strength will often prove to be cod-liver oil.

If, as sometimes happens, tartarized antimony fails to check or control the symptoms, a few small doses of calomel may be tried, for although mercury has no direct influence on the course of the disease, it often proves a powerful aid to other remedies, by modifying the secretion of the chylopoietic viscera. In many cases in which antimony fails to exert a beneficial influence, quinine is well borne from the first, and may be given in full doses in combination with digitalis and a few drops of hydrocyanic acid. By pursuing this course I have often been successful in arresting the course of an acute attack when antimony and other remedies had failed.

If, again, from whatever cause, the disease has produced extreme oppression of the breathing, with lividity of the lips and general surface, or other symptoms of approaching suffocation and collapse of the vital powers, our aim must be to sustain the failing strength, induce expectoration, and so relieve the pulmonary congestion. For this purpose nothing proves more serviceable than the inhalation of oxygen, but at the same time strong beef-tea, wine, brandy, and carbonate ammonia must be given internally, whilst repeated dry cupping between the shoulders, the application of mustard poultices and turpentine fomentations, and the use of various stimulating and vesicating ointments or liniments, are resorted to externally. In truth, there is no specific treatment for bronchitis; each case is a study of itself, and must be treated on general principles, due regard being had to the age, strength, and constitutional peculiarities of the patient.

Throughout the treatment the temperature of the room should be maintained at from 63° to 68° of Fahrenheit, and its atmosphere moistened, according to the sensations of the patient, by steam from a kettle, or by the evaporation of boiling water from a dish near the bed. Under ordinary circumstances the diet should at first be restricted to milk, whey, broth, beef-tea, gruel, and arrowroot; but if the attack be of an asthenic character, a more generous diet must be had recourse to, and stimulants may be cautiously employed.

Chronic Bronchitis.

Chronic bronchitis may result from any cause which excites and keeps up irritation or subacute inflammation of the bronchial mucous membrane. Sometimes it occurs in sequel of an acute attack, which has been neglected or imperfectly subdued; sometimes it is preceded by general ill health, and by a long succession of "colds" and coughs; sometimes it comes on gradually and insidiously, without any previous "colds" or feverishness, or any of the ordinary acute symptoms, and, sometimes, it seems to be a chronic ailment, dependent on the congestion of the bronchial mucous membrane, which is often associated with cardiac derangement. Further, it is a frequent accompaniment of tubercular consumption, for which it is liable to be mistaken, and it derives a special interest from the care required to form an accurate diagnosis between the two diseases.

Like the acute disease, it varies greatly in the character and severity of the symptoms. In one class of cases the affection consists of little more than slight occasional cough, with expectoration of a grayish or greenish, or yellowish-white muco-purulent matter. It commences with the approach of winter, ceases or diminishes as soon as warm weather sets in, is not attended with pain in the chest, and does not materially affect the health; indeed, the patient often feels that he would be perfectly well if he could but get rid of his cough. When it thus occurs, it constitutes the mildest form of so-called "winter-cough," and depends simply on passive congestion of the air-tubes with somewhat increased quantity and altered character of the bronchial secretion.

In another class of cases the cough is more severe, of more frequent occurrence, and attended with much more profuse expectoration. It is most troublesome early in the morning, and on first lying down in bed at night, and sometimes, though very severe at those times, it is comparatively quiet throughout the day. At one time expectoration is difficult, and the sputa, which are comparatively scanty, consist of stringy tenacious mucus, of a grayish or yellowish-white color, occasionally streaked with blood; at another it is easy, and the sputa are more copious, muco-purulent in character, of a yellowish-green color, having a faint unpleasant odor; at another, again, the sputa are profuse, almost wholly purulent, of a nauseous, and sometimes a fetid odor,[1] usually running together in one mass, but often remaining separate, and forming distinct, nummulated masses, which sink or float in water, according as they happen to be more or less aerated. There is oppression at the chest, but seldom much pain, except after a fit of coughing; the temperature of the body is scarcely raised; the pulse is not much hurried, nor is the respiration much more frequent than natural, but the violence of the cough often leads to vomiting, and its frequency to sleepless nights, and consequent exhaustion. As a natural result of the broken rest, there ensue loss of appetite, excessive debility, night sweats, and wasting. Indeed, in some instances, attended by frequent cough and profuse expectoration, the loss of flesh is so considerable as to suggest the existence of tubercular disease, and not a few cases of the form of disease now under consideration have been regarded as consumptive by the unwary practitioner. But the absence of any family tendency to consumption, the existence of the physical signs of bronchitis in all parts of the chest, the similarity of the signs on the two sides of the chest, the prominence of rhonchi, or sounds of vibration as contrasted with râles or sounds of bubbling, together with the comparative slowness of the pulse and respiration, the lowness of the temperature of the body, which is seldom above the normal standard, the absence of night sweats, and the gradual cessation of wasting which is observed after the first few weeks of the attack, combine to stamp it with distinctive characters, which, in most cases at least, are sufficiently legible to the experienced physician.

In another class of cases, termed dry catarrh by Laennec, there is considerable tightness and oppression at the chest, with cough almost more frequent and distressing than in the last variety, but seldom accompanied by the same debility and wasting. The expectoration is scanty, and composed of small, semi-transparent, roundish masses of viscid, pearl-like or starch-like mucus; and sometimes there is much difficulty of breathing, dependent,

[1] The fetid odor simulates that produced by gangrene of the lung. It resembles the odor of the butyrates of ethyl, and is probably connected with the presence of butyric acid, which has been detected in such cases in the sputa. See a case reported by Dr. Laycock in the "Med. Times and Gazette" for May, 1857, p. 480. Dr. Gamgee, whilst admitting the presence of butyric acid, refers the odor to the existence of unoxidized sulphur. See "Edin. Monthly Journ.," March, 1865.

I believe, on the temporary collapse of certain portions of the lung-tissue, consequent on obstruction of the bronchi by plugs of mucus. These cases, which are often met with in gouty habits, are apt to be regarded as referable to or connected with "stomach," and doubtless are aggravated by any irritation of the solar plexus occasioned by disorder of the stomach or bowels. They are generally associated with more or less emphysema, are dependent on passive congestion of the bronchial lining membrane, and are seldom marked by much febrile excitement.

In a fourth class of cases an opposite train of symptoms is observed. The cough, instead of being dry, irritable, and frequent, occurs in paroxysms, at considerable intervals, and is accompanied by profuse expectoration of a thin, watery, ropy fluid, which varies in opacity, but is usually somewhat transparent, resembling gum-water. This form of the disease, which has been styled bronchorrhœa, is generally observed in elderly persons, and sometimes leads to the expectoration of half a pint of fluid in the course of an hour. A patient under my care, a short time since, in the York Ward of St. George's Hospital, presented an excellent illustration of the fact. He rarely coughed more than once or twice in twenty-four hours, and then expectorated, within twenty or thirty minutes, more than half a pint of thin, semi-transparent ropy secretion.

In whatever form the disease occurs, it generally returns every winter, or, in old persons, lasts, with few intermissions, throughout the year. Further, whatever form of the disease prevails, an acute attack may supervene at any moment, and induce the symptoms of the ordinary disease. In all cases, but especially during the existence of the second variety, which is characterized by profuse secretion, the occurrence of such an attack is extremely dangerous; for in a patient whose breathing is oppressed by the accumulation of viscid secretion in the bronchi, such an attack is necessarily of a suffocating character, and in elderly persons, already exhausted by a long continuance of the chronic disease, it is always a question whether the strength will be equal to the long-continued cough and profuse expectoration which is necessary to free the air-passages and relieve the breathing. Indeed, there is no more frequent cause of death amongst the aged than this form of disease, which from its proving so fatal to elderly persons, has been termed "senile bronchitis."

The morbid changes to which chronic bronchitis gives rise, and the physical signs resulting from them, are collocated, for the sake of easy comparison. In the first instance those changes alone are described which accompany a simple uncomplicated attack; each of its more frequent complications is then described separately, viz., pulmonary collapse, dilatation of the bronchi, and emphysema. By this means it will be seen to what physical sign each morbid change gives rise, and the student will be thus enabled more readily to detect the existence of such complication, should it arise in the course of the disease.

Morbid Changes.

The mucous lining of the bronchi is swollen and congested, and of a deep venous red color, either generally or partially, in streaks or in patches. The more asthenic the form of the disease, the more livid or darker colored the membrane. Its surface is uneven, and often abraded, and its substance thickened; the longitudinal and circular muscular fibres of the bronchi are much developed, and the walls of the air-passages generally are hard and thickened. The bronchi are clogged with secretion, more or less viscid, varying from a semi-opaque, ropy, sero-mucous fluid to thick tenacious mucus, or to true pus. The more viscid and tenacious the secretion, the more pertinaciously does it adhere to their sides. In almost all fatal cases the right cavities of the heart are found distended with blood, and a clot often passes from the right ventricle into the pulmonary artery.

Physical Signs.

Inspection shows defective expansion movements of the chest and prolonged and labored expiratory movements.

Palpation.—The vocal fremitus is not materially altered; rhonchial fremitus is usually perceptible.

Percussion.—The sound is not materially altered, unless the accumulation of thick secretion gives rise to obstruction of any of the bronchi, with consequent impaired inflation of the lung, and temporary dulness on percussion as its result. This is to be distinguished from the dulness of pleuritic effusion by the continuance of vocal fremitus, and from that of pneumonic consolidation by the absence of tubular breathing.

Auscultation.—The respiratory sounds are comparatively deficient over the entire chest, except, perhaps, for a moment after free expectoration, when they may be heard loud, harsh, and coarse, where a moment before they had been almost inaudible. Unless the air-tubes have been just emptied by expectoration, the breathiug is accompanied by every variety of rhonchus, chiefly of a sonorous, but sometimes also of a sibilant character. Large coarse rales or bubbling sounds are also present, and exaggerated or supplementary breathing is heard in the unaffected portions of the lungs. The sounds in any portion of the lung are constantly varying, and may either cease for a time or have their character changed by cough and deep inspiration. Vocal resonance is not materially altered in most cases, but is sometimes bronchophonic, and when dulness exists on percussion may be almost wholly absent.

Bronchitis, complicated by Pulmonary Collapse.

A few years ago, Dr. Gairdner, of Edinburgh, pointed out[1] that a particular form of condensation of the lung, formerly regarded as the result of lobular pneumonia, but now acknowledged to be referable to pulmonary collapse, is a frequent and very serious complication of bronchitis. This fact before unknown or practically ignored, is now regarded as exercising an important influence on the issue of bronchitis, especially in old people and young children, and it may be well, therefore, to point out the causes from which it arises and the mechanism by which it is brought about. The causes of collapse may be briefly stated thus: First, weakness or insufficiency in the power of inspiration; secondly, the presence of some obstruction—usually thick tenacious secretion—in the bronchi; thirdly, inability to remove the obstruction by the act of coughing. These causes are obviously more likely to exist under conditions of asthenia than in persons who are strong and vigorous; and thus it is that pulmonary collapse is found to complicate bronchitis in young infants and in aged or weakly persons more frequently than when the disease occurs in the vigorous adult. It may happen slowly or suddenly, in small patches or over an extensive portion of the lung, and it is a frequent source of the sudden, but often temporary, accession of dyspnœa and oppressed breathing which is observed in the course of chronic bronchitis. Sometimes, when extensive, it is the immediate cause of death, and in fatal cases of bronchitis, at whatever age occurring, it is almost invariably an accompaniment of the attack. Its mechanism is very simple, and is admirably illustrated by an experiment of Messrs. Mendelsohn and Traube. They introduced a shot into the bronchus of a living dog, and the lung beyond the shot became collapsed and thoroughly emptied of air. A solution of gum injected into the bronchi produced the same result. The air gradually found its way out past the obstruction, and

[1] Dr. Gairdner "On Bronchitis," "Edin. Monthly Journal," 1850-51.

was expelled during expiration, but was prevented entering again during inspiration. So it is with a plug of tenacious mucus at the bifurcation of a bronchus. It acts the part of a ball-valve in a syringe; each expiratory blast may dislodge it so far as to admit of the escape of air around it, but not so far as to prevent its falling back into its old position, and thus closing the passage against the ingress of air during inspiration. This condition of things leads to the emptying and consequent collapse of the air-vesicles beyond the seat of obstruction, with attacks of dyspnœa more or less severe, according to the size of the portion of lung affected, and more or less enduring according to the period which elapses before the obstruction is removed.

The precise nature of the morbid changes and attendant physical signs are particularized below; but it must be remembered that they do not replace the changes and physical signs which mark an uncomplicated attack of chronic bronchitis, but are superadded to or associated with them, and occur in certain portions only of the lungs:

Morbid Changes.

Pulmonary collapse may be confined to certain lobules scattered through the lung, in which "lobular" form it has been mistaken for the effect of pneumonia, and was formerly miscalled "lobular pneumonia;" or it may be diffused over a considerable portion of one or both lungs, usually their posterior parts, and may give rise to the appearance which was formerly recognized under the title of "carnified lung."

When the mischief occurs in a lobular form the air-vesicles are generally emptied of air, the collapsed portions are small and sharply defined, being mapped out by the interlobular areolar tissue, and are not only hard and shrunken, but depressed below the general surface of the lung. In the diffused form the collapsed portions are larger, and the degree of collapse varies from partial emptying of the air-vesicles, and consequent diminution of the normal crepitus, to entire emptying of the air-cells and complete collapse. The collapsed portions are of a dark venous red tint externally, and of a deep plum-color internally; they are not crepitant under the finger, convey a feeling of solidity, sink when placed in water, and, unlike lung solidified by pneumonic hepatization, their cut surface is smooth, without a trace of granulation or morbid exudation. When the mischief is of recent occurrence, a small quantity of bloody serosity can be squeezed out, but even this contains nothing more than blood-corpuscles, epithelium, and other elements of normal lung-tissue, mixed now and then with a few pus- or mucus-cells; when it is of some standing the affected parts are shrunken, and no fluid can be obtained even by the firmest pressure. Unless the mischief is of such duration that the nutrition of the parts has become impaired and their tissue wasted, the collapsed portions usually admit of being restored in great measure to their normal condition by inflation through the bronchus. When atrophy and shrinking of the lung-tissue take place, vesicular emphysema usually occurs on the confines of the collapsed portions of the lung.

PHYSICAL SIGNS.

Inspection rarely furnishes any information except in young children, or when an extensive and superficial portion of lung is collapsed, in which case the corresponding portions of the chest cease to expand, and, when the mischief is seated in the lower portions of the lungs, may be even retracted during inspiration.

Palpation is never of any avail in the lobular form of the disease, and seldom of much use in the diffused form, unless the mischief be extensive. In that case it will reveal cessation of vocal and rhonchial fremitus.

Percussion never yields any evident result in the lobular form of the affection, and rarely even in the diffused form. If, however, a considerable portion of the lung is implicated, the percussion-sound is at one time completely dull, at another clear and resonant, according as the lung-tissue happens to be collapsed or inflated at the time of the examination. Rapid alteration in the character of the percussion-sound often accompanies this form of disease.

Auscultation reveals entire absence of the respiration over the affected parts, often followed in a few hours by breathing of the same character as exists in other portions of the chest, and accompanied by the same abnormal sounds.

In some instances, the suddenness with which the dyspnœa comes on in chronic bronchitis, and the equally sudden manner in which it passes off, is almost of itself sufficient to denote the existence of pulmonary collapse; and when the presumption which has thus arisen is verified by the occurrence of rapid variations in the percussion-sound, and of equally rapid alterations in the sounds of respiration—their presence at one time, their entire absence at another—the presumption amounts to a certainty. In a case which I saw some years since with Dr. Latham and Dr. Dundas, the rapidity with which changes such as these took place was most remarkable. The patient, a gentleman, aged fifty-four, would be breathing quietly in the morning, and would be suddenly seized in the afternoon with an attack of dyspnœa, threatening suffocation. These attacks occurred frequently during a period of three months prior to his death, and were invariably accompanied by complete dulness on percussion, and entire absence of breathing over a considerable portion of one of the lungs, usually the lower two-thirds of the right lung, where previous to the accession of dyspnœa the percussion-note had been clear and the respiration distinct, though accompanied by sonorous rhonchi and coarse râles. The attacks usually lasted from one to twenty-four hours, and were followed immediately by resonance on percussion, and by a return of breathing over the affected part. Sometimes, however, the attacks were of several days' duration, and then the change in the percussion-note and the return of breathing were more gradual. The same train of phenomena has been observed by Gairdner, West, and other observers, especially in young children and in aged persons, and are of extreme importance, as bearing upon the prognosis of the disease, the probability of an unfavorable issue in bronchitis being greatly increased by the occurrence of pulmonary collapse.

Bronchitis complicated by Dilatation of the Bronchi.

It has been already stated that dilatation of the air-passages takes place in all cases of bronchitis, and occurs more readily in children than in adults. In most instances, however, the dilatation is only temporary. The symptoms to which it gives rise pass off with the subsidence of the bronchial inflammation, and there is no reason for doubting that the tubes reassume their normal condition. But when the bronchial inflammation has been very severe, or of long duration, and has been accompanied by much spasm and difficulty of expectoration, the bronchial muscles sometimes lose their contractility, the air-tubes become more than ordinarily dilated, and the dilatation does not disappear with the subsidence of inflammation, but, on the contrary, becomes permanent. This occurs most commonly in bronchi of the third or fourth divisions, in which the cartilaginous plates are few or absent.

Dilatation of the bronchi is followed, as a natural consequence, by condensation of the surrounding lung-tissue, with obliteration of some, at least, of the air-cells, and a corresponding diminution in the air-containing capacity of the lung; so that, even when unaccompanied by bronchitis, this affection results in habitual shortness of breath. And when, as is much more frequently the case, it is attended by bronchial irritation and profuse muco-purulent secretion from the air-passages, it leads to symptoms, general as well as physical, which simulate those produced by tubercular disease of the lungs. It may be well, therefore, without further preface, to collate the morbid changes and physical signs by which this affection is usually accompanied.

Morbid Changes.

Sometimes one or more of the bronchi undergo dilatation almost throughout their whole extent, the enlargement being tolerably uniform, and often excessive; so much so, indeed, that tubes of the fourth or fifth division may exceed the main bronchus in size, and bronchi whose diameter in a healthy state corresponds pretty closely with that of a straw, may become as large as the distended finger of a glove. This form of dilatation is accompanied by thickening of the walls of the affected bronchi, and by hypertrophy of their circular muscular fibres, whilst the lining membrane of the tubes is of a dark red color, and secretes a thick, tenacious, muco-purulent fluid. Sometimes the dilatation is not so uniform; the bronchi are irregularly sacculated, and distended in different parts of their course, and present a series of alternate enlargements and contractions, the enlargements varying in size from that of a bean to that of a chestnut or even larger. Sometimes, again, the dilatation is almost confined to the extremities of the tubes, which form large, thin-walled, globular cavities.[1] In the two last-mentioned forms of dilatation, the walls of the affected tubes are atrophied and attenuated, instead of hypertrophied, and are lined by a smooth membrane, which secretes a glairy mucus, or more commonly a thick, yellowish, muco-purulent fluid. In cases of old standing,

[1] The most remarkable instance I ever met with of this form of dilatation occurred in the person of John Jenkins, who was admitted into the Hope Ward of St. George's Hospital, in November, 1843. Several of the bronchial tubes as they approached the lower portion of the lung became enormously dilated, and terminated in cavities large enough to contain a walnut. The membrane lining these cavities was of a deep brown color, and continuous with the lining membrane of the bronchi. In several instances these dilated cavities communicated with each other, and so enormous was the dilatation at the lower part of the right lung that the appearance was presented of a large cavity intersected by portions of condensed lung. For further particulars see "Hospital Post-mortem book" for December 28th, 1843.

the interlobular cellular tissue is infiltrated with a black or dark-colored indurated material.

Physical Signs.

Inspection.—The expansion movement of the chest is diminished, and the expiratory movement much prolonged. Occasionally retraction or falling in of the chest-walls occurs over the affected parts.

Palpation.—The vocal fremitus is usually increased, and rhonchial fremitus is often perceptible. The force of the vocal fremitus varies, not only with the character of the voice, but with the size of the dilatation, the degree of pulmonary condensation by which the affected air-tubes are surrounded, and with their freedom from secretion and proximity to the chest-walls. If the tubes are obstructed, the fremitus cannot be felt.

Percussion.—The resonance is always abnormal. If the affected tubes are deepseated, surrounded by much condensed lung-tissue, or obstructed by an accumulation of secretion, there is decided dulness on percussion; if free from secretion, distended with air, and close to the chest-walls, the resonance may be abnormally clear but shallow-toned. In some such instances, the cracked-pot sound may be elicited.

Auscultation.—The respiration at first is harsh and rough, accompanied, and often masked by, râles and rhonchi. When the disease is more advanced, it is either of a hollow, blowing character, and attended by gurgling, or else, if the tubes are free from secretion, it is of a clear, hollow, ringing, metallic quality. The vocal resonance may be diminished or altogether absent if the tubes are obstructed with secretion; of a clear bronchophonic or pectoriloquous character if they are unobstructed and free from secretion. In some instances, when dilatation is excessive, distinct splashing may be heard during cough.

It will be obvious, from a comparison of these morbid changes and physical signs, that the form of disease now under consideration is productive of alterations in the physical condition of the lungs in many respects analogous to those which occur in tubercular consumption, and is consequently accompanied by physical signs, many of which are identical with those which are met with in that disease. And when to this is added the fact that the expectoration in both cases is of a muco-purulent or purulent character, that in both instances spitting of blood may occur,[1] and that diarrhœa, night sweats, and considerable wasting may and often do take place, it will be readily understood how difficult in some instances must be the diagnosis between the two forms of disease. Indeed, at a first interview, it is sometimes impossible to arrive at a positive conclusion on the subject; but careful investigation continued for some weeks will seldom fail to elucidate the question. The whole course of the disease and of its physical signs is different in the two cases. In phthisis the physical signs of the disease are most marked in the upper part of the chest, and in their fully developed form are often confined to the apices of the lungs; in simple bronchial dilatation they are commonly most marked in the mammary and inferior scapular regions. In the former disease the signs of pulmonary excavation are usually preceded by dulness on percussion over the diseased part; whereas in the latter dulness is not produced until after the signs of excavation have been established. In the one hæmoptysis and diarrhœa are the rule, in the other the exception; in the former the progress of the general symptoms and physical signs is often very rapid, in the latter their progress is usually very slow; and although emaciation may be considerable, it seldom proceeds to the same extent as in tubercular disease. In the one the sputa are not only muco-purulent, but contain portions of the yellow, elastic areolar tissue of the lung, which can be detected by careful microscopic investigation; in the other no trace of this tissue can be discovered, however careful the search, and however profuse or purulent the expectoration. Therefore when these two forms of disease exist separately, it can rarely happen but that daily observation of the symptoms and physical signs will enable the careful practitioner to discriminate between them; but when, as sometimes happens, tubercular disease coexists with bronchial dilatation, it is impossible to form an accurate diagnosis, and it is probable that the dilated bronchial pouches would be confounded with tuberculous excavations. The physical signs and general symptoms of phthisis being present, it is difficult to conceive any combination of circumstances which could justify a positive opinion as to the existence of dilated bronchi.

Treatment of Chronic Bronchitis.

The treatment of chronic bronchitis is a subject at once of exceeding interest and unusual difficulty; for whether the disease

[1] Hæmoptysis is not common in these cases, but I have seen it occur to a slight extent in two cases in which subsequent post-mortem examination proved that no tubercle existed, and that there was no valvular disease of the heart.

be simple and uncomplicated, or whether it be attended by dilatation of the bronchi, pulmonary collapse, or vesicular emphysema, in either case its treatment must be regulated as much by the constitutional peculiarities of the patient, as by the precise stage of the disease, and the actual symptoms and physical signs which present themselves to our notice. It is the neglect of these circumstances which leads to the frequent want of success in the treatment of this class of affections. In every case the chronicity of the disorder is a point which must be constantly borne in mind, nor must it be forgotten that the disease is frequently connected with an enfeebled state of system; that even in the more robust and vigorous it has a tendency to exhaust the patient, and induce, "bad health;" that it frequently coexists with a weak and damaged heart; and that when it proves fatal the issue is commonly dependent on coagulation of blood in the right auricle and ventricle and on want of power to rid the air-tubes of the secretion which obstructs them, rather than on the occurrence of active inflammation. Therefore from the first it behooves us to be careful not to distress our patient unnecessarily. Bloodletting is never needed so long as the disease maintains its chronic character; and even when an accession of bronchial inflammation occurs it is still inexpedient, and in most cases dangerous, to have recourse to active depletion. The application of a few leeches to that part of the chest where fine crepitating or small bubbling rales show that active mischief is going on; or, better still, the abstraction of a few ounces of blood, not exceeding four or five, by cupping, is as much as can be safely ventured upon; and if this is followed by repeated dry cupping, it will be found to be as much as—nay, more than—is commonly needed. But even if it should fail to produce the desired effect, bloodletting must rarely be had recourse to again. It would not prove either a safe or an efficient remedy. The failure of strength is so rapid in these cases, and the accumulation of mucus and the interference with the circulation consequent on its failure are so dangerous in their effects, that recourse must be had to other and less lowering means for relieving the breathing. Dry cupping and counter-irritation are now the external remedies on which we must rely; and when combined with the administration of antimony, squills, ipecacuanha, digitalis, nitric ether, the sesquicarbonate of ammonia, and other expectorants, and, if the bronchial mucous membrane is dry and secretion is deficient, with the inhalation of an atmosphere charged with steam—to keep up, as it were, a constant fomentation of the irritated membrane—they seldom fail to lead to satisfactory results. A blister should be applied to the anterior surface of the chest, and should be followed by a poultice covered by oiled silk, while the posterior surface of the chest is subjected to the application of mustard-poultices, turpentine fomentations, stimulating liniments, or counter-irritant ointments. Wine and other more powerful stimulants must of course be given if the patient manifests any tendency to sinking.

When the disease is purely chronic, and uncomplicated by acute exacerbations, and is characterized by profuse muco-purulent secretion from the bronchi, bloodletting is never needed, and if employed is of no avail. The remedies here must have special reference to three objects, viz., 1st, to subdue the irritability of the bronchial mucous membrane and allay spasm; 2d, to promote free or easy expectoration; and 3d, to reduce the quantity of the secretion. Inhalations of various kinds, whether in the form of vapor or of atomized fluids, assist in fulfilling of the first-named indication; emetics, expectorants and antispasmodics facilitate the second; and the third must be attained by means of tonics and certain remedies which seem to exert a special influence in restraining discharge from the bronchial mucous membrane. The steam of water, either alone or charged with dilute hydrocyanic acid, ether, or chloroform, or else impregnated with the active principles of conium, hyoscyamus, or stramonium, may be mentioned as inhalations which sometimes prove of service in diminishing the irritability of the bronchial membrane, and so may water holding these and other ingredients in solution and pulverized by any of the various apparatus contrived for the purpose. Remedial agents so employed possess these special virtues beyond all remedies administered by the mouth, that they admit of almost constant use, and are applied directly to the seat of mischief. They should therefore be made use of in every case in which frequent and distressing cough, and the occasional presence, perhaps, of a few streaks of blood, denote the existence of much bronchial irritation.

Expectorants are always beneficial, as facilitating the ejection of the bronchial secretion, which, if allowed to accumulate and obstruct the air-passages, proves a source of serious danger. But it must even be borne in mind that tartarized antimony, which is of great value in the acute form of the disease, is seldom needed in the purely chronic stage of the disorder, and if administered is apt to prove hurtful rather than beneficial. There are here no active symptoms to subdue, no need to promote the flow of secretion; our object is only to render it less tenacious, and, by thus promoting its separation from the sides of the bronchial tubes, to facilitate expectoration. Therefore, although in certain cases marked by more than usual inactivity of the skin, ipecacuanha may be found useful in slightly nauseating doses, we ought not to have recourse to tartarized antimony unless the expectoration shall have become frothy, or thin, watery, and ropy, indicating the accession of more active mischief. Even under such circumstances it should be used very cautiously, and should be discontinued as soon as the more active symptoms have subsided. Generally there is little vascular excitement, and our reliance will have to be upon the more stimulating expectorants. Squills, ipecacuanha, the liquor cubebæ, naphtha, the compound tincture of benzoin, and balsam of Peru, the mistura ammoniaci of the Pharmacopœia, the decoctum senegæ, the sesquicarbonate of ammonia, combined, when no difficulty of expectoration exists, with small doses of paregoric or of morphia, constitute some of our more valuable remedies. In certain instances conium proves more directly sedative and antispasmodic than the salts of opium, and it possesses the recommendation of exerting its soothing influence without checking the expectoration as opium is apt to do. In other cases attended by much spasm, the ethereal tincture of lobelia and hydrocyanic acid afford excellent results; in others, again, which occur in persons of a rheumatic diathesis, iodide of potassium exerts a magical influence. In gouty persons, and in some even who have never manifested any positive symptoms of gout, colchicum, combined with small doses of carbonate of soda, gives rise to effects which are sought in vain from other remedies; and so again, several instances have come under my notice in which in successive attacks no remedy has appeared to control the cough until after mercury has been freely given with a view of stimulating the liver. In these latter cases the tongue has been coated, and the pulse frequent, the cough hard and frequent, and the expectoration very scanty, consisting of small, starch-like masses; but as soon as a free secretion of bile has been established the cough has changed its character and become loose, the expectoration free and easy, the pulse has fallen, and the dyspnœa has passed off. Some of these cases have occurred in gouty persons, but neither colchicum nor any other remedy has proved of service until after the action of mercury. In some such instances a hot air bath or an ordinary warm bath is extremely serviceable.

One word of caution should, perhaps, be added, respecting the administration of opium and its salts, in cases of chronic bronchitis, accompanied by profuse muco-purulent secretion. Its action in quieting the cough, and so checking expectoration, is apt to prove extremely dangerous to old and feeble persons; for with them a few hours' cessation from cough leads to an accumulation of secretion in the air-passages, which their strength hardly suffices to get rid of, and many a person has died asphyxiated in consequence of the incautious administration of the sedative. In the young and vigorous it may be often employed with advantage, as soon as all inflammatory symptoms have subsided; but extreme caution should be exercised in its administration to the old, the feeble, or exhausted. In the former it tranquillizes, allays irritation, and economizes strength; in the latter it induces lethargy and suffocation.

When the bronchial secretion is very profuse, and in spite of occasional expectoration tends to accumulate in and obstruct the bronchi, producing blueness or lividity of the face and lips, emetics prove the most useful of all remedies. Not only are they speedy and effective in their action, unloading the bronchi in a few minutes more thoroughly and with less distress to the patient than cough of some hours' duration, but, at the same time, they promote a free action of skin, and copious secretion from the liver. They thus fulfil a double purpose, and often set up a train of actions, which lead to the happiest results. So long as there is not undue vascular excitement, or any evidence of active inflammation in the lungs, no harm can accrue from their employment; and when their action is followed by the inhalation of oxygen to decarbonize the blood or stimulate the nervous system, the results which are obtained are surprisingly beneficial. The only question to be decided is as to the emetic which shall be selected. If the patient is young and vigorous, a grain of tartarized antimony and half a drachm of ipecacuanha may be had recourse to without fear; and if, at the same time, there be any heat or dryness of the skin, it will probably be found a more effective emetic than any other; but, in the majority of cases, I prefer an emetic of sulphate of zinc. Not only is it more speedy and less lowering in its action, but it seems in some way to restrain the amount of bronchial secretion. I have noted this effect in so many instances, that I cannot entertain any doubt on the subject, and therefore, unless there are special reasons for employing tartarized antimony and ipecacuanha, I always have recourse to sulphate of zinc.

It will sometimes happen that, notwithstanding the use of emetics and stimulating expectorants, and, in spite of the action of sedatives, the cough continues frequent, and the sputa remain profuse. The medicines do their duty in facilitating expectoration, but they fail to control the secretion from the bronchi. In these cases emaciation often occurs, and the strength fails. To remedy this condition the inhalation of steam charged with carbolic acid, naphtha, creasote, iodine or other stimulants, or of pulverized water holding these or other ingredients in solution, is often of the greatest service and should never be neglected. Tonics of various kinds are absolutely required, and quinine, bark, gallic acid, the mineral acids, sulphate of zinc, and cod-liver oil, are among the more valuable means at our disposal; and if the patient be pale and cachectic, iron may be given with advantage. In many instances no medicine proves so effective as sulphate of zinc in gradually increasing doses. Administered in combination with quinine, conium, and cod-liver oil, I have frequently seen it subdue an attack of bronchorrhœa in the course of a few days, which had been going on for weeks unchecked, in spite of remedies; and it is the more valuable, because its administration does not in any way interfere with the exhibition of other tonics or expectorants. Another astringent recommended by Dr. Alison, of Edinburgh, is tannic acid, in two- or three-grain doses. A generous diet must be had recourse to, and wine is usually needed.

In some instances, more especially when the skin is inactive, bronchorrhœa appears to be kept up by the extra work which devolves on the lungs in consequence of such inactivity. Common experience has proved, that a free action of the skin affords very great relief in such cases, and thus the profession in all ages have recognized the value of diaphoretics in aid of expectorants; but it has been reserved for us, in these later days, to ascertain and appreciate the assistance afforded by a full and continued action of the skin in many varieties of chronic bronchial affection. The introduction of the Turkish bath has furnished us with the means of exciting and sustaining full cutaneous action without materially weakening our patient; and I have met with several instances in which a bath on alternate days has speedily checked and ultimately arrested a profuse bronchial flux, which had long continued in spite of internal remedies.

Nothing, however, proves so useful in many of these cases as change of air, or change of climate. It is not the mere change from a colder to a warmer atmosphere, which exercises a beneficial influence; a uniform temperature may be maintained by artificial means, and the atmosphere may be artificially dried or moistened without any notable improvement in the cough, whereas forty-eight hours passed in another locality, even though the air may be colder and more bracing, will be often followed by an entire change in the train of symptoms. The cough will diminish, and the secretion will be reduced, probably, to one-half its former quantity. Indeed, I feel so strongly on this point, having seen immediate benefit result from change of air in numerous instances, that I do not hesitate to urge the removal of the patient as soon as it becomes evident that remedies are not exercising their usual salutary influence.

The only legitimate inference in such cases is, that some atmospheric condition of the locality in which the patient is residing, for the time at least, counteracts our efforts for his relief; and, under such circumstances, it is obviously our duty to recommend the adoption of a course which experience has proved to be often successful. Persons who are extremely prone to bronchitic attacks will do well to wear a Jeffrey's respirator during the winter months, and to seek a milder climate in the south of Europe; but, for most persons, a residence abroad is unnecessary, as I am satisfied that, in almost every instance, a climate may be found in our native island well suited to obviate and remove the customary winter cough. The question as to the precise localities to be selected will be discussed in a future chapter.[1]

Chronic Bronchitis complicated by Vesicular Emphysema.

Vesicular emphysema is a common accompaniment of bronchitis, and forms one of its most serious complications. It may exist to a slight extent without the concurrence of bronchial affection; but it rarely calls for medical interference, unless the bronchi are more or less affected. The patient may be more short-breathed than natural, the breathing being slow and labored —not hurried; he may be unable to walk quickly, to run up stairs, and otherwise to exert himself as much as usual; and he may even find it difficult to maintain a conversation without stopping frequently to take breath. But although he may be conscious of dyspnœa, and may even make complaint of being short-breathed or asthmatic, he seldom has recourse to medical advice, unless troubled with cough, the result of bronchial congestion. Then the emphysema becomes, indeed, a distressing malady, aggravating all the evils of the cough, permitting an accumulation of secretion in the air-passages leading to paroxysms of suffocative dyspnœa, and sometimes producing fatal consequences. Therefore, in a practical point of view, emphysema cannot be regarded apart from chronic bronchial affections.

Vesicular emphysema appears to be an hereditary disease,[2] and

[1] See chap. iv, of this treatise.

[2] Of 43 emphysematous persons whom I examined with a view to this inquiry, 26, or, in other words, 60.4 per cent., acknowledged an hereditary taint. Their ages varied from seven to sixty-five, and it will be seen from

may occur at any period of life, from infancy to extreme old age. It is most common, however, in persons beyond the middle period of life, and occurs more frequently in men than in women. Consisting of distension and rupture of the air-cells, with rarefaction and loss of elasticity of the lung-tissue, it has been attributed by some authorities to the effects of forcible *expiration*,[1] and by others to forcible *inspiratory* efforts.[2] Neither of these theories, however, appears to me to be exclusively correct. The pulmonary organs are so efficiently supported on all sides, that so long as they expand and contract freely, it is difficult to conceive how the deepest inspiration, or the fullest and most forcible expiration, can occasion rupture of their air-cells. But it is obvious that anything which produces contraction or imperfect expansion of certain portions of their tissue, while the adjoining portions are distended with air, must lead to inadequate support by the surrounding tissue of those portions of the lung in which the air is contained. Under these circumstances, the air-distended vesicles may give way under the strain put upon them, as the result of atmospheric pressure. Accordingly, emphysema is often found associated with pulmonary collapse, which leads to partial and irregular inflation of the lung. In these cases, it usually exists in the immediate vicinity of the collapsed portions, and may occur as an effect of forcible inspiratory efforts. So, again, when one lung has be-

the subjoined table that among the younger sufferers the influence of an hereditary tendency was most marked. Thus:

Age of sufferers at the commencement of emphysema.	Number of sufferers at the respective ages.	Predisposed by disease in either parent.	Percentage.
7 to 20	5	5	100.
20 to 35	8	6	75.
35 to 50	14	8	57.1
50 to 65	16	7	43.7

A very similar result has been obtained by Dr. Jackson, Jr., of Boston, U.S., who reports that no less than 18 out of 28 emphysematous persons had either a father or mother, or both, afflicted with emphysema, whereas 3 only out of 50 non-emphysematous persons sprang from an emphysematous stock. Again, he states that an hereditary taint was distinctly traced in each one of 14 persons, in whom the disease occurred from early childhood, but in 2 only of other 14 persons in whom the disorder did not manifest itself until the sufferer had attained to an advanced period of life.

[1] See paper by Dr. Jenner, "Med.-Chir. Trans.," vol. xl, p. 25.

[2] Amongst the advocates of this doctrine may be mentioned Laennec, Williams, Watson, Rokitansky, and others, and especially Dr. W. T. Gairdner, of Edinburgh, who has ably enforced his views in a series of papers published in the "Edin. Monthly Journal" for 1850.

come contracted, emphysema is usually found affecting the lung on the opposite side of the chest, the tissue of the sound lung having given way under atmospheric pressure during inspiration.[1] On the other hand, when, from the existence of viscid tenacious secretion in the air-passages, or from any other cause, air is retained in certain portions of the lungs after adjoining portions have been partially or completely emptied, a violent fit of coughing, as in chronic bronchitis or whooping-cough, may produce rupture of the air-cells, in the immediate vicinity of the empty portions of the lung during the act of expiration; and even when no obstruction exists in the lungs themselves, violent straining, as first shown by Dr. Jenner, may possibly lead to the same result at the apex and anterior margin of the lungs, which receive least support from the thoracic parietes. No portion of the lung is exempt from this affection. Not unfrequently the posterior and lower surface of the inferior lobes present large emphysematous patches; but the upper lobes and the anterior edges of those lobes are its favorite seat; and it is often met with in that situation when it exists in no other part. It is not a frequent accompaniment of phthisis, though the two diseases do sometimes coexist;[2] it is still less frequently associated with pneumonia; but it is a common sequel of obstruction of the bronchi from whatever cause arising. Whatever the disorder with which it is associated, it is always a consequence of pre-existing disease of the lungs or air-passages, and may be limited to a few vesicles, or to one or more lobules, or it may affect one or both lungs, or parts only of either lung. Giving rise as it does to serious impediment to the pulmonary circulation, it tends to cause the accumulation of blood in the right cavities of the heart, and so to produce hypertrophy and dilatation of their walls, with venous congestion of the head and face; and there are few advanced cases of emphysema in which the pulmonary symptoms are not greatly aggravated by the congestion thereby produced. In fact, the two forms of disease act and react on each other to a degree which, when emphysema is far advanced, renders the continuance of bronchial congestion

[1] A case illustrative of this cause of emphysema on a large scale has been recorded by me in the "Trans. Path. Soc.," vol. xi, p. 15. It is unique in its character, and the details will well repay perusal. Another, in which the same cause appears to me to have operated, has been recorded by Dr. Barker, in the "Med.-Chir. Trans.," vol. xxxiv, p. 131.

[2] For the details of a remarkable case in confirmation of the fact see "Trans. Path. Soc.," vol. xi, p. 15. The patient was under my care at St. George's Hospital, and though the symptoms of emphysema did not supervene until within a week of his death, the disease was developed to an extraordinary extent.

and labored breathing almost inevitable, and keeps up, even in the mildest weather, constant wheezing, and a distressing wearing cough. But the cardiac obstruction ultimately becomes a very serious malady, quite irrespectively of its action on the lungs. The right ventricle, unable to empty itself with its accustomed ease, contracts more forcibly on the contained blood, its walls yield to the gradually increasing pressure, and its cavity at length dilates to such an extent, that the auriculo-ventricular valves become inoperative. Fearful attacks of palpitation then occur, especially during the paroxysms of cough and dyspnœa, the blood is thrown back on the system, congestion of all the internal organs takes place, and œdema of the feet and legs is the natural result. Thus dropsical effusion is a frequent termination of this sadly afflicting malady.

When slight in degree, vesicular emphysema is seldom regarded as of much importance. It produces little inconvenience beyond shortness of breath on active exertion, and even in connection with chronic bronchitis it manifests itself chiefly by the amount and character of the dyspnœa it induces. The shortness of breath is out of all proportion greater than would be expected from the nature of the concurrent symptoms; but not only so, the respiration is effected with obvious difficulty, the breathing being slow, and the expiration much prolonged, and attended with considerable wheezing. In fact, the elasticity of the lung being seriously impaired, the power of emptying the air-vesicles and bronchi is necessarily diminished; the bronchial secretion, therefore, accumulates in the air-passages, and presents a still further obstacle to the egress of the air, and greatly aggravates the difficulty and distress of breathing. To make matters worse, there is a tendency to bronchial spasm in these cases, so that the cough and dyspnœa are generally more or less paroxysmal.

But after all, when it exists in a slight degree, emphysema is an annoyance, rather than a dangerous or distressing malady, and rarely arrests attention or calls for medical advice, except in connection with an attack of bronchitis. But it is otherwise when the disease is more fully developed. The suffering and discomfort are then so great that the patient is forced to seek advice, even when not oppressed by bronchial inflammation. He is not only short-breathed, and distressed by a constant sense of fulness and oppression at the chest, but the difficulty of breathing is aggravated by spasm, giving rise to frequent paroxysms of suffocative dyspnœa. During these attacks, which usually come on at night, he pants and often struggles for breath, loosens his clothes, throws open the door, and rushes to the open window for air; in short, he suffers severe paroxysms of asthma; is unable to lie down in bed; and often passes days and nights in a sitting posture. Even during the interval between these attacks, his sufferings are very distressing. There is so much obstruction to the pulmonary circulation, that the bronchi are always more or less congested, and cough and wheezing are present to a greater or less extent. Nor is even this the sum of his troubles. The least accession of bronchial irritation excites a vast increase of the symptoms, and irregularities of diet, and ordinary dyspepsia, produce the same effect. By giving rise to flatulence and abdominal distension, they occasion an upward pressure against the heart, interfere mechanically with its action, and thus aggravate the existing obstruction to the circulation; whilst, by creating an obstacle to the descent of the diaphragm, they prevent the already distended lungs from receiving the limited supply of air which would otherwise pass into them at each inspiration. The unhappy patient bears about with him abundant signs of his discomfort and suffering. Though he seldom loses flesh, his complexion is dusky and the vessels of the face are congested, the lips are dark-colored, the nostrils thick, and the face and neck full, or even swollen, from turgescence of the veins and enlargement of the accessory muscles of respiration. His expression is suggestive of anxiety and distress, the respiration is slow and labored, effected by the joint action of all the muscles of respiration; the shoulders are elevated, the back is rounded, and the slightest exertion aggravates his cough, and occasions distressing wheezing. The pulse is weak, and usually slow, except during the existence of bronchial inflammation; the impulse of the heart is often forcible, and a severe paroxysm of cough and dyspnœa will sometimes induce an attack of palpitation. In the more aggravated cases of this disease, attended by much difficulty of breathing, there is utter want of strength and incapability of bodily exertion; and the patient will often remark, that on the slightest exercise his breath and his strength seem to leave him instantly and simultaneously. The disease commences gradually, is chronic in its nature, is not attended by fever, but is usually accompanied by cough, with thin, frothy mucus, or watery expectoration. The appetite in most instances remains unimpaired, and the urine clear. The bowels are generally costive.

The following are the morbid changes and physical signs induced by this disease:

Morbid Changes.

These consist of dilatation, attenuation, and rupture of the air-cells; atrophy, rupture, and gradual disappearance of their septa; coalescence of their cavities, forming air-cells and bladders varying in size from that of a millet-seed to that of a hen's egg; compression and obliteration of the capillary bloodvessels; consequent dryness and anæmia of the emphysematous tissue, which is always pale and sometimes quite white. Oil sometimes exists in the walls of the air-cells; and it is probable that fatty degeneration and consequent loss of elasticity of the lung-tissue is one of the conditions which favors the production of emphysema.[1] The bronchial tubes are usually dilated in old-standing cases: their mucous membrane is pale, and their circular muscular fibres are highly developed. If emphysema is general, the lungs do not collapse when the chest is opened, but instantly protrude, as if too big for their bony case. Indeed, their volume is greatly increased, and leads to displacement of all the contiguous organs. They push in front of and displace the heart; they overlap in front of the chest at about the level of the second costal cartilages, and forcing themselves downwards, depress the diaphragm and the liver below it; and they exert so much outward pressure that their surface becomes indented, sometimes deeply, with impressions of the ribs, which form their bony prison. They are of a piebald color, and studded with irregular groups of air-bladders, which, though not very visible in the interior of the lung, form globular projections on its surface; they are dry, pale, or even white in parts, crepitate feebly, or not at all, under the finger; do not convey a feeling of elasticity, but feel soft and yielding, like a down pillow. Their specific gravity is low, so that they float high above the surface of water like an air-distended bladder. If the lung is inflated through the bronchi the spaces between these emphysematous bullæ are filled up, and the external surface is rendered smooth and even, showing that these projections result simply from the loss of elasticity in the tissue of the diseased parts, and the consequent non-escape of the air during the natural collapse of the lung. When emphysema is partial only, the morbid appearances correspond in character though not in degree with those observed when it is more generally diffused over the chest, and the amount of displacement of the thoracic organs is in strict relation to the extent and seat of the disease.

[1] See a paper by Mr. Rainey, "Med.-Chir. Trans.," vol. xxxi.

Physical Signs.

Inspection usually discovers alterations in the shape and movements of the chest. The chest-walls are abnormally rounded and prominent on the diseased side, or on both sides if both lungs are affected; or the bulging may be confined to certain parts corresponding to the diseased portions of the lung within. The infra-clavicular, mammary, and sternal regions are the parts in which bulging is most commonly observed, but the whole of the anterior surface of the chest is sometimes unusually prominent, and its antero-posterior diameter is much increased. The clavicles and scapulæ are elevated, and the former are often ill-defined, as in fat persons, in consequence of the filling up of the supra-clavicular space by the upward pressure of the emphysematous lung. The ribs are less oblique in their position than they should be, the intercostal spaces are wide, and sometimes well-marked; but in advanced cases, especially in old persons in whom the intercostal muscles become weak or paralyzed, the intercostal depressions are more commonly obliterated. In some few instances, in which the general symptoms of emphysema are well-marked during life, the lung is atrophied and no bulging or prominence of the chest occurs, either general or local. The motion of the chest-walls is remarkably small; the entire thorax is dragged upwards with the shoulders as one piece, but the expansion movement is diminished or almost *nil*. This arises from the fact that, in consequence of the loss of elasticity of the lung-tissue, expiration fails to empty the chest, and there is little room therefore for the introduction of fresh air, and consequently little dilatation of the chest during inspiration. The respiratory efforts are labored and powerful; the intercostal and all the accessory muscles of inspiration are seen in full activity, but the breathing, nevertheless, is chiefly abdominal, and in aggravated cases the lower part of the sternum and the cartilages of the lower ribs may even sink in during inspiration.

Palpation.—The vocal fremitus varies: sometimes it is normal; more commonly it either exceeds or falls short of the average standard of health. Not unfrequently the heart's action cannot be felt in the precordial region, and sometimes it is felt much lower than natural, even in the epigastrium.

Mensuration shows great increase in the peripheral size of one or both sides, and a remarkable increase in the antero-posterior diameter of the chest.

Percussion.—The sound is morbidly clear, and of a somewhat tympanitic quality; the area over which it is heard is more extended than natural, and owing to the very slight movement of the lungs is not materially affected either by forced inspiration or forced expiration; the precordial region often emits a clear resonance; the chest-walls are unusually elastic and resilient under the finger.

Auscultation proves that there is little or no air in motion in the emphysematous portions of the lung, for the inspiratory sound is either short and weak, or else altogether absent, whilst the expiratory sound, if not inaudible, is weak and remarkably prolonged—the ratio of the two sounds being as 1 to 3 instead of as 3 to 1, in consequence of the diminished elasticity of the lung-tissue and of the obstruction offered to the egress of the air by bronchial spasm and by mucus in the air-passages. Both sounds may be jerking and interrupted, and are often marked by sibilant or sonorous rhonchi, or by bubbling bronchitic rales. The vocal resonance varies greatly. Sometimes it is diminished or altogether absent; at others, it is loud and of a bronchophonic character. The latter I believe to be its natural condition in emphysema, and its diminution or absence to be referable to obstruction of the air-passages by spasm or mucous secretion. The heart's sounds are feeble or altogether inaudible.

Laennec speaks of a dry crackling as a result of sudden inflation of the emphysematous bullæ; and both he and Dr. Walshe invest the "prominent air-distended nodules" with the power of producing "a dry, grazing, friction-sound." I have never been able to satisfy myself as to the existence of either of these phenomena in emphysema; and even if their occasional existence be admitted, I should doubt their being correctly ascribed to the causes mentioned. I have never seen or known of an instance in which the lung-tissue, however damaged by emphysema, has been dry enough to crackle on expansion, and I do not think it possible that the smooth, soft surface of the attenuated lung could give rise to friction. The crackling I should attribute to the bursting of bubbles caused by the occasional passage of air through viscid tenacious secretion in the air-passages—the dryness of its character being referable to the viscidity and tenacity of the secretion—and the friction-sound when it exists—as it rarely does—to attrition of the two surfaces of the pleural membrane roughened by inflammatory products.

Pneumothorax is the only disease with which emphysema can be confounded. In both forms of the disease the affected side is rounded and prominent; in both, resonance on percussion indicates the presence of air in immense quantity beneath the chest-walls; and in both, feebleness or absence of respiratory sounds proves that the air is either in very slight motion, or is pent up in some cavity uninfluenced by the act of respiration. This cavity may be the pleura, or it may be the distended bladder-like pouch of an emphysematous lung. A distinction, however, may be drawn without much difficulty. Pure pneumothorax is extremely rare, if, indeed, it ever exists; and when accompanied, as it is almost invariably, by liquid effusion, its characters are well marked and distinctive.[1] But there are other means of distinguishing pneumothorax. It comes on suddenly, instead of gradually like emphysema; it produces more urgent symptoms of dyspnœa; it gives rise to greater bulging and to resonance of a more thoroughly tympanitic character than the ordinary resonance of emphysema; and whereas emphysema, when existing to such a degree as to simulate pneumothorax, invariably affects both sides of the chest, pneumothorax is necessarily confined to one side, as it would prove instantly fatal if it were to occur on both sides.

It is stated that subpleural emphysematous air-bags have been known to give way, and produce an attack of pneumothorax, but I have never chanced to meet with such a case.

Emphysema is naturally a progressive disease. Accompanied as it is by compression and obliteration of the capillary blood-vessels, and consequently by œdema and atrophy of the lung-tissue, it follows that each attack of bronchitis, and whatever puts a strain upon the weakened organ, is almost certain to be followed by some increase of the local mischief. Repair of the injured part is impossible, for its nutrition is hopelessly interfered with; and the utmost that can be expected is that, by carefully avoiding every act of straining, violent bodily exercise, and all other acts which are calculated to make an unusual demand on the inflation of the lungs, or on their power of resistance during respiration, we may in some measure prevent the *rapid* increase of the malady. But, slowly or quickly, it assuredly will increase, whatever steps may be taken to prevent it, until at length, as the function of the lungs becomes more and more impaired, the act of stooping, a recumbent posture, the slightest catarrh, an attack of dyspepsia with flatulent distension of the abdomen, or anything which even temporarily creates the slightest interference with the action of the heart and lungs, will suffice to induce a violent paroxysm of difficult breathing, with palpitation, and general distress.

If, then, emphysema be irremediable, what can be done to relieve the sufferings of our patient? Much good may be ef-

[1] See pp. 218–19 of this treatise.

fected by enforcing a mode of life calculated to guard against the various circumstances which aggravate the emphysema, or induce attacks of dyspnœa. Anything which puts the patient out of health is necessarily bad for him, and everything is beneficial which conduces to tranquillity of mind and body. Regularity of habits, moderation of diet, the avoidance of cold, entire absence from active exertion and mental excitement, are conditions essential to his well-being and comfort, and, when there is much disposition to catarrh, accompanied by bronchial spasm, removal to a warm climate during the winter months, and the use of Jeffrey's respirator whenever the external temperature is low and the air keen, are conditions of great importance. When these precautions are steadily observed, life may be prolonged for very many years even in aggravated cases.

But active medical interference is necessary in many cases of emphysema. This disorder is apt to be complicated by bronchitis, and in almost every instance in which a doctor's advice is sought, is attended by bronchial spasm, and bronchial congestion, with profuse secretion of mucus. The removal or mitigation of these symptoms is of the last importance to the patient, as tending, not only to promote his immediate comfort, but to prevent the further extension of the disease. The means to be employed for the purpose must necessarily vary, according as the attack is characterized by more or less extensive bronchitis and pulmonary congestion, or by spasm of the air-tubes. It is unnecessary to recapitulate what has been already stated respecting the treatment of bronchial inflammation, whether in its acute or chronic stage, inasmuch as what applies to cases of simple bronchitis holds good equally in respect to bronchitis complicated by emphysema. The only remarks which are called for in this part of the subject are, that extreme caution must be exercised in having recourse to depletory measures, and that, as soon as the more active symptoms have subsided, emetics, especially the sulphate of zinc, are extremely serviceable. Further, that with the view of relieving the passive pulmonary congestion, which is a distinguishing feature of these attacks, repeated dry cupping between the shoulders, the frequent employment of hot-water leg-baths, and the inhalation of oxygen, are useful expedients. If the stomach be loaded or otherwise out of order, or if there be flatulent distension of the abdomen, emetics, followed by warm carminative medicine and a gentle purge, will prove of the greatest service. If, again, there be no evidence of stomach derangement, but nevertheless the dyspnœa be considerable, the wheezing excessive, and expectoration scanty, it is fair to conclude that the symptoms are attributable chiefly to bronchial spasm, and means must be adopted accordingly. Nothing proves more useful in such cases than a full dose of opium combined with half a drachm or a drachm of sulphuric ether and twenty drops of the ethereal tincture of lobelia. Small doses of opium frequently repeated, which are of much service in ordinary catarrh, are of little avail to an emphysematous patient, whilst a full dose given as above directed will often act like a charm, subduing the cough at once, and speedily removing the wheezing and distress of breathing. The blueness or excessive venous congestion which accompanies a paroxysm of asthma resulting from emphysema is only a temporary effect of spasm of the bronchi; whereas when it occurs as a result of bronchitis unconnected with emphysema and bronchial spasm, it is the effect of obstruction caused by excessive loading of the air-passages with muco-purulent secretion. Thus it is that in the one case lividity of the surface is not a bar to the administration of opium—nay, rather calls for the employment of sedatives and antispasmodics—whilst in the other, as before stated, it contra-indicates the use of sedatives, which have the effect of checking expectoration and of so increasing the loading of the bronchi. If opium fails to afford relief, belladonna, conium, stramonium, the ethereal tincture of the Cannabis Indica, and hydrocyanic acid may be tried, and each in turn may be found of service. Even when all medicines taken internally fail to afford relief, belladonna, camphor, and stramonium will sometimes act beneficially if introduced into cigarettes or put into a pipe and smoked. The seed-pods of the Datura tatula smoked in this manner are more powerfully antispasmodic than even the pods of the Datura stramonium, and I have known them afford relief after stramonium had completely failed. The inhalation of ether and chloroform occasionally proves serviceable, as do also the fumes of burning bibulous paper which has been steeped in nitre. The inhalation of very dilute carbonic acid has also been reported as conducive to the alleviation of bronchial spasm. On general grounds its administration is contraindicated, and, although its immediate action may be sedative, its effect can scarcely prove otherwise that noxious. I am bound, however, to add, that my objections to its use are purely theoretical, for I have never employed it, and have not had any personal experience of its inhalation.

Interlobular Emphysema.

Interlobular emphysema is the name given by Laennec to a form of disease in which air is poured out from a ruptured air-vesicle into the cellular tissue which exists between the lobules of the lungs. In this form of complaint the air is not necessarily confined to the spot where it was first effused, but may pass into the cellular tissue of the mediastinum, and thence into the cellular tissue of the chest and neck. The disease is seldom connected with vesicular emphysema, and, if not produced by a puncture of the lung, occasioned by a broken rib or some similar accident, it usually arises from rupture of one or more air-cells under the influence of violent straining efforts, as during childbirth, in defecation, in coughing, and in lifting heavy weights. It may be recognized during life by the suddenness of its attack. Something is felt to give way in the chest during the act of straining; shortness of breath immediately ensues, and may be followed in a short time by an emphysematous condition of the neck or chest. If the air finds its way into the subcutaneous cellular tissue, it may be felt and heard crackling beneath the finger, and such an occurrence, coupled with the history of dyspnœa occurring suddenly after an act of violent straining, cannot fail to lead to a recognition of the disease. The idiopathic disease is said to be more common in childhood than in adult life, but under all circumstances it is so uncommon that I am not able from personal observation to speak confidently as to its treatment. Theoretically, however, nothing can be of much avail, unless the air has found its way into the subcutaneous cellular tissue, in which case a few punctures will allow the air to escape, and thus will lessen the pressure which results from the deeper-seated emphysema. Therefore, if the dyspnœa be excessive, this plan ought to be adopted; but if otherwise, it is better not to interfere, inasmuch as the mischief tends to spontaneous cure. The air which has escaped is gradually reabsorbed, and the difficulty of breathing ceases coincidently with its absorption.

If the air should fail to pass into the cellular tissue of the neck or chest, the diagnosis of this affection is by no means easy, for under these circumstances it is almost impossible to distinguish this form of disease from ordinary emphysema. The suddenness of the attack forms the only other clue to the nature of the mischief; and as dyspnœa may occur suddenly from a variety of causes, this cannot be depended on as a means of diagnosis.

In this, as in the former variety of emphysema, Laennec thought that a "dry crepitant rhonchus with large bubbles" is to be heard, and that the subpleural air-pouches give rise to friction-sound; but I have never been able to hear those sounds, and it is almost certain, theoretically at least, that emphysema in either form cannot lead to their production. In this opinion I am confirmed by many careful observers of large experience.

Sir Thomas Watson "doubts exceedingly whether" this large dry crepitation "really occurs at any time," and Dr. Herbert Davies does not hesitate to remark that Laennec's statement "is not confirmed by later observers." Dr. Walshe indeed states it as his opinion that "it is *quite possible*" that both these sounds may be produced "if the surfaces of the interlobular spaces become prominent through distension;" but inasmuch as his statement implies that he has never met with a case in support of his theoretical "possibility," his opinion cannot be allowed much weight in opposition to so much extended practical observation as has been brought to bear on the subject at issue.

Interlobular emphysema may be easily recognized after death. Bubbles of air of various sizes pervade the surface of the lung, elevating the pulmonary pleura, and producing more or less the appearance of foam. These bubbles, instead of being stationary as are the bladder-like projections formed by the dilated air-cells in vesicular emphysema, can be moved to and fro through the cellular tissue by pressure of the finger. Further, the lobules of the lung may be seen mapped out and sharply defined by small bubbles of air in the surrounding cellular tissue; and sometimes, if the emphysema be extreme, the interlobular cellular tissue is thoroughly inflated, and the divisions between the lobules are of considerable width. Pneumothorax is said to have been produced by the giving way of one of the subpleural air-pouches.

Bronchitis complicated by Plastic Exudation into the Bronchi.

A complication which is sometimes met with in bronchitis, and is sufficiently remarkable to merit special notice, is the formation of plastic casts of the bronchial tubes. Just as in croup a plastic exudation takes place into the larynx and extends downwards towards the bronchi, so in this form of disease fibrinous matter is poured out into the smaller air-tubes and their ramifications, and extends upwards towards the primary divisions of the bronchi. The exudation rarely reaches the larger bronchi, and never implicates the trachea and upper air-passages. Indeed, the disease bears no affinity to croup, but is essentially a form of bronchial inflammation, connected probably with some peculiar diathesis which leads to the outpouring of concrete albuminous and fibrinous matter. It may occur in either sex and at any period of life; and although most frequent in delicate persons of a strumous or a consumptive constitution, it has been known to arise in persons of robust frame and healthy appearance, born of healthy parents, and who, neither before nor subsequently to their seizure, have exhibited any symptom of tubercular disease. Whatever its precise cause, this form of disease is exceedingly rare, and is of interest to the pathologist as well as to the practical physician. Not only are its symptoms of great severity and little under control, but its cause and true nature deserve further investigation.

The patient, possibly a healthy-looking person, is attacked with an irritative, hacking cough, which is not accompanied by much expectoration. After a period, varying from half an hour to several days, there is a considerable and rather sudden accession of difficulty of breathing, sometimes, indeed, of an alarming character, and attended by a distressing sense of constriction across the chest. The cough increases in severity, small fragments of white fibrinous matter mixed with ordinary bronchitic expectoration or with blood-tinged mucus are spit up, and at length, during a violent paroxysm of cough, one or more white fibrinous casts of the bronchi are ejected, and the cough and difficulty of breathing in great measure pass off. The disease may then subside, or may continue for weeks or even months, marked from time to time by severe accessions of cough and dyspnœa, invariably followed and ultimately relieved by the expectoration of fibrinous casts of the air-tubes. Indeed, in one patient who came under my care at St. George's Hospital, and the particulars of whose case I communicated to the Pathological Society,[1] these attacks continued for several months, and recurred during a period of many years. The general health does not usually suffer much, and during the intervals between the attacks the patient's breathing may be unaffected. Sometimes even during a paroxysm of the disorder there is no pain at the chest, nor is there necessarily any heat of skin, or coating of the tongue, or loading of the urine, and the pulse, though frequent, may be soft or even weak. But the symptoms vary as much in kind and degree as they do in duration; for whilst, as already stated, there may be almost entire absence of febrile excitement and a very slight amount of mucous expectoration, there may be, on the contrary, excessive febrile distress and profuse muco-purulent bronchitic secretion, more or less tinged or mixed with blood. The secretion, however, is rarely tinged with blood, unless acute bronchitis or pneumonia be present; and perhaps the symptoms most generally met with are those which accompany that variety of chronic catarrh, which is attended by scanty starch-like expectoration.

In former days the concrete masses which are expectorated

[1] "Pathol. Trans.," vol. v, p. 41.

in these cases were falsely termed bronchial polypi, and are described as such by Dr. Warren.[1] They vary from mere fragments to large pieces of from one to four inches in length, and may be either tubular or solid. Their ejection may be preceded, and is often accompanied by spitting of florid blood. Sometimes the mucous expectoration is merely streaked with blood; at others the quantity of blood is considerable, and escapes in gushes; and this has led to the supposition that only the former variety of casts are the result of exudation, and that the latter are mere decolorized fibrinous coagula, resulting from ordinary hemorrhage, and moulded in the smaller air-passages.[2] But I am not inclined to coincide in this opinion. Without denying the possibility of fibrinous coagula forming in the bronchi as the result of pulmonary hemorrhage, it is undoubtedly the fact that such an occurrence is rarely observed even when hæmoptysis is very profuse; and inasmuch as when these plastic casts of the air-tubes have been met with in any case, they have been so usually not once only, as if formed by accident, but repeatedly, at intervals of days or months or even years, it is obvious that some special circumstances ought to be adduced in order to afford the slightest pretext for referring them to a cause which, to say the least, very commonly exists without giving rise to them. Moreover, long-continued observation at the bedside has led me to conclude that the white fibrinous casts now under discussion are invariably the products of true exudation. In two instances I have met with them without any previous or coexistent spitting of blood; and in the remarkable case which I reported to the Pathological Society, and which I have since had under my observation, the casts were sometimes rejected so soon after the accession of cough and difficulty of breathing, that it is morally certain they would have exhibited some trace of the coloring matter of the blood if they had been attributable to bronchial hemorrhage. Nevertheless, though some of them were solid and others tubular, and though in several instances their expulsion was preceded or else accompanied by more or less hæmoptysis, yet they were invariably white, without a trace of coloring matter. Further, I would remark that, however solid these casts appeared to be, they were always found on examination to consist of concentric laminæ, evidently deposited, or rather exuded at different periods, in successive layers; and if placed under the microscope they were seen to consist of amorphous granular matter, intermixed, *not* with blood-globules,

[1] "Med. Trans.," vol. i.

[2] See a paper by Mr. North in "Med. Gazette," vol. xxii.

but with mucus-corpuscles, compound granular cells, oil-globules, and ovoid cells containing dark coloring matter, such as exist in ordinary bronchial mucus.

In most of these cases of plastic exudation the prognosis is favorable. In two of the instances which I have met with complete temporary recovery took place; and in the case of the widow whose bronchial casts I brought before the Pathological Society, there has been complete temporary recovery after each of the repeated attacks which have occurred during a period of sixteen years. This patient was last in St. George's Hospital in 1858, when she again recovered thoroughly, after paroxysms of cough and difficulty of breathing, which for some weeks threatened her very existence. In an able report on the casts which I exhibited at the Pathological Society, Dr. Peacock has given an analysis of thirty-four cases, of which he has collected records; and of these no less than twenty are said to have recovered their health entirely; whilst in the ten who died, it appears, on investigation, that death was attributable either to phthisis, pneumonia, low fever, or profuse hæmoptysis. In no instance, as far as can be ascertained, did death arise directly from the presence of these plastic casts; on the contrary, there appears to have been little danger to life; for however urgent the cough and difficulty of breathing prior to and during the expulsion of the concretions, the natural efforts sufficed in every instance to expel them; and the urgency of the symptoms passed off as soon as free ingress was again afforded to the air.

The physical signs of this form of disease are peculiar. The formation of fibrinous concretions tends to obstruct the bronchi, and consequently to deprive certain portions of the lung of air. The natural result of this is weakness, or entire absence of breathing over the affected portions of the lungs. The percussion-note varies in its degree of resonance or dulness, according as more or less air is pent up in the superficial portions of the lung; but if one of the larger air-tubes becomes implicated, whilst as yet the portions of lung beyond it, and supplied by it, remain unaffected, collapse of those portions may take place, and complete dulness on percussion may ensue. This might lead the inexperienced practitioner to suspect the existence of pleurisy or pneumonia; but the rapidity of the disappearance of the respiratory sounds, the speedy occurrence of dulness on percussion, the absence of friction-sound, and of ægophonic resonance of the voice, combine, with the non-occurrence of acute catching pain and other general symptoms, to distinguish these cases from pleurisy; whilst, in like manner, they may be distinguished from pneumonic consolidation by the rapidity of their occurrence, by the absence of tubular breathing, and by the non-existence of those general symptoms which usually mark the inroad of inflammation of the lungs. The ordinary signs of chronic bronchial inflammation are commonly present during these attacks; but acute bronchitis or pneumonia may supervene, in which case the physical signs will of course undergo a corresponding change.

I know of nothing which can be relied upon to afford decided relief during the paroxysms of dyspnœa which precede or accompany the expulsion of these plastic casts of the bronchi. Calomel and opium, tartarized antimony, salines, alkalies, and blisters, have been fairly tried without effect; and I have seen venesection and sedatives employed in vain. Lime-water has been said to possess the power of dissolving these plastic exudations, and as, by means of a pulverizing apparatus, we can now introduce fluids into the bronchi by inhalation, it may perhaps be worth while to make trial of it. At present I am unable from personal observation to speak as to its effects when thus employed, but experiments lead me to doubt its efficacy. Judging from the effect produced on diphtheritic exudation, I should have more confidence in the use of a solution of permanganate of potash.

When the more active symptoms have subsided, our aim must be to prevent the recurrence of plastic exudations; but unfortunately our means are very inadequate to the end we have in view. Though I have had the opportunity of watching one of these cases for many months, I have failed to satisfy myself that any class of remedies possesses the power of controlling the formation of these bronchial casts. The only medicine which appeared to prove serviceable was tartarized antimony, in moderate doses, persevered in for a period of several weeks; and even that, on several occasions, proved utterly inoperative.

But although, when the disease has once arisen, we cannot arrest its progress, I am convinced that we may do much to shield our patient from danger, by sustaining the general health, and warning him to take every precaution to avoid congestion and inflammation of the bronchial mucous membrane. Quinine, iron, and cod-liver oil should be given whenever his strength begins to flag: he should be taught to shun exposure to cold and damp, to wear one of Jeffrey's respirators whenever he is likely to be subjected to variations of temperature, and on the slightest symptom of bronchitic irritation to remain indoors, in a warm atmosphere, inhaling the steam of boiling water, perseveringly, for hours. By these precautions I have reason to believe, that in one of the instances which have come under my observation, an attack was warded off on several occasions. Any bronchitic or pneumonic symptoms must be treated in the usual manner, without reference to the plastic nature of the secretion by which they are accompanied.

Epidemic Bronchitis, or Influenza.

The forms of bronchitis hitherto described are referable to endemic influences, and derive their characters less from any peculiarity in the causes by which they are excited, than from the constitutional condition of the patient at the time of the attack, the pre-existent state of the respiratory organs, or the physical changes to which the disease accidentally gives rise. It is otherwise, however, in regard to epidemic catarrh or influenza. Its occasional prevalence in summer as well as in winter, and the facts that it occurs at long intervals in an epidemic form, pursues an erratic course, always spreading from east to west, but not necessarily affecting contiguous places at the same time;[1] that it disappears from the different localities within or about the same period—from four to six weeks—and that it produces fever, more or less severe, with disturbance of the nervous centres and great general prostration—all point to some specific poison as its proximate or essential cause. Indeed, the peculiarities by which its progress is marked are obviously attributable to the nature of its exciting cause, and there are few circumstances within the range of our profession which, in a practical point of view, require more careful consideration.

The symptoms of influenza differ considerably in their character in different cases. In some persons catarrh is the prominent symptom; in others fever exists, almost unaccompanied by catarrh, whilst in the aged and weakly it sometimes happens that reaction does not ensue after the preliminary shivering, and the patient dies in a few hours in a state of collapse, without the supervention of any of the ordinary symptoms of the disease.

An ordinary attack, however, greatly resembles common catarrh, and differs from it chiefly in the suddenness of its invasion, the rapid development of its symptoms, and the degree of depression by which it is accompanied. It is ushered in by chilliness, seldom amounting to actual shivering, with aching pain and soreness of the limbs, headache, and muscular and nervous pros-

[1] Those who desire to make themselves acquainted with the history of influenza, and especially with the course which the various epidemics have pursued, will do well to consult Sir Henry Holland's "Medical Notes and Reflections," and Dr. Robert Williams's admirable work on "Morbid Poisons."

tration. In some instances sore throat is a prominent symptom, almost from the very outset of the attack; in others a running at the nose and eyes is most strongly marked; whilst, in a third class of cases, disturbance of the alimentary canal, with loss of appetite, vitiated taste, nausea, or even vomiting, may prove to be the earlier and more troublesome symptoms. There is extreme uneasiness, general restlessness, and nervous prostration. Usually the skin is warm and moist, though, sometimes, towards evening, rather hot, and dry; the face at one time is pale, at another flushed; the eyes are streaming; there is tenderness of the scalp, and aching of the jaw-bones; the alæ of the nose are red, and excoriated by acrid discharge from the nostrils; and the voice is altered, as in common "cold." At the same time the patient complains of intense frontal headache, of more or less nausea and loss of taste, of soreness and discomfort at the epigastrium, and of constriction and rawness across the chest, with frequent hard, tearing cough. If his throat be examined, it will be found to be of a dusky red color, but not ulcerated, and seldom much swollen; the pulse is always weak, and usually, but not always, frequent; the tongue is moist, and coated with a white or a yellowish-white fur; the bowels are either costive, or else disturbed and relaxed; and the urine in most cases is clear and acid. Notwithstanding the pain and soreness at the chest, and the frequency and severity of the cough, it often happens that the physical signs of pulmonary mischief are not of a well marked character. Sometimes, indeed, little more can be discovered than indistinctness of the respiration, with some prolongation of the expiratory sound; but, commonly, the signs of ordinary bronchitis are present, and sonorous and sibilant rhonchi are audible in all parts of the chest. In some cases pneumonia occurs between the third and sixth days of the attack, and, in others, pleurisy is set up. Seldom, however, is the inflammatory mischief of a sthenic character. If the lung-tissue is affected, there is seldom or never any rusty-colored expectoration, and whether pneumonia or pleurisy be present it is rarely attended by pain in the side, or, at least, by any urgent dyspnœa, or by the characteristic pulse of inflammation. Indeed, the mischief occurs in a latent form, and if not discovered by a careful physical examination of the chest, will run on unheeded, until, at length, blueness of the face and lips, with clammy perspirations and coldness of the surface, will indicate extensive pulmonary obstruction and failure of the powers of life.

In all cases of influenza muscular prostration and dejection of spirits are prominent features of the attack, and are infinitely greater than would have been expected from the local symptoms of the disease. The patients, if at all advanced in life, and not very vigorous, are often obliged to be supported while being subjected to a stethoscopic examination, and may even faint if kept long in an erect posture. Even the younger and more robust feel utterly exhausted, and incapable of bodily or mental exertion.

The various epidemics of the disease have differed somewhat in the character of the symptoms produced. That of 1782 was marked by the extreme severity of the frontal headache and pain in the temples, by the amount of fever which accompanied it, and by the frequency of delirium. "The nights were passed in disturbed and unrefreshing sleep, frequently with delirium, which in general did not continue long. In some cases, however, it appears to have been the most alarming symptom of the disorder."[1] In the epidemic of 1837 there was seldom any delirium, but there was a great tendency to pneumonia of an extremely asthenic type; so that Sir Henry Holland described it as "scarcely maintaining true inflammation, yet simulating the character of it." So again, in some instances, the chief strain has been on the digestive organs; and excessive nausea, with frequent vomiting, tenderness at the epigastrium, and diarrhœa, have been the more prominent symptoms.

The attack usually lasts from four to ten days, and terminates in profuse sweating or diarrhœa, followed in favorable cases by gradual subsidence of all the symptoms. But not unfrequently we find in its wake an abiding languor and debility, with cough and chronic bronchial irritation. So often does this occur, that it has become a matter of common remark that influenza is apt to lay the seeds of consumption. Nor are facts inconsistent with this common observation; for, although doubtless in many instances the epidemic disorder has done little more than favor the development of already existing tubercular mischief, yet it cannot be doubted that many strong and healthy persons have justly referred their ill health and consequent tubercular disease to the entire derangement of system and great vital prostration resulting from an attack of influenza.

To the aged and weakly influenza proves a very fatal disease, whilst amongst children and persons under forty years of age it rarely gives rise to fatal consequences, unless the patient is weakly and out of health at the time of the attack, or is reduced or weakened by injudicious treatment. But putting these considerations aside, and looking merely at the actual results, it cannot be doubted that influenza is one of the most fatal scourges;

[1] "Med. Trans.," vol. iii, p. 69.

in proof of which it need only be stated, that the French bills of mortality for 1837 prove that no less than 4800 persons succumbed to the complaint during the two months ending the 7th of March; and that, in the epidemic of 1557, 2000 persons are said to have died of it in a small town near Madrid, during the month of September.

In all epidemic disorders it has been proved beyond dispute that active treatment is seldom of much avail, and is often decidedly injurious. The same holds good in respect to influenza. In the milder cases it is seldom necessary to do more than confine the patient to the house, prescribe the free use of tepid diluents and entire abstinence from animal food and fermented liquors, employ a gentle purgative to unload the bowels, and a saline draught to promote the action of the skin and kidneys, and if necessary, apply a mustard poultice or some stimulating liniment to the chest, with the view of producing counter-irritation and relieving the cough. Even in the most severe examples of the disorder it is seldom judicious to have recourse to active or depressing treatment; for whatever measures are adopted, the disease will usually run on unchecked in its course.[1] Therefore, as we cannot arrest the disease, our object should be to prevent the occurrence of any untoward symptoms rather than to waste time in vain attempts to subdue it or cut it short. So long as the symptoms are confined to headache, pains in the limbs, slight sore throat, hoarseness, and cough, the simple measures above recommended will usually accomplish all that can be done for our patient's relief. But should severe sore throat occur, should the bronchitis prove unusally troublesome, or should pleurisy or inflammation of the lungs supervene, more active medical interference will be necessary in order to obviate serious mischief. And here it becomes necessary that sound discrimination should be exercised, and that a lesson should be learned from past observation. The supervention of pneumonia or pleurisy in these cases is sometimes accompanied by so much fever and vascular excitement as to suggest the propriety of free bloodletting. But experience has shown that these symptoms of excitement are speedily followed by great depression even in the younger and more robust, and that bleeding proves extremely dangerous, if not a fatal measure. "Experientia enim hoc comprobavit, omnes fere mortuos esse, quibus vena aperiebatur."[2] Even local bleeding by cupping or leeches is always

[1] In this I am confirmed by many authors. See Williams on "Morbid Poisons," vol. ii, p. 683.

[2] This is a statement of Linnertus respecting the epidemic of 1580: *vide* "Cyclopaedia of Practical Medicine."

a doubtful expedient. Seldom indeed, as far as my experience has gone, is bloodletting admissible in these cases. It is clearly inadmissible in the weakly and the aged; and I believe that even in the younger and more vigorous it is not needed for the relief of the symptoms. Assuredly it lowers the vital powers, and, according to my observation, it fails in relieving the inflammatory symptoms; whereas I have repeatedly traced the gradual disappearance of the signs of consolidation under the use of dry cupping, blistering, and turpentine fomentations, aided by internal remedies. Indeed, I am satisfied that by these means we may accomplish all that can be effected with safety to the patient. Nevertheless, some practitioners, as Dr. Robert Williams,[1] report, that "in general, when the patient was young, and the affection of the lungs limited to bronchitis, the substance being as yet unaffected, leeches to the chest, or cupping to a moderate amount, as ten or twelve ounces, were borne extremely well, and the symptoms were relieved." I am constrained, therefore, to admit that cases may occur, in which bloodletting, if judiciously employed, may prove of service; but at the same time I feel bound to caution the student and young practitioner, who has not had much experience in these cases, to exercise the greatest caution in the abstraction of blood. I have known rapid sinking follow bloodletting to the amount of ten ounces; and Dr. Walshe reports having seen "successive fainting fits" follow the application of a dozen leeches, in "a previously robust and healthy person."

Tartarized antimony and calomel, which are strikingly beneficial in sthenic pneumonia, are of questionable utility in influenza. The former is said to have proved successful in the hands of M. Hortloup, at the Hôtel-Dieu, and Dr. T. Davies and others have spoken highly of the latter; but I have seldom seen any marked improvement follow their administration, and in many instances a contrary effect has been observed. The patient's strength has failed rapidly under their influence, without any corresponding amelioration in the chest symptoms.

In short the remedies on which alone I place much reliance in the acute stage of this complaint are tepid diluents freely employed, saline diaphoretic medicines, ipecacuanha and gentle purgatives, assisted if necessary by opiates, hyoscyamus, or other sedatives, and by dry cupping, blistering, and turpentine fomentations. If the patient exhibits the slightest symptoms of depression, quinine or bark, with sesquicarbonate of ammonia, should be given from the first, and wine or brandy allowed in

[1] Loc. cit., p. 682.

moderate quantity; and if there be any difficulty of expectoration, the tincture of senega or some other stimulating expectorant should be added. In all cases, quinine should be substituted for the saline as soon as the first fury of the attack is past, and should be combined with squills and nitric ether or spirits of chloroform. During convalescence the salts of iron may be used advantageously.

In mild cases the patient may be allowed white fish, light puddings, and strong beef-tea throughout the attack; but in the severer forms of the disease, when the digestive organs are disturbed, the diet should be restricted, in the first instance, to beef-tea, broth, gruel, or barley-water.

In all cases, whether slight in character or severe, much care should be taken during convalescence. Tonics should be administered suited to the constitutional peculiarities of the individual; and if the slightest cough remains, cod-liver oil will often prove useful. But other means than mere internal remedies should be resorted to if the patient does not speedily recover. Change of air, under these circumstances, is imperative, and if not delayed too long, will seldom fail to effect our object; but if our warning be disregarded, and no efficient means are taken to get rid of the languor, lowness of spirits, want of sound refreshing sleep, and cough, slight though it may be, which often follow this disorder, permanent ill-health will ensue, and pulmonary consumption will probably terminate the patient's existence.

Bronchitis, associated with "Hay Fever," or "Hay Asthma."

There is yet another variety of catarrh, which has been supposed to be connected with emanations from the *Anthoxanthum odoratum*, or sweet-scented vernal grass, and hence has been termed hay asthma, or hay fever. Its symptoms are peculiar. and resemble those produced in some persons by the powder of ipecacuanha, and sundry other substances. They resemble in some measure the symptoms which would be produced by a combination of catarrh and asthma. The patient is seized with pain in the head, especially over the frontal sinuses, itching at the eyes and nose, with copious watery defluxion from the nostrils, frequent paroxysmal sneezing, pricking sensations in the throat. excessive irritation of the air-passages, with cough, usually dry, but sometimes accompanied by thin, watery, mucous expectoration, soreness, oppression, and tightness of the chest, shortness of breath, and difficulty of breathing. Few persons are susceptible of the disease; but those who have once experienced an attack, are apt to suffer from it year after year, in

May or June, when the grass is ripe, and haymaking is going on; nay, more, they will often experience an attack if brought into contact with the dust of dry hay at other times of the year. This, in short, constitutes a peculiarity of the disease, and together with the suddenness of its invasion, and the rapid development of its symptoms, serves to excite a suspicion of its true character.

As might be expected, from the nature of the exciting cause, the duration of the hay asthma does not usually exceed a month or six weeks. Sometimes, indeed, on removal from the source of irritation the symptoms subside in a few hours. Nevertheless, the complaint is so troublesome and distressing, that sufferers will have recourse to any expedient which seems to offer a prospect of relief. The most effectual remedy is avoidance of the source of irritation, as by taking a sea-voyage, or by removal to the sea-side, out of reach of the smell of hay. This is the course very commonly adopted by those whose time is at their own disposal; but it is not in the power of every one to leave home; and to such persons it becomes necessary to administer medicines which shall enable the system to resist the influence of the matter which is irritating it. For this purpose, many remedies have been proposed. Mr. Gordon[1] reports two cases effectually cured by quinine and sulphate of iron, aided by the use of the shower-bath; Dr. Elliotson has seen relief from the inhalation of an atmosphere more or less charged with chlorine; Dr. Gream has spoken highly of the efficacy of nux vomica; Sir Thomas Watson has recommended arsenic; Dr. Walshe, the use of creasote inhalations: but these and all other remedies which have proved of service in certain instances, have utterly failed in others, and we are yet in want of a trustworthy remedy for this disease. In two well-marked instances which came under my own observation, all the remedies already named had been employed in vain, and I was fortunate enough to alleviate my patient's sufferings by means of sulphate of zinc administered internally in gradually increasing doses, whilst zinc lotion was applied to the eyes, and a lotion composed of eight grains of sulphate of zinc and an ounce of glycerine to the lining membrane of the nostrils. Still more recently I have subdued the symptoms by the internal administration of five-grain doses of quinine, and by causing the patient to inhale atomized water holding zinc in solution; and one lady in whom zinc proved of no avail was benefited by the use of arsenic given internally, and inhaled in the same manner. In both these instances relief

[1] "Med. Gazette" for 1829, vol. iv.

was obtained so speedily and effectually, that I am warranted in recommending a further trial of this method of treatment.

Spasmodic Asthma.

Spasmodic asthma is an affection dependent on spasmodic contraction of the bronchi, which gives rise to difficulty of breathing of a paroxysmal character. It has been termed indifferently asthma, spasmodic asthma, and bronchial asthma; but the term "spasmodic asthma" should alone be applied to the disease under consideration, inasmuch as there is a form of bronchial asthma, to which allusion will be presently made, in which there are strong reasons for doubting whether the symptoms are not attributable to paralysis rather than to spasm of the bronchial muscles. In spasmodic asthma the difficulty of breathing, which is accompanied by loud wheezing, may come on either suddenly or slowly, at any period of the day or night. Very commonly it commences during sleep, and after lasting for some hours, is terminated by expectoration. In this case it has been termed "humid" asthma; whereas, when the fit is brought to an end without expectoration, the case is said to be one of "dry" asthma. In either case the phenomena which constitute a fit of the disease are much as follow: The patient probably wakes soon after midnight with a sense of tightness or constriction across the chest. The inability which he feels to expand the chest while he remains in a recumbent posture induces him to start up at once in bed; and a common posture for him to assume is that of leaning forward, with his elbows on his knees, and his head supported by his hands. In this position he will sometimes remain for hours, gasping for breath, his countenance meanwhile being anxious and distressed, the face red and congested, or else pale and rather livid; the eyes prominent and staring, the skin covered with a clammy perspiration, the extremities cold; the pulse small, feeble, and sometimes irregular. He coughs with difficulty, so "short" is his breath, and so feeble his efforts; and if he attempts to speak, he can only articulate a few words at a time. The bowels are often relaxed, and act hurriedly, yet imperfectly, as if under the influence of spasmodic action; and the urine, which, at the commencement of a paroxysm, is copious, pale, and watery, becomes scanty and high-colored towards its close. If he finds relief in the posture above described, the patient will maintain it until the breathing becomes less labored, and the paroxysm begins to pass off: but if he fails to do so, and experiences, as he often does, a desire for fresh air, he will open the door of his room, or throw open his window, and remain at it for hours, even in cold weather. When, at length, the fit begins to subside, he is much exhausted by the fatigue he has undergone, and generally falls asleep.

But his troubles do not cease with the termination of the paroxysm. The repose which follows is but a brief and imperfect respite from suffering. Though he may consider himself quite well during the intervals between the paroxysms, he is short-breathed, and unequal to active exertion; he is incapable of maintaining a continued conversation, and pants and wheezes, if he attempts to stoop or to run up stairs. Nay, more, he is in constant dread of a fresh attack; for experience has taught him that, day after day, or for many nights in succession, the asthma will return with undiminished violence, and that when at length it ceases to recur, it will leave him weak and thoroughly exhausted.

It has been already stated that asthma is a spasmodic disease. It may be stated further, that it is an hereditary complaint; in other words, that the tendency to it may be transmitted from parent to child, through several successive generations. Like all hereditary disorders, it may occur at any period of life, and instances are not wanting of its existence in infancy and early youth. One of the most frightful examples of it I ever met with was in the person of a boy thirteen years of age. Indeed, it often dates its origin from the diseases of childhood, from the straining efforts incident to whooping-cough, and the severe bronchitis which accompanies measles. Nevertheless, the most common period for its full development is between the ages of twenty-five and fifty-five; and it occurs more commonly in men than in women.

Valentin found that contraction of the rings of the trachea is induced by irritation of the par vagum; and there cannot be a doubt that the spasm of the bronchi, on which asthma depends, is excited by irritation of the filaments of that nerve connected with the lining membrane of the air-tubes. Further, it cannot be doubted, that the tendency to spasm, or, in other words, to be affected by certain causes of irritation, depends upon a constitutional peculiarity—an obscure and ill-defined condition of the nervous system—and that according as this peculiarity varies, so do the causes differ which lead to the production of asthmatic symptoms. Most asthmatic persons are affected by certain states of the atmosphere, not necessarily connected with its dryness or humidity, nor yet with its precise temperature, but probably having some relation to its electrical condition, or to the presence or absence of ozone, or of other matters, some not improbably of vegetable origin, which have hitherto escaped detection. One class of asthmatic persons can breathe freely in a close, damp valley; another on the summit of a hill. To one the fresh air of the country is unbearable; to another, the smoky atmosphere of a town. Nay, more, so subtle and incomprehensible are the disturbing influences which give rise to the disease, that mid-London air is sometimes the most congenial atmosphere, and mere removal from one house to another, or to a neighboring street in the same town, has been known to put an end to a patient's sufferings. Many persons are influenced by circumstances which interfere with the action of the heart or disturb the pulmonary circulation; and others by the occurrence of mental excitement, or whatever tends to disturb the nervous system. Thus, a fit of anger, sudden distress, exposure to intense cold or to a storm of wind, active exertion, flatulent distension of the abdomen, and derangement of the liver and bowels, are one and all, in certain instances, productive of asthma.

Thus, then, it will be seen that whilst in every case the disease is dependent on spasm of the bronchial tubes, its exciting cause may differ widely in different instances. In one case the spasm may be of centric origin, or arising primarily from excitation of the nervous centres: in another, its source may be eccentric, that is, it may originate in an impression conveyed to the par vagum from the surface of the air-passages, or from the coats of a disordered stomach. In either case it may be unaccompanied by organic mischief, or, in other words, may be purely spasmodic—a form of disease very rarely met with—and the patient may enjoy most perfect health during the intervals between the paroxysms, in which respect it resembles the disease already described under the title of hay asthma. But more commonly, it is associated with structural changes within the chest which impart to it a much more serious character, and render it impossible for the patient to obtain more than partial relief. Emphysema of the lungs, and organic disease of the heart and large vessels, are among its most frequent, or almost constant concomitants. Thus it happens that the sufferings of asthmatic patients, though aggravated in paroxysms, are persistent to a greater or less extent. The asthma may be subdued for a time, but the pulmonary or heart affection will remain, and lead to shortness of breath and urgent distress on the slightest occasion. A little extra exertion, unwonted excitement, or an attack of indigestion, is almost certain to aggravate the symptoms and induce a paroxysm. Indeed, the symptoms by which an attack is usually preceded are, loss of appetite, flatulence, acid eructations, drowsiness after meals, chilliness, irritability of temper, and excessive languor; symptoms which denote derangement of the digestion and circulation, and impairment of the nervous function.

Those persons who are affected with pure spasmodic asthma may and do sometimes outgrow their complaint, or get rid of it by change of habits or locality; but those in whom it is associated with disease of the heart or lungs experience year by year an aggravation of their symptoms. The spasm, which characterizes a fit of the disorder, tends naturally to increase the mischief in the chest, and that in turn reacts, and renders the paroxysms of asthma more frequent and more distressing. Nevertheless, in most instances the disease does not prove speedily fatal; on the contrary, with due care and attention, the short-breathed sufferer from asthma will often drag on his miserable existence for years after his more vigorous companions have disappeared from the scene of their earthly labors.

It has been asserted by some authors that asthma and consumption never coexist, the one being antagonistic to the other. The statement, however, is erroneous; for although the two diseases are not commonly associated, the one having nothing in common with the other, occurring usually at a different period of life, and being excited by entirely different causes, it is nevertheless true that there is no incompatibility between them, and that many asthmatic patients have died consumptive. Three cases of the kind have come within my own knowledge, and few persons who have enjoyed extensive opportunities for observation can have failed to meet with similar examples. The utmost that can be truly stated is that asthma does not predispose to consumption, nor tubercular disease to spasmodic asthma.

The general symptoms of the disease are so characteristic that its physical signs become of little practical importance. They consist of excessively labored respiration, with elevation of the chest-walls, deficiency or almost entire absence of thoracic expansion, and occasional retraction or falling in of the lower ribs during inspiration. At the same time there is almost entire absence of healthy respiratory murmur, which is replaced by sonorous or sibilant rhonchi. The only conditions under which vesicular respiration can be heard during the paroxysm are, if the patient has held his breath as long as possible, or has exhausted his chest by talking as long as he can without drawing breath; in both of which cases, as was pointed out by Laennec, the spasm is temporarily relaxed, a quiet inspiration takes place, and air can be heard entering freely into the lung.

In the treatment of asthma there are three points for consideration, namely, 1st, whether the disease is complicated by organic mischief in the chest, and if so what is likely to relieve that mischief; 2dly, how to subdue the spasm and arrest the paroxysm; 3dly, how to avert an attack or prevent its recurrence.

The first point is one which cannot be discussed at the present time, inasmuch as it involves the whole treatment of pulmonary and heart affections. It is obviously, however, of the greatest importance to determine whether there be organic mischief in the chest and to ascertain its precise nature, as the treatment of asthma, especially the preventive treatment, would be modified materially by considerations arising in connection therewith. Thus, if the disease were uncomplicated and purely spasmodic, we might safely permit our patient to take moderate exercise, provided he sedulously avoids those localities or those particular influences which experience has proved to be pernicious to him. Our efforts in such a case would have to be directed almost wholly to three points, viz., 1st, to discover a locality in which the patient can reside with comfort; 2dly, to invigorate his system, and render him less susceptible to the influences which excite the bronchial spasm; 3dly, to regulate his diet and mode of living so as to obviate, as far as may be, those disorders of the stomach which always aggravate and in some instances appear to prove the exciting cause of the bronchial spasm. The first object can be attained only by repeated trials; the second, if attainable at all, is so by the frequent use of the shower-bath, by the exhibition of sulphate of zinc, quinine, arsenic, iron, and other tonics, combined from time to time with stomachic medicines and alteratives; the third can be arrived at only by strict inquiry into the previous habits of the individual, and into his experience relative to different articles of food—ordinarily a generous though carefully selected diet is advisable, but instances are not wanting in which the quantity of food must be greatly restricted if beneficial results are to be attained.[1] Nothing, however, will countervail the influence of an atmosphere which does not suit the patient; so that until some locality be discovered in which he can breathe freely, the best remedy we can prescribe is repeated change of residence. In his search for such a spot we may often assist him by suggesting a trial of places which differ widely in their atmospheric conditions. But additional measures are needed when the lungs are emphysematous, or the heart and large vessels more or less diseased. In such cases active exertion must be avoided, and so must all influences, of whatever nature, which tend to accelerate the heart's action or disturb the respiration.

[1] See a case recorded by Mr. Pridham in "British Med. Jour." for 1860.

Moreover, care must be taken to mitigate or ward off by medicinal agency any symptoms arising from the structural changes within the chest. Without due attention to these points our efforts to afford relief will be futile, and our patient will be subject to a constant recurrence of asthma.

When the paroxysm has once commenced, we can shorten or mitigate our patient's suffering only by remedies of an antispasmodic nature. Perhaps it would be more correct to say "by remedies which in the particular case in question prove antispasmodic in their operation," for relaxation of the spasm is in certain instances produced by means which do not usually exert an antispasmodic influence. Thus in cases which are connected with gout and rheumatism, colchicum and iodide of potassium prove valuable antispasmodics; strong coffee does so in others, and ipecacuanha, and many nauseants might be mentioned as sometimes efficacious. But in most cases our chief reliance must be placed upon sedatives and antispasmodics of another class. Notwithstanding the congestion which often accompanies the earlier attacks of asthma, it is seldom advisable to have recourse to bloodletting; and so great is the prostration in advanced cases, that venesection is never admissible. Opium and the salts of morphia, belladonna, hyoscyamus, stramonium, and Indian hemp, the ethers, and the ethereal tincture of lobelia, digitalis, and hydrocyanic acid, make up the catalogue of our more valuable remedies, and great relief they oftentimes afford, though the efficacy of each of them varies in different cases. In several instances I have seen exceeding comfort derived from a mixture of opium and sulphuric ether internally, aided by the inhalation of chloroform; in others, the addition of belladonna to the mixture has appeared to exert a magical effect; and in a third class of cases, the administration of lobelia has given entire relief after the other remedies had failed. Indeed, lobelia is one of the medicines of which I entertain the highest opinion. One lady, somewhat advanced in years, who had been a martyr to asthma during the greater part of her life, assured me that she had been entirely cured of her complaint by two drachms of the ethereal tincture, taken in two doses at an interval of half an hour. She took these doses six years ago of her own accord whilst stopping in Leicestershire, at a considerable distance from any medical advice; the asthma subsided immediately, and since that time has not returned. Stramonium and other remedies had previously failed to relieve her, and so had small doses (ten to fifteen minims) of the ethereal tincture of lobelia.

Some authors have suggested doubts as to the safety of administering full doses of lobelia; and inasmuch as lobelia is a potent remedy, there are fair grounds for supposing that very large doses might give rise to disagreeable, if not dangerous effects. But I have employed it in full doses (twenty to thirty-five minims) in scores of cases of chronic bronchitis and asthma, and have never yet observed the slightest inconvenience, except occasional nausea, from its administration. I am justified, therefore, in stating, that under proper medical supervision it may be safely given in the doses above mentioned; and I can unhesitatingly assert, that in some cases at least it will be found extremely serviceable.'

The Datura stramonium, however, is the remedy which appears to be most efficacious in checking the spasm, and relieving the asthmatic patient. Taken internally, in the form of tincture or extract, or cut into small pieces, and put into a pipe and smoked, the herb will often act like a charm, arresting the spasm, inducing expectoration, and restoring the patient to comparative comfort within a short space of time. Indeed, so speedy and so complete is the relief it affords, that asthmatic persons will often declare that they have little dread of a return of their complaint so long as they are provided with stramonium; but, like other remedies, it sometimes fails, and, if the paroxysm is strongly established before recourse is had to its assistance, the patient may find himself unable to smoke, and unable, therefore, to avail himself of its virtues.

The Datura tabula is another species of the same genus of plant which possesses valuable antispasmodic properties. Ten years ago I gave it with success in two cases in which stramonium had failed, and many instances are now on record in which it has proved superior to that drug. In this country, however, it has not been imported in any quantity, and up to the present time has been comparatively little used. But from what I have seen and what I have heard of its effects from others, I am satisfied that it well deserves a trial.

In some instances a combination of stramonium, belladonna, and camphor may be employed advantageously in the form of a cigarette; in some the addition of arsenic is serviceable; in others, tobacco fumes, and the fumes arising from burning blotting-paper, or from the burning of bibulous paper, saturated with nitre, will give relief; in others, the inhalation of ether and chloroform will serve to *check* the spasm, though it will seldom, if ever, put an end to the disease; and in others, again, the inhalation of sedatives and antispasmodics dissolved in water, pulverized by means of Siegle's apparatus, is productive of a degree of relief which is sought in vain from other measures.

The Cannabis Indica has been lauded as a remedy, but I have

not been fortunate in my experience of its effects; digitalis proves useful in those cases especially in which the disease is complicated by affection of the heart; the prunus virginiana is often useful as a sedative, and hydrocyanic acid is a remedy, which often acts beneficially in conjunction with soda and other antacid medicines, when the stomach is irritable and out of order. I have never known it arrest a paroxysm, but on several occasions I have seen a patient's sufferings relieved by its administration. Iodide of potassium in full doses (ten to fifteen grains) has been known to act magically in certain instances. Nitric acid, in large doses, repeated at short intervals, is another remedy which deserves a trial. I am not aware that it has ever been prescribed as a remedy for asthma; but judging from the influence it exerts over the spasm of whooping-cough, it is fair to conclude that it may produce a similar effect in asthma. Its action is harmless, since the merest infants take it with impunity, and, even if it fails to relieve the spasm of asthma, it is likely to prove serviceable in preventing its recurrence; and this, after all, is often the most important point to which it is possible to attain in these cases.

Various other means have been adopted in obstinate cases in the vain hope of obtaining relief; amongst which I may mention emetics, cold dash, or the shock of cold water, galvanism, hot leg-baths, and the influence of an intensely hot atmosphere as in a Turkish bath. Each and all of them are said to have proved useful in certain instances, and possibly they may have done so; but it is notorious, that in many instances they have failed altogether, and in some have appeared to act prejudicially. The truth appears to be, that they are not equally adapted to every case, and that they each prove useful or prejudicial according as they are judiciously employed. As curative agents they are not entitled to our confidence, but in certain instances they may prove useful adjuncts to other and more specific treatment.

Baths of compressed air have been reported to be extremely serviceable; and if the reports by M. Bertin, of Lyons,[1] be correct exponents of their average utility, they ought to be had recourse to in every instance. The few instances, however, in which they have been employed, which have come within my own knowledge, lead me to doubt whether the favorable results obtained by M. Bertin will usually reward our efforts in this direction.

[1] "Gaz. Méd. de Lyon," Feb. 1, 1861.

Paralytic Asthma.

When describing spasmodic asthma, I stated that certain instances of asthma occur in which there is reason to believe that the symptoms are dependent on paralysis of the bronchial muscles, rather than on the existence of spasm. Thus, asthmatic persons are sometimes met with in whom the difficulty of breathing is obviously connected with expiration. They inspire with comparative ease, but labor in vain to empty their chest. There is comparatively little wheezing in such cases, but not the less difficulty in expiration. In some such persons the habitual dyspnœa is doubtless connected, in part at least, with the existence of emphysema of the lungs; but, in several cases which I have watched during life, and have had the opportunity of tracing to the dead-house of St. George's Hospital, the difficulty of breathing has been excessive in relation to the emphysema discovered after death; and in others little emphysema could be detected, although the dyspnœa, the deficiency of the expansive movement of the thorax, and the imperfect emptying of the lungs during expiration, combined with extreme clearness and resonance of the chest on percussion, had appeared to indicate its existence. In these cases occasional exacerbations of the symptoms may arise; but the difficulty of breathing is more constant and the attacks are less paroxysmal than in spasmodic asthma.

Whatever may be the cause of the symptoms in these cases, they are obviously unlike those produced by spasm of the air-tubes, as typified in spasmodic asthma; and as Laennec has traced distension of the lungs with air to paralysis of the vagi nerves, and as direct experiment has confirmed the result of his investigation, it seems fair to conclude that the cases under consideration are referable to this cause. The absence of all evidence of bronchial spasm and the existence of air-distension of the lungs do not admit of a doubt, and in the absence of emphysema, paralysis of the bronchial muscles seems the only assignable cause. Possibly, the existence of this condition of the respiratory apparatus may serve to explain the beneficial effects which, in some cases of asthma, have been observed to follow the use of galvanism, cold dash, exposure to a cold atmosphere, and other similar influences.

Hæmic Asthma.

Before quitting the subject of asthma, it may be well to draw attention to certain cases in which a morbid condition of the blood is productive of much apparent distress and difficulty of breathing, irrespective of any lesion of the heart or lungs. The patient in these cases seldom complains much of shortness of breath; for although the frequency and apparent difficulty of his breathing are great, the dyspnœa does not really distress him. It resembles the labored breathing of breathlessness from over-exertion, rather than that arising from true asthma. Ordinarily, the chest expands freely, there is little or no lividity of countenance, but nevertheless, the breathlessness continues. The careful and experienced physician cannot possibly be misled by such cases, but they may mislead, and often have misled the student and the unwary practitioner into a belief in the existence of organic disease of the thoracic organs. On this account they are deserving of special notice.

The characteristic features of this form of complaint are, that the frequency of the respiration is not so great as the distress to which it apparently gives rise would lead one to anticipate; that in most cases it is not accompanied by lividity of countenance, or by any marks of imperfect expansion of the chest; that the ratio naturally subsisting between the pulse and the breathing is not seriously disturbed, and that an examination of the lungs and heart fails in affording any evidence of mischief calculated to explain the symptoms.

In truth, this variety of hurried breathing originates in a morbid condition of the blood, which is either impoverished or insufficiently oxygenated, and thus is less stimulating than it ought to be, or else being charged with materials which it ought not to contain, exerts a morbid or depressing influence on the thoracic organs, and on the nervous centres whence they derive their power. Thus it is met with in certain cases of anæmia, and after excessive hemorrhage, as also in gout, Bright's disease, and other blood disorders. The precise cause in any particular instance must be determined by the history and symptoms of the patient, but if care be taken in the investigation, a satisfactory conclusion may be easily arrived at.

The treatment must of course be varied according to the nature of the disease from which each case originates, but in every instance the patient should be well supported during the interval which must elapse before the quality of his blood can be improved, or the morbid materials with which it is charged can be eliminated and got rid of. In most cases the preparations of iron are of essential service.

Whooping-cough or pertussis, is characterized by slight catarrhal fever, followed by a peculiar spasmodic cough, of which the paroxysms occur at uncertain intervals. It is met with sporadically at all seasons of the year, but sometimes prevails epidemically, thus leading to the inference that its cause must exist at all times diffused through the atmosphere, though varying at different periods in quality and intensity. Common report stamps it as infectious; and the evidence on this point is so strong and conclusive, that although Laennec[1] and some few members of our profession have hesitated to admit the fact, we may fairly state with Frank, that "nostro ævo nemo amplius de naturâ contagiosâ coqueluche dubitat." It may occur at all ages,[2] but is most common in childhood, for the reason that children are extremely susceptible to it, and few pass through early life without experiencing an attack.

The history of its introduction in the island of St. Helena[3] proves that the disease may be propagated by means of fomites; and observations which have been made in several instances of the same sort, as also in regard to the communication of the disease from infected persons, have led to the conclusion that its period of latency is about five or six days. It may coexist with many other disorders, and is often seen concurrently with small-pox and measles,[4] and though usually occurring once only in the course of life, it has been met with twice in the same person.[5] Some pathologists have ascribed it to cerebral irritation, and others to an affection of the pneumogastric nerve, but neither supposition is consistent with many acknowledged facts. Even if it be admitted that the peculiarity of its character is attributable directly or indirectly to nervous irritation, it is nevertheless true, as already stated, that the cause of that irritation—the essential or proximate cause of the disease—is an epidemic influence, a specific poison, infecting the whole system, and acting not only on the nerves, but on other parts of the body. Practically, therefore, if my opinion is correct, the disorder may be regarded as consisting essentially of bronchitic irritation,

[1] "Traité de l'Auscultation," vol. i, p. 156.

[2] Sir Thomas Watson in his "Lectures" (ed. i, vol. ii, p. 63) alludes to a child born with whooping-cough.

[3] See Williams on "Morbid Poisons," vol. i, p. 302.

[4] Williams, loc. cit., vol. i, p. 303.

[5] Dr. Heberden relates a case in point; and there are few men of large practical experience who have not met with similar instances.

usually not very severe, excited by some epidemic influence, and accompanied by reflex spasm of the air-passages.

The course of the disease may be divided into three stages: 1st, the catarrhal; 2dly, the spasmodic; 3dly, the terminal, or stage of convalescence.

The *first stage* is marked by the usual symptoms of "common cold"—sneezing, running at the eyes and nose, hoarseness, oppression at the chest, irritation of the air-passages, with dry or almost dry and often paroxysmal cough, impaired appetite, disordered bowels, and catarrhal fever. This stage lasts from three or four days to ten days or a fortnight, and is not distinguished by any special features, so that it is practically impossible to determine whether, to use a nurse's phrase, the disorder may not "turn" to whooping-cough. In mild cases, the feverishness, running at the nose and eyes, and loss of appetite are almost absent.

The *second stage* commences as soon as the cough becomes distinctly paroxysmal, and is accompanied or followed by the characteristic whoop. The disorder now appears to consist essentially of bronchitic irritation, with violent reflex spasm of the air-tubes, which at uncertain intervals, by day and night, produces abrupt paroxysms of spasmodic cough. On the approach of the fit the child instinctively grasps his nurse or lays hold of a chair or table, and thus prepares, as he best may, to support the shock of the cough by which his whole frame is about to be shaken. As soon as the paroxysm has fairly commenced, the respiration becomes labored to the greatest extent; the expiration consists of a succession of forcible, short, spasmodic coughs, which continue, without any intervening inspiration, until the air in the lung appears to be almost wholly expelled, the eyes seem about to start from their sockets, and the little patient turns red or even black in the face. Then the spasm momentarily ceases, and a prolonged and labored inspiration takes place, accompanied by a loud cooing or whooping noise, denoting partial closure of the rima glottidis. As soon as the inspiration is fairly completed, another spasmodic expiratory effort commences of the same character as the former; and a complete paroxysm, which consists of a succession of these alternate violent inspiratory and expiratory acts, may last from half a minute to a quarter of an hour. When the paroxysms are much prolonged, not only does the congestion of the head and face become excessive, but blood may burst from the distended vessels of the lungs, mouth, ears, and nostrils; convulsions may take place, and the urine and the contents of the bowels may be discharged involuntarily. Each fit is brought to an end by the occurrence either of vomiting or free expectoration, or by vomiting and expectoration taking place simultaneously, or sometimes, though rarely, through mere exhaustion, without the occurrence of discharge of any kind. The matter rejected from the stomach is of a glairy nature, semi-transparent and tenacious. That which comes from the bronchi is scanty, thin, and often streaked with blood in the earlier stages of the disease; more abundant, semi-opaque, ropy, or of a muco-purulent nature when the disease is more advanced. The former variety of bronchial secretion is expelled with difficulty, the latter with comparative ease; and thus it happens that the earlier paroxysms of cough are commonly more severe and of longer duration than those which accompany the later stages of the complaint. As soon as the paroxysm has subsided, the congestion of the head and face passes off, the pulse becomes quiet, and the respiration tranquil; so that in the course of a few minutes the little patient is again lively and apparently quite well, resumes his play or any other occupation which the cough may have interrupted, and takes his meals with a relish, as if nothing were the matter with him. This, at least, is the usual course of events in favorable cases; but in severe or unfavorable cases the patient remains feverish, pale, and exhausted, and shows a disinclination to food. In such instances there is too much cause for expecting some untoward complication, and good grounds for anxiety as to the issue.

The frequency of the paroxysms is usually found to vary according to their severity, the fits being most frequent when they are most severe. In ordinary cases, they return every hour and a half or two hours; but in some instances there may be only two or three fits in a day, whilst in others the paroxysms take place every twenty minutes or half an hour. They may occur spontaneously or without any obvious exciting cause; but more frequently they are occasioned by the act of swallowing, shouting, or laughing, by a fit of anger, or, in short, by any cause which influences the excito-motory system, and occasions reflex spasmodic action. After the third or fourth week the paroxysms usually diminish in frequency and severity, and about the eighth week the symptoms are so much mitigated that the third stage, or the stage of convalescence, may be considered to have commenced. Sometimes, however, the second stage may terminate at the end of three or four weeks, or, on the other hand, if improperly treated, it may continue for many months.

The third stage of the complaint is marked by diminution and ultimately by cessation of the symptoms. The paroxysms become milder, the intervals longer, the cough loses its convulsive character, the whoop gradually ceases, the expectoration loses its ropy appearance and becomes simply catarrhal, vomiting no longer occurs, and the patient's health improves. The average duration of this stage is from a fortnight to three weeks, but under unfavorable circumstances or inappropriate treatment it may persist for many months, the characteristic whoop remaining almost up to the last. Even after it has entirely ceased, the patient is for some time liable to a recurrence of spasmodic whooping in the event of his "catching cold" and having a cough.

In uncomplicated cases an examination of the chest does not afford much information. During the intervals of rest there will be found to be good resonance on percussion; and if the sounds of respiration are not altogether natural, they are only mixed with or replaced by rhonchi and moderate-sized râles—sounds which denote the existence of catarrh. Towards the close of a violent paroxysm the resonance of the chest becomes temporarily impaired, in consequence of the forced emptying of the air-cells; at the same time the respiratory sound is entirely suspended, and is not perceptible in any part of the chest. During the brief snatches of inspiration which occur during the spasmodic fits of coughing, the sound of breathing may be sometimes heard, either natural, or mixed with wheezing; but during the prolonged whooping inspiration little vesicular breathing is audible. The air may be heard to rush into the trachea and larger bronchi, occasioning rhonchi, and if there be much fluid in the tubes, bubbling râles may also be heard; but the air does not permeate the structure of the lungs, and therefore the natural respiratory murmur is not audible. This arises from the conjoint action of two causes, viz., spasmodic narrowing of the glottis, and spasm of the bronchial tubes; and as the action of these causes is limited to the duration of the cough, vesicular breathing is re-established, and the respiratory murmur may be heard as soon as the fit has passed off.

When the disease is mild and unattended by local inflammation, it seldom gives rise to fatal consequences; but when it is severe it is apt to be complicated by affections of the chest or head, which impart to it a very serious character, and often lead to fatal results. Acute capillary bronchitis, or pneumonia, or pleurisy may supervene, and may carry off the patient in a few days. The air-cells may be ruptured by the violence of the cough, and the patient may die suffocated by the effusion of air into the cellular tissue of the chest; the pleura may give way, and pneumothorax may result; or convulsions, or apoplexy may occur. Even if temporary recovery takes place, dilatation of the bronchi or emphysema may result from the long-continued straining cough, or the cerebral structures may not recover from the violent mechanical congestion to which they have been so long subjected. In this case the patient begins to squint, and may die hydrocephalic. It may be stated, however, that this chronic mischief is most apt to occur in very young children, and that immunity from it is obtained in proportion as the age is more advanced. Another complication which is apt to arise, and seriously increases the danger of the disease, is infantile remittent fever, with a disordered condition of the bowels. Oftentimes, when this train of symptoms sets in, the motions are relaxed, and are either dark-colored and offensive, or else, pale, consisting of little else than mucus. It is difficult to determine whether these symptoms are attributable to the action of the morbific agent from which the whooping-cough originates, or whether they are not the results of a disorder engrafted on or arising coincidently with it. My own opinion, however, inclines to the former view, for the reason that the disorder of the bowels is observed only in severe cases, and on more than one occasion has appeared to me to vary with the severity of the other symptoms of whooping-cough.

The post-mortem appearances correspond with the symptoms observed during life. In uncomplicated cases little more is discovered than a congested condition of the mucous lining of the air-passages, together perhaps with enlargement of the bronchial glands and slight effusion in the pia mater or into the ventricles of the brain, as a consequence of the long-continued congestion resulting from the cough. Even these morbid appearances are sometimes absent. But more frequently fatal cases are found to have been complicated by capillary bronchitis, pneumonia, or pleurisy, the usual pathological results of which are manifest after death. Pulmonary collapse—the so-called lobular pneumonia of former writers—is often present, as pointed out many years ago by Dr. Alderson.[1] In some cases there is considerable effusion of serum, and occasionally even of blood in the brain, and in cases complicated by bowel affection during life the stomach and alimentary canal have been found congested and Peyer's glands enlarged.

The character and duration of whooping-cough are apt to vary greatly, even when the disease occurs in an uncomplicated form. Some persons[2] have contended that it may "never put on any other form than that of a common catarrh;" and although

[1] "Med.-Chir. Trans." vol. xvi, p. 91. [2] Cullen and others.

it appears to me that the grounds on which this opinion rests are not sufficient to warrant such a positive statement, and that we should do wrong to recognize as pertussis any catarrh which is unaccompanied by the characteristic whoop, still it cannot be denied that the complaint is often exceedingly mild, and is sometimes limited to catarrh of two or three days' duration, with a few paroxysms only of spasmodic cough and stridulous inspiration. More commonly, if unchecked by proper remedial agents, it persists for three or four months, and may run on in extreme cases for six or eight months, or even longer.

The treatment of whooping-cough is a subject which deserves very serious attention. It is commonly supposed that the disease has a definite course to run, which will be longer or shorter according to the severity of the attack, and which is not to be checked or arrested by remedial agents. The advocates of this doctrine, amongst whom may be named the major part of the profession, confine their efforts to the prevention or subjugation of any untoward symptoms, and to the mitigation of the severity of the cough. They give saline medicine and a slight sedative with ipecacuanha, or, if necessary, a little tartarized antimony; occasionally an emetic is administered; the bowels are regulated by means of a gentle purge; and stimulating embrocations are employed down the spine.

Now, there cannot be a doubt that in the majority of cases this plan of treatment, if judiciously carried out and aided by a careful avoidance of cold and by due regulation of the diet, will suffice to conduct the patient safely to the termination of the disease. But it is equally certain that the average of cases will persist, under this treatment, for a period of from eight to sixteen weeks; that some at least will continue for a much longer period; and that in some at least a fatal result ensues. It behooves us, therefore, to inquire whether some more efficacious plan may not be discovered.

In the year 1847 I was led to try the effect of sulphate of zinc, in gradually increasing doses, with the view of checking the spasmodic cough. Regarding the complaint as consisting essentially of bronchitic irritation, usually not very severe, accompanied by reflex spasm of the air-passages as its prominent and more important symptom, it occurred to me that sulphate of zinc might control, if it did not put a stop to the whoop; nor were my expectations disappointed, for the remedy succeeded admirably, and I therefore adopted it in all cases of whooping-cough which subsequently came under my care. In five only of fifty-nine cases in which I had the opportunity of testing its virtues did it fail in giving marked and speedy relief. In one of these five—a male adult—it utterly failed; in two it could not be tolerated by the stomach; and in the remaining two it did not materially lessen the paroxysms in the course of three weeks, and therefore other medicines were substituted for it. In twenty-five of the remaining fifty-four cases, the severity of the paroxysms was greatly lessened by the end of a fortnight, and the whooping had ceased before the expiration of a month. In eleven it continued five weeks, in nine of them six weeks, in six it did not cease before the end of seven weeks, and in three it continued into the eighth week. In no instance did the disease persist after the expiration of the eighth week.

These results are sufficiently striking, and contrast most favorably with the ordinary plan of treatment. Indeed, they seem to indicate the exercise of a directly remedial influence, and to suggest the idea which has more than once forced itself upon me, that the disease may be kept within very narrow bounds, and in many cases may be almost limited to the duration of an ordinary attack of bronchitis.

Thus, then, until the end of the year 1858, when I discovered how readily children tolerate large and increasing doses of belladonna,[1] I trusted exclusively to sulphate of zinc. Since that time I have combined the zinc with belladonna, and have obtained even more satisfactory results. The whooping has been controlled in a remarkable degree, the severity of the disease has been much diminished, and, indeed, if further observation verifies my hitherto limited experience of this mode of treatment, it may be fairly stated that zinc and belladonna, judiciously employed, are capable of reducing the average duration of an uncomplicated attack of whooping-cough to three weeks or a month.[2]

My plan of proceeding is as follows: During the catarrhal stage of the complaint, the patient is kept in a warm and equable atmosphere; the diet is limited to milk and beef-tea or broth; the bowels are regulated by mild aperients; salines and ipecacuanha or antimony are prescribed; and mustard poultices, or even blisters, are applied, if necessary, to the chest. Leeches are rarely employed, and bleeding from the arm is never practised. As soon as the whoop declares itself, a draught is given every three or four hours, containing a grain of sulphate of zinc and a sixth of a grain of extract of belladonna to two drachms of syrup of orange, in from two to six of water, and an additional grain of the sulphate of zinc, and an additional sixth of a grain of

[1] See a paper of mine in vol. xlii of the "Med.-Chir. Trans."
[2] See "Lancet" for July 28, 1860, p. 85.

belladonna, are added to each dose daily, or every alternate day, until the quantity taken daily amounts to from six grains to a drachm of zinc, and from two to six grains of the extract of belladonna, according to the age of the patient.[1] To children under a twelvemonth old I have never administered more than ten grains of the zinc and two grains of the belladonna, daily, which were given in doses of a grain and a quarter of the zinc and a quarter of a grain of belladonna every three hours; whilst for children of eight or ten years of age I frequently prescribe half a drachm, or two scruples of the zinc and six grains of belladonna. If the dose be gradually and cautiously increased, the medicine will not occasion sickness; and as it neither heats nor excites the patient, it may be given as soon as the true nature of the complaint is ascertained. Its administration, however, need not preclude the exhibition of other remedies; and if there be feverish heat of skin and persistent quickness of breathing, indicating inflammation of the lung, or if the bronchial flux be great, and oppresses the breathing, it will always be prudent to have recourse to auxiliary measures. In the former case it is my practice to administer the vinum antimonii in doses varying with the patient's age; in the latter, if sickness does not occur spontaneously, to order a mixture containing the vinum ipecacuanhæ, of which a dose is to be taken every evening sufficient to cause vomiting. Nothing unloads the air-passages so thoroughly, promotes easy expectoration, and gives so much relief as free vomiting, and nothing, I believe, conduces more directly to the safe and favorable progress of the disease. Nevertheless it is essential to guard against the production of sickness by the sulphate of zinc, as, if nausea is once excited by its agency, the stomach will thenceforth refuse to tolerate it, and its use will have to be abandoned. Therefore it is prudent to administer it alone, in the manner recommended, and if necessary, to give the vinum antimonii as a cough-drop, with a little extract of conium or syrup of poppies.

Various remedies have been recommended by authors, and some have been vaunted as specifics. Amongst the former may be mentioned the whole class of sedatives and antispasmodics, including opium, henbane, hemlock, and lettuce, assafœtida,

[1] The only precautions necessary to be observed in prescribing full doses of belladonna to children, are—1st, that the dose of the medicine be *gradually* increased; and, 2dly, that the quantity given daily be administered in divided doses at intervals of not less than three hours. The mere occurrence of dilatation of the pupils need not be considered a bar to its exhibition. For full information respecting the action and mode of administering full doses of this drug, reference may be made to my paper on the subject published in vol. xlii of the "Med.-Chir. Transactions."

valerian, musk, and camphor, chloroform, and hydrocyanic acid. Amongst the latter, alum and nitric acid. The former are sometimes useful, if judiciously administered, just as they would be under similar circumstances in any other disorder; but they certainly do not materially hasten the period of convalescence, and do not deserve any special notice in connection with this disorder. Of the latter, I know little, except by repute. Alum is a very popular remedy, and appears, from the concurrent testimony of many intelligent observers, to mitigate the disease and shorten its duration. Nitric acid, in like manner, has many advocates, who assert that the disease yields rapidly to its influence. Dr. Gibb in this country, and Dr. Arnoldi, of Montreal, may be named as among those who have spoken of it in the highest terms, and who assert that, if given in full doses (half a drachm to two drachms daily), it subdues the disease in the course of three weeks. I am unable, from personal observation, to bear witness either for or against it in more than a very limited number of cases; but in two of these its beneficial effects were well marked, and the reports which have reached me of its efficacy are so encouraging as to induce me to consider it worthy of further trial. If the statements made respecting it are borne out by more extended observations, it must be regarded as a valuable method of treatment. Bromide of ammonium has also been recommended by Dr. Gibb, and garlic, cochineal, cantharides, digitalis, and many other remedies, have attracted attention at one time or another; but they have not obtained the confidence of the profession, and there is not sufficient testimony in their favor to lead to their employment in preference to the medicines already mentioned. Counter-irritation to the chest and spine has been much extolled by many writers, and whilst there are few practitioners who do not employ stimulating embrocations during the spasmodic stage of the complaint, some[1] have gone so far as to recommend that one or two blisters be applied to the nape of the neck. Certain of the applications which have been recommended, such as the tartar emetic ointment and croton oil liniment, are scarcely justifiable in early childhood, and do not, I believe, give greater relief than the less formidable compound camphor and soap liniments, or turpentine fomentations. The liniment which has appeared to me most serviceable consists of equal parts of the chloroform and the belladonna liniments of the British Pharmacopœia; but I doubt very much whether any endermic treatment has a material in-

[1] See a paper by Mr. Holl in the "Med. Times and Gazette," April 30, 1859.

fluence over the duration of the disease. It may, and I believe it does, mitigate the paroxysms; but I question whether it renders them less frequent, or in any way prevents their recurrence.

Chloroform inhalations have been proposed with the view of shortening the paroxysms and lessening their severity; and there cannot be a doubt that chloroform so employed does have the desired effect. But may not its constant use be productive of mischief? If employed at each recurring paroxysm, it is difficult to conceive how it can fail to give rise to unpleasant symptoms. Indeed, it appears to me that this species of inhalation cannot be regarded as a *curative* agent, and that its use should be restricted to very bad cases, in which it may be employed occasionally with the view of checking the spasm temporarily, and so of securing sleep and economizing the strength of the patient. A few whiffs taken when a paroxysm is commencing allay the spasm in a remarkable degree, and by shortening the duration and violence of the cough contribute materially to prevent exhaustion.

Mr. Atcherley, of Liverpool, reports[1] that he has seen great benefit derived from the inhalation of nitrous fumes generated in the room which the patient occupies, by the deflagration of bibulous paper steeped in a solution of nitrate of potash. The paroxysm appears to be thereby shortened, and the spasm lessened. I have not had the opportunity of testing its virtues in whooping-cough, but it often proves serviceable in spasmodic asthma; and as the remedy is simple, and can be employed without difficulty day and night, it must, if efficacious, prove extremely valuable.

The inhalation of the vapors disengaged during the distillation of coal-gas is another remedy, which comes to us with strong recommendations from France. It is said to be productive of excellent results at all periods of the disorder, and to be perfectly devoid of danger, even to the youngest patient. Personally I have had no experience of this treatment, for inquiries which I instituted at our largest gas-works led to the discovery that the mode of distilling coal-gas in France is very different from that usually employed in England, and that whereas the products of the distillation in France consist principally of ammoniacal gases and tar vapors, the products of the process generally employed in England are of a totally different description, and are quite unfit for respiration. Knowing, however, how closely carbolic acid is allied to some of the products of coal-tar distillation, it occurred to me to try the effects of air charged with that remedy, inhaled through one of Godfrey and

[1] See "Med. Times and Gazette," Feb. 26, 1859.

Cooke's inhalers. At present I have had so few opportunities of testing its action, that I hesitate to recommend its employment too strongly; but in the few instances in which I have made use of it the symptoms of the disease have been modified so speedily and so favorably, that I shall certainly give it a further trial.

The inhalation of atomized fluid charged with various medicinal substances is another mode of medication which may be advantageously practised in many cases of whooping-cough. At present my experience of its influence on the course of the disease is too limited to justify any opinion as to its efficacy, but in three cases in which a solution of carbolic acid combined with belladonna was pulverized by Siegle's apparatus, and the atomized fluid inhaled, such speedy alleviation of the symptoms ensued, that I am inclined to regard it as one of the most potent means we possess of combating this distressing malady.

The application of a strong solution of nitrate of silver to the pharyngeal mucous membrane, and to the orifice of the glottis, is another remedy which has been thought to lessen the tendency to spasm. In some instances undoubtedly it exercises a beneficial influence, but in many others it fails in affording the slightest relief, and therefore, whilst on the one hand I would not join with those who indiscrimately deprecate its use, I would not on the other have recourse to its employment except in appropriate cases. When, as often happens, the throat is relaxed, and the mucous lining of the pharynx is congested and covered with unhealthy secretion, the stimulating effect of the application induces a more healthy action, and lessens the irritability of the parts; but when the throat is in a healthy state, local stimulation is not needed, and fails to mitigate the severity of the spasm. In short, its use should be restricted to those cases in which it would be applicable if bronchitic irritation existed irrespective of whooping-cough.

Towards the close of the disease, when all febrile symptoms have subsided, quinine and iron are often useful adjuncts to sulphate of zinc, and change of air has been found of the greatest service; indeed, under ordinary treatment it appears in some instances to be indispensable for the entire removal of the tendency to whooping, and the restoration of health. A second change is sometimes beneficial, even without reference to the nature of the locality.

One word must be added in respect to the general management of the patient. It is too much the habit to disregard those precautions which are usually taken in cases of bronchitis, and the little patient is not only not dieted and not confined to his room, but is even permitted to go out of doors during the catarrhal stage of the complaint. Hence, I am satisfied, the frightful mortality from whooping-cough which stands recorded in the reports of the Registrar-General. In all cases during the continuance of catarrhal symptoms the diet should be restricted as in ordinary bronchitis, and in winter-time it is essential not only that the patient should not leave the house throughout the whole course of the disorder, but that the temperature of his apartment should be at least 60°, and kept so as uniformly as possible, and that he should be protected from cold by wearing flannel next the skin. Even in summer-time he should be strictly guarded from exposure to cold throughout the catarrhal stage of the complaint, and should not be permitted to leave the house until the severity of the complaint be overpast. So important is this precaution to the patient, that change of air should not be sought until towards the close of the disease.

Bronchitis secondary to Gout, Fever, and other Constitutional and Blood Disorders.

There are few disorders to which the human frame is liable which do not lead occasionally to the production of bronchitis. The altered and poisoned blood of typhus fever, of measles, and other eruptive disorders, of gout and rheumatism, albuminuria and diabetes, of scurvy, secondary syphilis, and many other maladies, can hardly fail, sooner or later, to produce irritation of the bronchial mucous membrane. Hence arises a morbid condition of the bronchial secretion, with cough and expectoration—in other words, bronchitis. The precise nature and extent of the pulmonary affection varies with the condition of the blood; so that, whilst in one instance excessive pulmonary congestion is a marked feature of the case, the mischief in another is limited to mere irritation and slight congestion of the mucous lining of the air-passages. In either case, however, the pulmonary mischief is the consequence of a specific cause of irritation, to the removal of which our chief efforts must be directed. It is vain to bleed and to give tartarized antimony in the hope of getting rid of the congestion ; the only rational and successful treatment is the employment of counter-irritation, and the exhibition of slight sedatives and expectorants, to mitigate the severity of the cough and facilitate expectoration, and of tonics to support the general health during the time which must elapse before the primary disease can be got rid of. The remedies calculated to subdue the primary disease are those which exert the greatest influence on its secondary effects. Thus in bronchitis connected with a syphilitic taint, iodurated inhalations and iodide of potassium, or biniodide of mercury in conjunction with cinchona, sarsaparilla, and cod-liver oil, often prove our most serviceable agents; whereas, when bronchial irritation occurs in conection with gout, our chief reliance must be on vapor or hot alkaline baths, with colchicum, soda, magnesia, and similar remedies. Practically, therefore, in the cases under consideration, very little treatment, save counter-irritation, is required specially for the relief of the cough; our chief aim must be to remove the primary disease and invigorate the patient, and meanwhile to keep up counter-irritation by means of mustard poultices, turpentine fomentations and blisters. Soothing, or possibly stimulating expectorants in the form of a cough-drop, are sometimes useful, but even a linctus is not always needed.

Narrowing or Obstruction of the Bronchi.

Before quitting the subject of disease of the bronchial tubes, it may be well to point out the nature of the causes which produce narrowing or complete obstruction of the bronchi, and to trace out the pathological effects and physical signs to which such obstruction gives rise.

Narrowing or obstruction of the smaller bronchi is a phenomenon of common occurrence, and may depend on causes operating within the air-tubes, or on causes which exert their influence from without. Amongst the former may be mentioned thickening of the mucous membrane, such as is met with in many forms of bronchitis, the retention and accumulation of viscid tenacious mucous secretion, plastic exudation on the surface of the membrane, as often occurs in diphtheria, and the depósit of tubercle or cancer; amongst the latter may be named the pressure of tuberculous and cancerous deposits, the contraction of plastic exudation-matter, whether in the lung-tissue itself, or on the surface of the pulmonary pleura, and the pressure of enlarged bronchial glands, and of aneurismal and other thoracic tumors. The larger the tubes, the less liable they are to become wholly obstructed; but masses of enlarged bronchial glands, and aneurismal, encephaloid and other tumors in the thorax, do sometimes exert sufficient pressure to obstruct even a main bronchus, and frequently give rise to their perforation by ulceration.

Whatever the immediate cause of complete obstruction, the result is invariably the same, namely, collapse of that portion of the lung to which the obstructed bronchus leads. Thus obstruction of the main bronchus will occasion collapse of the entire lung, whilst obstruction of one of the smaller tubes will be followed by collapse of correspondingly limited extent. In either case, emphysema commonly results. In the one it occurs in the opposite lung, which expands under the influence of atmospheric pressure to assist in filling up the space left unoccupied by the collapsed lung;[1] in the other, it arises in the immediate vicinity of the collapsed portions of the lung under the operation of the same exciting cause.

The physical signs of obstructed bronchi are—first, dulness on percussion over the seat of obstruction and over the collapsed portion of the lung, unless the dulness in the latter situation be masked by emphysema, which commonly arises in its immediate vicinity; secondly, weakness or deficiency of the respiratory sounds, with greatly prolonged expiration, and occasional sonorous and sibilant rhonchi; thirdly, entire absence of respiration over the collapsed portions of the lung; fourthly, deficiency or entire absence of vocal resonance and vocal fremitus. Extreme prolongation of the expiratory sound over a limited space is perhaps the most characteristic sign of a partially obstructed bronchus.

Difficulty of breathing, sometimes accompanied by stridor, is the general symptom which principally attracts attention, and that varies in severity, according to the extent of the obstruction and the rapidity of its occurrence; but the rapidity with which the obstruction takes place appears to exercise a greater influence in the production of dyspnœa than the extent of the obstruction; so that the sudden obstruction of a moderate-sized bronchus will induce a paroxysm of labored and suffocative breathing, whilst obstruction of even a large tube occurring gradually, though necessarily attended by considerable frequency of respiration, will not be accompanied by any striking accession of dyspnœa, or notable distress of breathing. The former is well exemplified in many cases of bronchitis, the latter in cases of intrathoracic tumors. The effect of partial obstruction and narrowing of the tubes in producing prolonged expiratory murmur is seen fully developed in chronic bronchitis.

The treatment of obstructed bronchial tubes must depend on the nature of the obstructing cause, and must be in strict relation to it. Practically, however, the whole subject of obstructed bronchi acquires its interest from the light which it may throw upon the diagnosis of intrathoracic tumors; and unfortunately in these cases remedial agents are usually of little avail. In a diagnostic point of view the importance of the subject can hardly

[1] See a case which I reported in the "Trans. Path. Soc. Lond.," vol. xi, p. 15.

be overrated, as will be seen by reference to the portion of this work where disease of the bronchial glands is discussed,[1] as also to the section on aneurismal tumors.

CHAPTER IV.

ON PULMONARY CONSUMPTION.

Pulmonary consumption is the commonest and most fatal malady to which the human race is subject. Occurring at every age and in every rank of life, it selects the most beautiful and most gifted as its victims, and entails months or years of weary suffering. What need then for further incentive to a careful study of its history—to a diligent inquiry, as to whether something may not be done to prevent its occurrence, retard its development, or arrest its progress?

With a view to the attainment of precision in our ideas, it will be desirable to commence by an inquiry into the nature of consumption, and into the part which is played by *pulmonary* disease in producing the fatal termination of the disorder.

Briefly, then, it may be stated, that consumption is essentially a constitutional malady, and is intimately connected with perverted nutrition and imperfect sanguification. When the blood has become vitiated up to a certain point, an unorganized material, which has been termed tubercle, is deposited in various parts of the body. This material is not deposited indifferently in all parts of the body; it affects certain parts in preference to others, and none so frequently as the lungs.[2] Hence, the disease is often designated by the terms "phthisis pulmonalis," "pulmonary consumption," "tubercular disease of the lungs"—terms which, though expressive of one of its commonest and most salient features, ignore its constitutional origin, and lead to the erroneous impression that it is a purely local disease of the lungs. Practically this is of little importance, provided the nature of the malady is understood; but inasmuch as the lung af-

[1] See pp. 410–12 of this treatise.

[2] Thus, of 566 tuberculous patients examined in the dead-house of St. George's Hospital during the ten years ending December 31, 1850, no less than 517 had tubercles in the lungs. In 49 only were tubercles absent from the lungs. See St. George's Hospital "Post-mortem and Case Books," and the analysis of their contents in the "Decennium Path.," chap. iii.

fection is one only of the local results of a great constitutional disorder—a complication which, though of frequent occurrence, and a serious aggravation of the patient's danger, is not an essential part of the disease, or necessary to its fatal termination—it follows that the more appropriate titles are "consumption," "phthisis,"[1] or tuberculosis, which do not convey the erroneous impression of the disorder being of pulmonary origin.

In what, then, does the constitutional derangement consist? It is not easy to give a definite answer to this inquiry, but a few points may be indicated, which will serve to throw some light on the subject. Chemistry has shown that the blood in consumption contains an excess of water in relation to its solid constituents, that it exhibits a relatively increased proportion of albumen, is deficient in red corpuscles, is less alkaline than in health, and is otherwise altered in character,[2] while the general tendency of combined chemical analysis, microscopical research, and clinical observation, has been to prove not only that the nutrient fluid is vitiated or impoverished, but that its vitality is below the healthy standard. The same inherent defect or infirmity is traceable throughout the entire organization of the body. The structural peculiarities which characterize the disorder are indicative of imperfect cell-formation and extreme delicacy of the tissues; whilst the functional derangements in their ever-changing variety afford evidence of the weakness of the different organs, and of their liability to become disordered on the slightest exciting cause. So that, although consumption is attended by local mischief, which produces corresponding local symptoms, there is not in the whole catalogue of human ailments, a malady which more strictly deserves to be classed among constitutional disorders.

But what is the starting-point of this derangement? A variety of theories have been advanced on the subject. Some persons have referred the disturbance to imperfection of the primary processes of digestion;[3] some to secondary mal-assimilation and mal-nutrition of the tissues; some to a morbid condition of the lymph;[4] some to a specific poison in the blood;[5] some to want

[1] Signifying a wasting or consuming away.

[2] For the detailed results of the analyses of the blood of consumptive patients by Becquerel and Rodier, Andral and Gavarret, Fricke, Lehéritier, Nicholson, Simon, Karl Popp, Elsner, and others, see Ancell on "Tuberculosis," pp. 8–10 and 639–642.

[3] Bennet on "Pulmonary Tuberculosis."

[4] See Simon's Lectures on "General Pathology," 1850.

[5] "Thoughts on Pulmonary Consumption," by Dr. Madden; also M. Villemin, "Gazette Hebdomadaire," 1865–6.

of power in the organic nervous force;[1] some to imperfect respiratory action; some to deficient oxygenation of the blood; and others, again, to a variety of causes, which it is needless to enumerate.[2] But, in truth, these theories, however ingenious, have one fault in common. They are all too exclusive, and hence inconsistent with many acknowledged facts! Even were it not so, they would only serve to throw us one step further back in the line of causation. Admitting, for the sake of argument, that each or any one of them may be correct, we still must ask, what is the cause of the deviation from health to which the theory points? or, in other words, what is the essence of the disease?

It has been already stated that the blood is vitiated, altered in character, and of low vitality; that the tissues, therefore, are delicate, and the organs weak and easily deranged. Pathological chemistry and microscopical research enable us to go even a step beyond this, and afford ground for believing that in a portion of the recently formed blood—the "young blood," as Mr. Simon terms it—the proteiniform and oleaginous principles are not perfectly elaborated, and consequently are less organizable than those which are formed in health, more apt to become granular, more prone to aggregation into masses; that the fatty matter is of a lower grade than healthy fat, and approximates to the nature of cholesterine; and that these deviations from the healthy type, both of the blood and of the blastema from which the tissues are formed, produce the various observed anomalies in the growth and development of the body, and give rise to products which constitute the pabulum of tubercle. Beyond this we are unable at present to proceed, for there are no trustworthy data for our guidance. Nay, more, it seems improbable that we shall ever arrive much nearer to the truth than the point here indicated. The revelations of science may ultimately throw additional light on the precise character and constitution of the blood at different stages of the disease, but they can never bring to light the hidden power which regulates molecular aggregation and cell development, modifies the character and action of the corpuscles of the blood, determines the transformation of its various constituents, and stamps the different tissues with their respective characters. This vital or formative power, which operates in accordance with fixed laws impressed upon us by our Creator, must ever be beyond man's compre-

[1] Dr. Copland on "Consumption."

[2] For a detailed account of various theories which have been advanced respecting consumption, see Ancell, lib. cit., pp. 548-579.

hension; and all that can be surmised respecting it is, that so long as certain conditions exist, it promotes the elaboration of healthy blood, and leads to healthy assimilation, healthy nutrition, healthy excretion—in a word, to the maintenance of health; that any alteration in these conditions, by giving a fresh direction to, or otherwise interfering with, the exercise of this power, modifies the results which it produces in our organization, leads to altered chemical action, and so to changes in the character of the blood; and that this derangement, which issues under certain circumstances in the formation of tubercle, constitutes the essence of consumption. Why disturbance of the system should tend in certain instances only to the production of the peculiar chemical changes whereby tubercle is generated, is one of the mysteries which we shall probably ever fail to unravel. The fact is akin to many others of a similar nature, and may be illustrated by the variations in complexion, visage, form, stature, and constitution observed amongst children of different families, and even amongst children born of the same parents. It is explicable only by supposing that there is an inherent difference in the formative power implanted in each individual, and a corresponding difference in the abnormal chemical combinations to which, when deranged, this power gives rise, and thus it is more prone in some persons than in others to promote those peculiar abnormal combinations of the constituents of the blood which result in the formation of tubercle. Observation has led to the belief that, in some instances, repression of the skin's action may be the cause which first upsets the balance of the machine; in some impaired respiratory action, or the respiration of an unrenewed or vitiated atmosphere; in some defective action of the liver; in some imperfect digestion or primary mal-assimilation; in some secondary mal-assimilation; in some an inherited deficiency of organic nervous force; in some a derangement or exhaustion of nervous energy, whether produced by mental or physical causes; in some the introduction of a morbid poison, such as that of syphilis, into the blood; and, in others, disturbance of the uterine function, or the occurrence of any morbid action which serves to depress the system, and derange the general health. But, whatever the disturbing cause may be, the conditions which are invariably antecedent to the formation of tubercle, are defective vital formative power, an impoverished blood, imperfect assimilation and general mal-nutrition: these constitute the essence of consumption. This brings us to a consideration of the predisposing and exciting causes of the disease.

It will be obvious, from what has been already stated, that

amongst these must be classed a great variety of agencies. The inheritance of a scrofulous or tuberculous tendency, long-continued exposure to cold and wet, improper or insufficient diet, irregularities of living, insufficient exercise, unhealthy and sedentary occupations, the constant inhalation of a vitiated atmosphere, the depressing influence of an unsuitable climate or locality, the indulgence of vicious habits, excessive sexual intercourse, over-protracted suckling, and long-continued grief and mental anxiety, are amongst the causes which most powerfully predispose to the inroad of consumption; whilst irritation and congestion of the lungs, whether produced mechanically as in the case of the Sheffield fork and needle grinders, or through the agency of influenza, catarrh, and other diseases which, in the absence of a predisposition to consumption, would pass away without producing the development of that disorder, may be regarded as its more common exciting causes.[1]

Without going into the operation of these causes, the influence of which has been fully discussed by several authors, I feel bound to remark on some few points on which erroneous views are generally entertained.

Common observation has stamped consumption as an hereditary disorder, and amongst the public the opinion is entertained that in a vast majority of cases the disease is traceable to an hereditary taint. "We have no consumption in our family," is considered a conclusive reply to the inquiry, whether certain symptoms may not be attributable to that disease. On the other hand, a large portion of the profession has been unwilling to admit the extended operation of an hereditary tendency; and statistics have been published by various observers to prove that this influence does not operate in more than 24 or 25 per cent. of the cases met with in practice. Thus, the records of the Brompton Hospital[2] are supposed to show an hereditary tendency in 246 only out of 1010 cases, or in other words to prove its influence in only 24.4 per cent. But the truth I believe lies between these two extremes. Without denying that consumption may be acquired, or that it is acquired in many instances through the agency of causes conducive to nervous exhaustion, an impoverished blood, and feeble health, I do not hesitate to assert that an hereditary predisposition exercises a far more extended influence than is indicated by the Brompton Hospital report. The 24.4 per cent.

[1] For evidence respecting agencies such as these in exciting the disease, see Sir James Clark's work on "Consumption," and Mr. Ancell's treatise on "Tuberculosis."

[2] See first report of the Brompton Hospital for Diseases of the Chest.

which is there referred to relates only to a predisposition acquired from an avowedly consumptive parent, and does not include the many cases in which the tendency has been inherited from a parent who, though he may not have exhibited at the time the report was drawn any decided symptoms of consumption, has come nevertheless from a consumptive stock, and eventually may have died consumptive, or who again, though not himself consumptive, may have lost brothers and sisters of consumption, and may have inherited and transmitted a consumptive disposition.

The great discrepancies observable in the opinions expressed by authors relative to the hereditary transmission of the disease, are explicable only by assuming a difference in the mode in which the inheritance of the tendency is calculated;[1] and, as the matter is one which has important bearings on many social questions, I have taken considerable trouble to arrive at a correct conclusion. The result will be seen by an inspection of the subjoined table, which shows that if the inquiry be limited to the direct transmission of the disease from either parent, the proportion which the cases of inherited disease will bear to the whole mass of cases of consumption, will be about 25.7 per cent.; whereas, if the existence of the disease in either grandparent be considered as evidence of the transmission of the disease to the grandchildren, the proportion will rise to 43.6 per cent.; and if the predisposing influence exhibited by the death of uncles or aunts from consumption be included in the calculation, the proportion will rise to 59.5 per cent. Thus the various conflicting statements may be reconciled, and a trustworthy insight obtained as to the bearing of an intermarriage with a consumptive family.

[1] Thus, M. Louis names one-twelfth as the proportion of cases in which a consumptive disposition is inherited; M. Piorry one-fourth as the proportion of his hereditary cases; the Brompton Hospital about one-fourth; Dr. Cotton rather less than two-fifths; M. Briquet two-fifths; Dr. Copland nearly one half; M. Portal two-thirds; M. Rufz somewhat more than three-fifths; M. Ruysch four-fifths.

TABLE exhibiting the existence or non-existence of an hereditary tendency in 385 cases of consumption. The cases were taken indiscriminately from hospital and private practice, the only cases excluded being those in which the family history could not be ascertained with tolerable certainty. In every instance in which a parent died consumptive the tendency is referred to that head; in cases in which the parents escaped, but a grandparent died consumptive, the tendency is referred to that head. In no instance is an hereditary transmission assumed in the collateral relationship of uncles and aunts, unless two or more uncles or aunts in the branch of the family had fallen victims to the disease.

	Males.	Females.			
Father consumptive,	27	21	99 or 25.7 per cent.	168 or 43.6 per cent.	229 or 59.5 per cent.
Mother do.,	24	16			
Father and Mother do.,	6	5			
Grandfather (paternal) do.,	11	10	69 or 17.9 per cent.		
Grandmother do. do.,	10	8			
Grandfather (maternal) do.,	8	9			
Grandmother do. do.,	8	5			
Uncles or aunts (paternal) do.,	14	15	61 or 15.8 per cent.		
Uncles or aunts (maternal) do.,	17	15			
No consumptive taint in either of the above relations,	87	69	156 or 40.5 per cent.		
Total,	212	173			

It is obviously impossible to obtain any data for an opinion as to the proportion of persons who, though born of consumptive parents, pass through life without exhibiting any symptoms of the disease. The experience of life insurance offices shows that it is small, and this conclusion must be indorsed by all who have had an extended sphere of observation. Whole families are sometimes swept away, and even under favorable hygienic circumstances, a few members only ordinarily escape. M. Roche has gone so far as to assert that the children of consumptive parents almost necessarily prove victims of the disease; and if the statement were restricted to those persons whose parents were both consumptive, my own experience would have led me to concur with M. Roche. But the proportion of cases in which a transmitted tendency to the disease is developed in the children is not so large when one parent was healthy. In that case, as far as my observation has gone, the disease is developed sooner or later in about three-fifths of the offspring, and is especially prone to be developed in the children of that sex to which the consumptive parent belonged; so that the daughters of a consumptive mother will often fall victims to the disease whilst their brothers escape, and the sons of a consumptive father whilst their sisters escape. Whether the sex of the grandparent from whom the disease is transmitted has any influence in determining a proclivity to the disorder in the corresponding sex amongst the grandchildren, is a point which I have been unable as yet to ascertain, but the extraordinary facts lately elicited in Norway and Sweden respecting the hereditary transmission of leprosy[1] renders such an influence extremely probable.

Another fallacy which commonly prevails, is respecting the age at which consumption occurs. Amongst the non-professional public, youth is considered the harvest-time of consumption, and middle age the extreme limit of the period within which the whole crop is garnered; and even in the ranks of our own profession there is too frequently a disposition to regard the disease as occurring almost exclusively in early or middle life. But nothing can be further from the truth, or more utterly at variance with modern statistics. The post-mortem records of St. George's Hospital show that a large percentage of persons who die over sixty years of age are affected with pulmonary consumption,[2] and the valu-

[1] See "British and Foreign Med.-Chir. Review," for April, 1858, pp. 332–346.

[2] The subjoined table exhibits the number and age of the patients who died in St. George's Hospital during the ten years ending December 31, 1850, distinguishing those in whom tubercle existed in the lungs, and the percentage which those cases bear to the total number of cases examined. See "Decennium Path."

—	Number of patients examined.			Tubercles in the lungs.				Percentage of pulmonary tuberculosis.		
	Males.	Females.	Total.	Males.	Females.	Sex unknown.	Total.	Males.	Females.	Total.
From birth to 15 inclusive,	94	60	154	20	17	0	37	21.2	28.3	24.0
From 15 to 30,	377	259	636	128	82	1	211	33.9	31.5	33.1
" 30 to 45,	472	179	651	120	35	0	155	25.4	19.5	23.8
" 45 to 60,	299	139	438	66	11	0	77	22.0	7.9	17.5
Above 60,	109	58	167	10	1	0	11	9.1	1.7	6.5
Age unknown,	74	37	111	14	10	2	26	18.9	27.0	23.4
Total,	1425	732	2157	358	156	3	517	25.1	21.3	23.9

able reports of the Registrar-General confirm in a striking manner the extended operation of a tuberculous disposition, and prove that, in relation to the number of persons living at the respective ages, the mortality from consumption does not vary materially between the ages of fifteen and seventy, but is actually somewhat less between the ages of ten and forty than it is between the ages of forty and seventy.[1]

A still more prevalent, and a most mischievous error relates to the predisposing agency of cold and wet, and atmospheric vicissitudes. Nothing is more common than to hear consumption attributed to the effect of wet and cold, the inclemency of the season, or the variableness of the climate in which the person is resident; and nothing, unhappily, is more consistent than the practice which is based on this supposition. Incased in warm clothes, and shut up in heated rooms from which every breath of fresh air is carefully excluded, the unfortunate victim to this prevalent superstition becomes rapidly enervated and falls an easy prey to his disorder. But careful observation, which is amply corroborated by statistical records, proves incontestably

[1] The subjoined table is constructed from the Registrar-General's returns for 1847, in which the deaths from consumption in all England and Wales at the several epochs of life are given in connection with the number of persons of each specified age living in the middle of that year:

Years of age.	Mortality from phthisis.			Estimated population in 1847.			Ratio of the deaths from phthisis to the estimated population.
	Males.	Females.	Both sexes.	Males.	Females.	Both sexes.	
Under 5	2642	2559	5201	1,112,027	1,136,766	2,248,793	1 in 432.3
5—10	780	876	1656	1,020,042	1,022,576	2,042.618	1233.4
10—15	910	1432	2342	942,534	915,115	1,857,649	793.1
15—20	2294	3232	5526	836,845	865,094	1,701,939	307.9
20—25	3521	3899	7420	774,542	888,360	1,662,902	224.1
25—30	2983	3683	6666	657,138	722,120	1.379,258	206.9
30—35	2373	3094	5467	604.706	646,972	1,251,678	228.9
35—40	2212	2545	4757	465,847	482,871	948,718	199.4
40—45	1847	1903	3750	466.338	486,047	952,385	253.9
45—50	1534	1404	2938	345,472	349,380	694,852	236.5
50—55	1261	1111	2372	328,848	351,469	680,317	286.8
55—60	1025	874	1899	202,956	217,199	420,155	221.2
60—65	749	715	1464	223,801	247,912	471,713	322.2
65—70	515	505	1020	129,211	149,231	278,542	273.1
70—75	253	246	499	111,365	129,120	240,485	481.9
75—80	128	109	237	59,546	69,089	128,635	542.8
80—85	41	26	67	33,295	42,745	76,040	1134.6
85—90	9	10	19	10,770	14,881	25,651	1350.0
90—95	2	3	5	2,662	4,333	6,995	1399.0
95 and upwards,	—	—	—	622	1,175	1,797	—
All ages,	25,083	28,234	53,317	8,368,914	8,755,174	17,124,088	1 in 321.0

that the pure air of heaven which God has provided for us to breathe, and the variations of temperature, to which in His all-wise providence He has seen fit to subject us, are not so noxious or productive of ill health as man in his ignorance has oftentimes asserted. No climate is more variable than ours, and none certainly is more healthy, as proved beyond dispute by the bills of mortality. Even if considered solely in its relation to consumption, England will bear comparison with any other equally civilized country. The subjoined statistics are decisive on this point,[1] whilst modern research has tended to show that the vicious habits of civilized life, the confined atmosphere of the dwellings, the want of sufficient active out-door exercise, the various depressing passions, and the exhaustion of mind and body resulting from the anxious struggles and vicissitudes of life, have far more influence than wet and cold in preparing the way for the inroad of consumption. Indeed, those classes of the community who lead an active out-door life, though necessarily exposed to wet and cold, are precisely those who are least subject to consumption.[2] Even when the disease is already developed or far advanced, the pernicious effects resulting from over-caution in respect to exposure to atmospheric vicissitudes are often painfully apparent. Nowhere is this more strikingly

[1] The following table, constructed as the result of his researches by Dr. Caspar, of Berlin (see "British and Foreign Review," July, 1847), exhibits the ratio of deaths from phthisis to the deaths from all causes in the following localities:

In Berlin, during	10 years,	there was	1 death	from phthisis to	5.7 deaths	from all causes.
Paris,	4	"	1	"	5.5	"
London,	2	"	1	"	6.2	"
Hamburg,	3	"	1	"	4.6	"
Stuttgard,	10	"	1	"	4.7	"
New-York,	11	"	1	"	5.	"
Philadelphia,	7	"	1	"	7.7	"
Baltimore,	8	"	1	"	6.7	"
Boston,	7	"	1	"	5.9	"

whilst from other sources we learn (see "Bullet. de l'Académie," Août, 1839), that at the civil hospitals in Rome 1 in 3.4 of the deaths from all causes is due to phthisis; in Naples 1 in 2.33.

[2] Thus Verhaegle reports that the deaths from consumption, in relation to the deaths from all causes, amount to 19 per cent. amongst the population of the interior, whilst the deaths from the same cause among seamen at Ostend are limited to 4.6 per cent. (see "Edinb. Med. Journ.," May, 1859, p. 1063); and Dr. Christison, to the same effect, tells us "that in Glasgow 385 persons in every 100,000 of the population die annually of consumption, whilst in Edinburgh 283 is the number; in the Highlands, 179; in the Lothians, a fine agricultural country, only 125; and in Berwickshire, where there are no towns with over 3500 inhabitants, there are only 105 deaths from consumption in 100,000;" and further, that the natives of the Western Isles of Scotland are almost entirely free from the disease. So, again, Dr. Livingstone, the African explorer, has shown that the native tribes of South Africa, who live an out-door life and know nothing of town employments and the confinement of civilized life, are free from the curse of consumption.

exemplified than at our large hospitals, where the consumptive out-door patient who, ill clad and often ill fed, in hot weather and in cold, on wet days as on dry, has come to the hospital twice a week for medicine, notoriously lives far longer than his brother who, more fortunate in being well fed and protected from the inclemency of the weather, is shut up in the equably heated wards of the hospital, and thus loses the bracing, invigorating stimulus of the fresh breezes of heaven.

Without further discussing the predisposing causes of consumption, we will pass on to the consideration of tubercle, the material which forms the characteristic anatomical element of the disorder.

Tubercle is a substance formed by a retrograde metamorphosis of an exudation from the blood. It is a material of imperfect organization and low vitality, unfit for the construction of new tissues, and therefore purely excrementitious. Sometimes it is deposited in one organ, sometimes in another, sometimes in all the organs of the body; but in every instance it is a foreign body, and damages the organs in which it is imbedded. In appearance it varies greatly, the varieties observed being referable in part to differences of structure, but partly also to the age and extent of the deposit. It may exist in two distinct forms. In the one it produces minute, firm, semi-transparent granules of a bluish-gray color, very different in appearance from ordinary tubercle, and known as "gray miliary tubercles." These may be sparsely scattered through the lungs, or may crowd their entire tissue. In most instances they are isolated, and sometimes are so minute as to be scarcely discernible by the naked eye, but more commonly they are about the size of a millet-seed, and occasionally, when several of them coalesce, they may form a mass as large as a small pea. Usually firm and of a semi-cartilaginous hardness, they are occasionally softer and less resistant, and admit of being crushed between the fingers, whilst not unfrequently they are intermixed with black pigmentary matter, or else are surrounded by it.

The other and more common form of deposit is that known under the title of "crude yellow tubercle." This form of tubercle is quite opaque, and varies in color from a dirty white to a drab or a bright buff, and in consistence from that of firm tough cheese to that of diffluent cream cheese. In many instances it exists in roundish or irregularly-shaped, isolated masses; in others, a large portion of the lung is infiltrated with it; and in others, again, it forms isolated masses in one part of the lung, and in another infiltrates large portions of the lung-tissue. The isolated deposits may vary in size from that of a millet-seed to that of a hazel-nut or a large walnut,—the larger masses being formed by the aggregation of the smaller deposits, which give rise to compression and atrophy of the intervening tissue. Thus the size of the tuberculous masses is determined by the extent of lung-tissue implicated in the mischief, and their shape by the form of the spaces in which the deposits take place. Not unfrequently the deposit is soft and pultaceous in one part of the lung, firm and tough in another, extremely friable in yet another; or, if the deposit be old, it may have undergone calcareous or cretaceous transformation, and may prove of stony hardness. In another class of cases, the tuberculous matter becomes mixed up with dark-colored carbonaceous deposit, which not only colors the tubercle, but gives a bluish-black tinge to the pulmonary tissue.

Opinions have varied as to the precise relationship subsisting between these two forms of tubercular deposit. Laennec and his followers regarded the gray miliary granulations as nascent tubercles, the germs of crude yellow tubercle, whilst others have maintained that the two forms of deposit are essentially distinct, and not merely different stages of the same deposit. Even in the present day pathologists are not agreed on the subject.[1] My own opinion is, that they both originate in perverted nutritive action consequent on a tendency to tubercular disease, and though intimately allied in their nature, are each probably connected with a somewhat different condition of the system.[2] The frequent occurrence of extensive deposits of gray miliary granulations without any appearance of yellow tubercle, and of yellow tubercle without any trace of gray granulation, is of itself sufficient proof that the two forms of deposit are independent of each other, or, at least, that the gray granulation is not a stage through which yellow tubercle must necessarily pass. On the other hand, it is certain that both varieties occasionally coexist in the same lung, and even in the same parts of an affected lung; nay, more, specks of yellow tubercle may be sometimes observed imbedded in the substance of the gray granulations, as if the gray granulations may undergo degeneration and may be ulti-

[1] See Laennec (loc. cit.), Louis, Bayle, Rokitansky, Henle, Vogel, Lebert. A concise review of the different opinions entertained on the subject is given by Ancell, loc. cit., pp. 127–132.

[2] Rokitansky lends the weight of his authority to this view, but M. Villemin, whose experiments on the inoculation of tubercle have thrown considerable light on the causation of tuberculosis, draws no distinction between the two forms of deposit. Indeed, if the results he obtained are not due to some peculiarity in the animals on which he operated, it would appear that Laennec's suggestion is correct, and that the gray miliary is in many instances the earliest form of tubercle.

mately transformed into yellow tubercles. Further, the differences observed between these two varieties of tubercle disappear in great measure when recourse is had to chemical analysis and microscopical research. The microscope has shown that tubercle, whether "gray miliary" or "crude yellow tubercle," consists of granular matter, elementary molecules, and fat globules, and of so-called tubercle corpuscles, which are imperfectly formed cells of irregular and often angular shape, containing granules, but without any distinct nucleus. The relative proportion of these constituents varies greatly in different forms of tubercle, and to these variations are attributable the differences observed in the appearance of that deposit. In recent and firm tubercle, and especially in the gray miliary tubercle, the cell element prevails and fatty matter exists in the smallest possible quantity; whilst in soft yellow tubercle of low vitality, and in old tubercle which is undergoing degeneration and softening, the fatty and granular matters are in much larger quantity, and the cells fewer in number, if not altogether absent.[1] Epithelium, pus-globules, and portions of yellow elastic fibrous tissue, the products of disintegration of the lung, may also be visible. When calcareous or cretaceous transformation has taken place, earthy granules are seen of irregular form and size, and crystals of cholesterine are frequently present; whilst if pigmentary matter exists, it is seen as black molecules infiltrating the tubercle, or else is collected round it into small, irregular black masses.

Chemistry, equally with the microscope, has failed to point out any very great difference between "gray miliary" and "crude yellow" tubercle. Chemists are agreed in regarding tubercle as consisting of some modified protein compound or compounds with fat and earthy salts;[2] they disagree only as to

[1] It has been suggested by Dr. Cotton (loc. cit., p. 16) that these differences in the relative proportion of the constituents of tubercles are due not only to the age of the tubercle and the precise stage of its degeneration, but to the "degree of phthisis" which has produced it, and that to this, in a great measure, are attributable the varieties in the course and duration of the disease. I am strongly inclined to believe that the influence here referred to does modify the character of tubercles, but it will not wholly serve to explain the difference between the "gray miliary" and the "crude yellow" or scrofulous tubercles, which are probably connected with different states of constitution.

[2] Scherer gives the following as the result of an analysis of tubercles after the salts and all foreign matters had been carefully removed, viz.:

Carbon,	53.888
Hydrogen,	7.112
Nitrogen,	17.237
Oxygen,	21.767

This corresponds with the formula $C_{43}H_{35}N_6O_{13}$. Hence, tubercle may be

the precise character and relative proportion of these constituents in the specimens they have severally examined. The albuminoid matters, however, appear to exist in largest quantity in firm and recent tubercle; the fatty matters and the earthy salts—which consist principally of carbonate and phosphate of lime, with a small portion of some salt of soda—in old tubercle, or in tubercle which has undergone transformation subsequent to its deposition.[1]

Tubercle possesses one feature by which it is distinguished from most other modifications of deposit in the lungs, namely, that it is reproducible by inoculation. M. Villemin has shown[2] that when the anatomical elements of tubercle are introduced into any part of the body, miliary tubercles are developed in the lung, and he has thus succeeded in establishing an important distinction between true tubercle and syphilitic, fibroid and other nodular deposits which often occur in the lungs, and which not only give rise to the symptoms of phthisis during life, but are too often regarded after death as of a tubercular nature.

Tubercle, when once deposited, may long remain very nearly *in statu quo*, or may slowly or speedily undergo transformation. The precise nature and rapidity of the change will depend in part on the original constitution of the deposit, and partly on the existing condition of the patient's system. The gray miliary tubercle does not ordinarily break up or undergo softening; it loses its transparency, and ultimately shrinks into a dense, hard, tough mass. In certain instances, however, in which it becomes the seat of yellow tubercular deposit, or undergoes transformation into yellow tubercle, it loses this character, and is liable to soften and break up; in others it becomes the seat of slight earthy deposit.

Yellow tubercle, on the contrary, is prone to soften, the process of disintegration commencing sometimes in the centre of the tuberculous mass, but more commonly at its periphery.[3] Softening, however, does not necessarily occur in yellow tubercle. It usually takes place sooner or later if the patient's health con-

regarded as protein ($C_{48}H_{36}N_6O_{14}$) from which five atoms of carbon, one of hydrogen, and one of oxygen have been removed. (See Simons's "Chemistry," translated by Dr. Day, p. 479.)

[1] For full particulars respecting the chemical analysis of tubercle by Thénard, Lassaigne, Hecht, Boudet, Güterbock, Scharlau, Vogel, Lehmann, Lehéritier, Scherer, Preuss, Wood, Simon, Glover, and others, see Ancell, loc. cit., pp. 148–151. Also an analysis by Wright, "Med. Times," vol. ii, pp. 418–419.

[2] "Gazette Hebdomadaire," 1865-6.

[3] For a detailed explanation of the modes in which softening occurs, see Lebert, loc. cit.

tinues to fail; in which case the tissue entangled in the tuberculous mass or immediately surrounding the softened tubercle becomes inflamed and ulcerates, the softened scrofulous matter, mixed with portions of the disintegrated lung-tissue, finds its way into the air-passages, excites cough, and is then expectorated and got rid of, leaving cavities or vomicæ in the lungs. But if the patient's health improves, and the tendency to the disease is arrested, the tubercles may remain *in statu quo*, or may ultimately be absorbed, or may undergo calcareous or cretaceous transformation. After softening has taken place and the disintegrated tubercle has been removed by expectoration, the cavities which remain in the lung may cicatrize under the influence of returning health and increased vital power.

Many persons have argued against the absorption of tubercle, and have boldly asserted that its absorption is impossible. But there are no valid grounds for this statement. True it is that the general health and the tone of the system rarely improve so far as to render absorption of tubercle probable, and it is equally true that from its nature and position tubercle does not easily admit of absorption; but experience at the bedside and observation in the dead-house leave little room for doubt that its absorption may take place under favorable circumstances. Its gradual disappearance from the glands of the neck and other external parts of the body admits of ocular demonstration, and the stethoscope affords presumptive proof of its occasional absorption from the lungs; whilst its almost entire absence from the textures surrounding the cicatrices of old-healed vomicæ, and its occasional transformation into cretaceous or calcareous matter—which implies the disappearance of the liquid and animal matter of tubercle—complete the chain of evidence which induces me, in common with Andral, Carswell, and many recent pathologists, to regard its absorption as possible.

Tubercles may be deposited on the free surface of the mucous membrane of the air-cells or bronchi, or in the pulmonary tissue beneath it—a fact which has an important bearing on the causation of certain physical signs.[1] Very generally they are deposited in isolated masses, but in the infiltrated form of tubercular deposit they block up the air-cells and smaller bronchi, just after the manner of inflammatory exudation. They may be few in number, or extremely numerous, and may be distributed through every part of the lungs, though generally the seat of their de-

[1] Thus the deposit of tubercle in the submucous tissue gives rise, as I believe, to the more marked degrees of prolonged expiration, and leads to the remarkable phenomenon of dry clicking. (See pp. 142–145 of this treatise.)

posit varies with the age of the patient and the form of the malady. In cases of acute phthisis, and especially when the deposit is of the gray miliary variety, tubercles are apt to be disseminated through all parts of the lungs, and in children the same fact is frequently observed; but in chronic cases and in adults the deposits usually commence in the summit of the lungs, and become scarcer or altogether absent in the lower lobes. In a diagnostic point of view, this fact is of the utmost importance; but it must ever be borne in mind that exceptions are numerous, and that every possible variation may be met with. In most instances both lungs are affected to a greater or less extent, but the tubercle is deposited earliest, and the tubercular deposit is larger in amount, and in a more advanced stage of softening at their apices than in their middle or lower portions; so that when large, empty cavities exist at the summit of the lungs, smaller and more recent vomicæ are found lower down, and crude unsoftened tubercle or healthy lung-tissue at the base. But sometimes the middle or base of one or both lungs is alone or principally affected, and at others one lung may escape altogether, while the entire structure of the other is infiltrated with tubercle, or broken up into vomicæ. Sometimes, again, both lungs may be partially or extensively implicated, but the tubercles may have undergone degeneration and softening in the one lung, whilst they are hard and present all the characters of recent deposits in the other.

Andral, Carswell, Louis, and others, have asserted that tubercles are more commonly found in the left than in the right lung—a statement, which, if correct, would have been of diagnostic value. But careful observation in the wards and in the dead-house of St. George's Hospital has convinced me that this opinion has no valid foundation, and the statistics diligently collected from our register by Dr. Chambers show that the statement commonly promulgated on the subject is at variance with the facts revealed by the scalpel.[1]

A few points connected with vomicæ require further notice.

[1] Thus among 517 consumptive patients examined in the dead-house of St. George's Hospital during the ten years ending December 31st, 1850, there were 171 in whom tubercle was found in its crude, unsoftened state. Of these 32 had tubercle in the right lung and not in the left; 21 in the left and not in the right; 117 had tubercle in both lungs; whilst of 343 patients in whom the tubercular matter had softened and formed vomicæ, 56 had vomicæ in the right and not in the left lung; 59 in the left and not in the right; and in 224 vomicæ existed on both sides. In 2 cases the records are imperfect, and, without defining the side, state only that vomicæ existed on one side and not on the other; and in 2 others the description is omitted, in consequence of the imperfection of the notes kept. For further details see "Decennium Path.," chap. iv, pp. 47, 48.

In the first place, they may vary greatly both in size and number. Sometimes, though rarely, they occur singly; at others a whole lung is riddled with them; sometimes they are no larger than a pea, whilst in certain instances they are as large as a walnut or an orange, or are even capable of containing a pint or more of fluid. The larger cavities are formed by the coalescence of smaller vomicæ, and are, therefore, irregular in shape, and usually intersected by bands of tissue which have escaped disintegration. The tissue surrounding them is generally much solidified, partly by the presence of crude tubercle, and partly by an albuminoid exudation, the result of the inflammatory process which has accompanied the deposit and subsequent softening of the tubercular mass.

Sometimes, though rarely, the large bloodvessels which traverse those portions of the lung which are converted into vomicæ are laid open by ulceration during the process of softening, giving rise to copious and fatal hemorrhage. But almost invariably the pressure of the tubercle leads to obliteration of the vessel and coagulation of the blood which it contains long before its coats yield to ulceration, so that hemorrhage from that source is extremely rare; and when spitting of blood does occur in connection with vomicæ, the blood generally issues from minute congested vessels on their ragged pulpy walls. Occasionally, though very rarely, a pervious vessel may be found imbedded in the bands which intersect the vomicæ,[1] and sometimes an obliterated vessel, which has escaped ulceration, will itself form one of the bands.

Those portions of the bronchial tubes which are included in the tubercular mass become obstructed by the deposit of tubercle, so that during the earlier stages of a vomica there is seldom much communication between the cavity and the main air-passages; but, after a time, the obliterated bronchi yield with the surrounding lung-structure to the process of ulceration, and are thus broken down and expectorated. Consequently, the patulous, ulcerated extremities of one or more pervious air-tubes of considerable calibre are almost sure to be seen terminating abruptly, as if cut across, in the walls of a large vomica, and they form the openings by which air finds admission into the cavity, and the outlets by which the matter which is secreted from its internal surface, together with the disintegrated tissue of the lung, escape into the main bronchi and windpipe.

As long as a tendency to disintegration continues, the in-

[1] Louis examined 123 cases in which vomicæ existed, intersected by bands, and in five instances only did he discover a pervious bloodvessel in the bands.

ternal surface of the vomicæ remains pulpy and ragged; but if the general tone of the system improves, and a tendency to repair ensues, their internal surface becomes smoother, and ultimately is lined by a sort of false membrane. Sometimes this membrane is extremely vascular, and secretes a large quantity of puriform fluid; at others it becomes dense and cartilaginous in appearance, and shows no disposition to contract, so that the cavity may remain open for years.[1] But more generally, when the reparative process is established, the fibrinous membrane contracts, the cavity closes, and the surface of the lung immediately over the seat of the vomica becomes puckered. Not unfrequently a portion of calcareous or cretaceous matter is found entangled in or immediately adjoining the cicatrix.

It would naturally occur to any one unacquainted with pathological phenomena, that ulceration occurring in the pulmonary tissue would be likely to make its way through the pleural membrane, and thus establish a direct communication with the cavity of the pleura. Such is indeed the case sometimes, though rarely—so rarely that it occurred ten times only (five times on the right and five times on the left side) amongst 168 consumptive patients whom I carefully examined throughout the period of their residence in St. George's Hospital.[2] And the reason of its not occurring more frequently is at once revealed by post-mortem examination; for when tubercles are seated, as they often are, near the surface of the lung, they are apt to excite adhesive inflammation of the pleural membrane—not extensive inflammation attended by copious effusion, but dry, adhesive

[1] This fact was remarkably exemplified in the case of a young friend of mine, who was seen during life by the late Dr. Chambers, and by Dr. C. J. B. Williams. Those gentlemen detected a vomica at the apex of either lung, and when the patient died nine years afterwards, the cavities were found patulous, about the size of a small walnut, and lined with a dense, smooth, vascular membrane.

[2] This is a considerably higher percentage than that deduced by Dr. T. Chambers from the post-mortem records of St. George's Hospital. He gives 19 cases of pneumothorax (11 on the right side and 8 on the left) amongst 514 cases of tuberculosis. But, in truth, those numbers do not adequately represent the frequency of tuberculous perforation of the pleura. A certain proportion of the patients in whom this complication occurs recover sufficiently to leave the hospital; and in a considerable proportion of those who die in the hospital the effused air is absorbed, the two surfaces of the membrane become adherent, and the lesion escapes notice in the post-mortem record. Further, pneumothorax rarely takes place except towards the later stage of consumption, and a large proportion of the 515 cases analyzed by Dr. Chambers were not, strictly speaking, fatal cases of consumption, but were rather cases in which death occurred from some other causes, though tubercles to a greater or less extent were found in the lung. If the inquiry were limited to *fatal* cases of consumption, the proportion in which pneumothorax occurs would be found infinitely larger than that which I have given above.

pleurisy, usually corresponding in position and limited in extent to the area of the tubercular deposit, and causing the costal and pulmonary pleura to cohere. The risk of a mixed aeriform and softened tuberculous effusion into the pleural cavity is thus avoided, and partial or local adhesion of the pleura is the only consequence. This, however, is of very constant occurrence, and is most common and most extensive at the apices of the lungs, where the tubercles are usually most numerous. Louis informs us that it was absent in one only out of 112 consumptive patients whom he examined after death, and it was present in forty-six out of fifty fatal cases of phthisis which I noted with a special view to this inquiry; two out of the remaining four cases being examples of the gray miliary deposit. It was also present in 392 cases out of 514 cases of pulmonary tubercle (not all fatal cases of consumption) which were examined in the dead-house of St. George's Hospital during the ten years ending December, 1850.[1] Thus, partial or local pleurisy may be regarded as a frequent, if not a constant, complication of phthisis pulmonalis; perforation of the pleura and pneumothorax as a comparatively uncommon event. But it must not be understood that the pleurisy which accompanies phthisis is always dry and limited in extent. Commonly, indeed, it is so, and still more generally the inflammation, if acute, is limited to one side of the chest; but occasionally it exists on both,[2] and in a considerable proportion of casos in which it is limited to one side it is diffused over a large portion of the pleural membrane, and is accompanied by a profuse outpouring of fluid, which is a serious aggravation of the patient's danger, and hurries him to an untimely grave.

Other complications are also of frequent occurrence. Amongst these may be mentioned atrophy of the lung-tissue, resulting from the cutting off of a due supply of blood by the pressure of tubercle on the bloodvessels; emphysema, which was observed in forty-one, or in other words, in 7.9 per cent. of the phthisical cases examined in St. George's Hospital during a period of ten years;[3] and still more frequently bronchitis and pneumonia. The morbid condition of the blood in phthisis, and the local irritation which attends tuberculization of the lung, are apt to produce not only congestion, but active inflammation of the bronchial mucous membrane, or of the surrounding lung-structure, and thus the phenomena which have been already discussed in the sections

[1] "Decennium Path.," p. 257.

[2] Thus it was present on both sides in no less than 29 of the tubercular cases already alluded to as having been examined in the dead-house of St. George's Hospital during the ten years ending December 31st, 1850.

[3] "Decennium Path.," p. 63.

devoted to bronchitis and pneumonia are met with during the course of consumption. Pneumonia, however, is less commonly excited by tubercular deposit than congestion and inflammation of the bronchial mucous membrane, which are prone to occur at all stages of the disease.

Another complication which is apt to arise, especially in children, is tuberculization and enlargement of the bronchial glands;[1] and another, which is extremely frequent, more particularly in adults, is ulceration of the mucous lining of the upper air-passages. The epiglottis, the internal surface of the larynx, and the tracheal and bronchial mucous membrane are all liable to be implicated in the ulcerative process, and when so affected, give rise to symptoms which will be fully described hereafter.

But consumption is not merely a disease of the lungs and air-passages; it is essentially a constitutional disorder, and, as stated at the commencement of this chapter, it may leave its traces in other parts besides the chest.[2] The brain and its membranes are

[1] For a full description of the physical signs and general symptoms produced by this affection, see pp. 410–12 of this treatise.

[2] This is seen by reference to the subjoined record of the post-mortem examinations of 566 tuberculous patients at St. George's Hospital during the ten years ending December, 1850, as recorded in the "Decennium Pathologicum." The first table shows the seat of tubercle in 517 cases of pulmonary consumption; the second exhibits their seat in the 49 cases in which tubercles did not exist in the lungs.

TUBERCLE IN OTHER ORGANS CONJOINTLY WITH THE LUNGS.

Seats of tubercular deposit in cases of pulmonary consumption.	In 37 cases, from birth to 15 inclusive.	In 211 cases from 15 to 30 inclusive.	In 155 cases. from 30 to 45 inclusive.	In 77 cases, from 45 to 60 inclusive.	In 11 cases above 60.	In 26 cases of unknown age.	In 517 cases at all ages.
1. Intestinal canal,	10	71	37	14	—	9	141
2. Mesenteric glands,	17	37	26	9	1	3	93
3. Kidneys,	6	46	26	4	1	1	91
4. Peritoneum,	6	23	8	4	—	1	42
5. Bronchial glands,	10	13	6	2	—	3	34
6. Nerve centres,	10	15	3	—	—	—	28
7. Spleen,	6	12	4	—	—	1	23
8. Liver,	5	7	1	—	—	—	13
9. Bladder and prostate gland,	—	6	3	2	—	1	12
10. Organs of generation (male and female divided below),	—	5	1	2	—	1	9
11. Heart and pericardium,	2	3	—	—	—	—	5
12. Male organs of generation,	—	3	1	—	—	1	5
13. Female organs of generation,	—	2	—	2	—	—	4
14. Axillary and cervical glands,	1	—	2	1	—	—	4
15. Anterior mediastinum,	—	3	—	—	—	—	3
16. Pancreas,	—	—	1	—	—	—	1
17. Cranial bones,	—	—	1	—	—	—	1

prone to be the seat of tubercle; so are the liver, spleen, and kidneys: so also are the cervical and mesenteric glands, which in children are seldom exempt. The glandular structure of the bowels may be implicated in the same way, and small hard masses of yellow tubercle may be seen projecting from the surface of the intestine;[1] or the tubercle may soften, and cause ulceration, and lead eventually to perforation of the peritoneum, and to peritonitis resulting from the escape of fæcal matter into the cavity of the abdomen; or tubercular deposit may take place in the peritoneum, and occasion chronic inflammation and ascites. Piles, and fistula in ano, are other of its common accompaniments;[2] fatty and amyloid degeneration of the liver is another, and œdema of the lower extremities, resulting from coagulation of blood in the veins, and consequent obstruction to the circulation, is another which occasionally proves very distressing.

CASES IN WHICH THE LUNGS WERE FREE FROM TUBERCLE.

Seat of tubercular deposit.	From 1 to 15.	From 15 to 30.	From 30 to 45.	From 45 to 60.	Above 60.	Age doubtful.	In 49 cases at all ages.
1. Kidneys,	1	4	5	4	1	—	15
2. Bronchial glands,	1	5	3	—	—	—	9
3. Peritoneum,	—	3	1	2	—	1	7
4. Abdominal glands,	1	2	1	1	—	1	6
5. Parts of generation,	—	1	—	1	1	2	5
6. Liver,	—	2	2	1	—	—	5
7. Intestinal canal,	1	—	4	—	—	—	5
8. Nerve-centres,	2	2	1	—	—	—	5
9. Spinal bones,	—	3	1	—	—	—	4
10. Parietal pleura,	1	2	—	—	—	—	2
11. Cranium,	—	2	—	—	—	—	3
12. Hip,	2	—	—	—	—	—	2
13. Other joints,	1	1	—	—	—	—	2
14. Cellular tissue and muscles,	—	1	1	—	—	—	2
15. Pericardium,	—	1	1	—	—	—	2
16. Shaft of femur,	1	—	—	—	—	—	1
17. Spleen,	—	1	—	—	—	—	1
18. Dura mater,	—	1	—	—	—	—	1
19. Nervus abducens,	—	—	1	—	—	—	1
20. Axillary glands,	—	1	—	—	—	—	1
21. Cervical glands,	—	1	—	—	—	—	1

[1] Louis states that he met with tuberculous ulceration of the smaller bowels in five-sixths of the phthisical cases he examined after death, and ulceration of the larger bowels was almost as frequent.

[2] The statistics deduced from the post-mortem records of St. George's Hospital throw considerable doubt on the commonly received opinion as to the almost invariable connection between fistula and phthisis, and afford grounds for believing that in a large proportion of cases fistula in ano is unconnected with tuberculosis. For full details see "Decennium Pathologicum," chap. vi, p. 75.

The physical signs of pulmonary consumption, no less than its general symptoms, correspond in great measure with the extent and condition of the tubercular deposit; and as there are three well-marked stages of tubercle, viz., those of deposition and induration, of softening, and of excavation, it will be desirable to discuss the physical signs of the pulmonary disease as they usually present themselves at each of these stages.

1st stage, or stage of deposition and induration.—Inspection of the chest affords very little information unless the tubercular deposit be large in amount, or be confined to one side. In the former case an abnormal rapidity of the respiratory movements will be observable, and in both the former and the latter a decrease of expansion in the infra- and supra-clavicular regions on the affected side, and, possibly, a little flattening of the upper part of the chest-walls on that side. When the deposit takes place equally on both sides, the flattening which it occasions at this early stage may escape notice.

Palpation will sometimes detect deficient expansion in the infra- or supra-clavicular regions, even when inspection has failed to do so. It also gives us notice of consolidation of the lung-tissue, by making us aware of increased vocal fremitus on the affected side; the increase, however, is seldom great, and on the right side cannot be relied upon as evidence of disease. At the left apex it is a sign of considerable importance.

Mensuration will sometimes detect a slight diminution in the antero-posterior diameter of the infra-clavicular regions, and also a decrease in the local expansion movement.

Percussion is always a fallacious guide in these cases. In most instances miliary tubercles do not perceptibly affect the resonance of the chest; and even if they exist in quantity sufficient to produce a slight modification of the percussion-note, they usually do so on both sides of the chest, so that the alteration of sound fails to impress the ear or to convey any definite information. Even when considerable masses of crude tubercle exist at the apex, dulness is not an invariable consequence. If they are seated superficially, the sound on percussion is dull, and the sense of resistance to the fingers is increased; but if, as often happens, some healthy or some emphysematous lung-tissue intervenes between the consolidated lung and the chest-walls, or if haply the superficial air-cells become distended with air in consequence of any obstruction offered to its egress by the pressure of tubercle on the smaller bronchi, not only will no dulness occur, but there will be an abnormally clear though shallow resonance over the affected lung.[1] These alterations in the percussion-note

[1] See chap. v, pp. 63–67, of this treatise.

will be most perceptible in the clavicular and infra-clavicular regions; and if doubts arise as to the existence of dulness on either side, recourse should be had to percussion, practised during full inspiration, and then during deep expiration. In the former case, if tubercles exist, the increase of resonance on the affected side will be slight as compared with that on the healthy side, whilst in the latter case the dulness will be far greater on the affected side than over the healthy, unaffected lung.[1] In all cases it is the difference of the percussion-note on the two sides of the chest, rather than the actual quality of the sound, which is indicative of subjacent disease.

Auscultation may afford very valuable assistance, though in many instances it will fail altogether to do so, and in others, its indications will not suffice to establish the precise nature of the disease. Thus, when gray miliary tubercles, or a few isolated, yellow tubercles, exist in a quiescent state in the lungs, they will often fail to produce any appreciable auscultatory signs. Even when they are present in somewhat larger quantity, the utmost auscultatory disturbance produced is slight deficiency and harshness of the respiration, with prolongation of the expiratory murmur, and jerking irregularity of the respiratory rhythm—signs which are not peculiar to phthisis, but are met with in bronchitis and other forms of disease. Therefore, it is only when they occur persistently, and are confined to one side of the chest, that they can be regarded as indicating the existence of tubercle. To these, however, in some instances, must be added dry clicking, which is always suggestive of tubercular deposit, and imparts a dangerous significance to the jerking respiration and prolonged expiratory murmur. These signs are usually developed earliest, and are most marked in the supra-scapular, the supra-clavicular, and the infra-clavicular regions.

When local pulmonary irritation is excited by the tubercular deposit, or by the condition of blood which has led to the deposit, the ausultatory signs of catarrh are superadded, viz., weakness or deficiency of the healthy vesicular murmur, which is replaced by coarseness of respiration, with small bubbling râles and sonorous and sibilant rhonchi. In this, however, there is nothing pathognomonic of phthisis; and it is only when these symptoms of bronchitis are persistent, and are almost confined to the apices of the lungs, being inaudible below the second intercostal space, and especially when they are confined to the apex of *one* lung, that their existence can be regarded as suspicious. Sometimes,

[1] For detailed precautions respecting the performance of percussion in these cases, see chap. v, pp. 56–60, of this treatise.

however, these sounds are accompanied by the grazing or crumpling sound of pleural friction, which renders their true character apparent.

When the deposit of tubercle is more extensive, and includes within it bronchial tubes of considerable diameter, the evidence is much more conclusive. If the tubes are filled with tubercle, there is almost entire absence of vocal resonance and of respiratory murmur over the affected part, and not unfrequently exaggerated breathing in the parts immediately adjoining; whilst if the air-passages remain pervious, hollow-sounding respiration and increased vocal resonance are met with. Further, if the deposit be extensive, and not merely limited to the apices of the lungs, the pulmonary consolidation leads to the transmission of the sounds of the heart and large vessels to a far greater extent than in health; so that, if the right lung be principally affected, the cardiac sounds may be more audible at the right than even at the left apex. But many conditions of the intervening lung-tissue serve to prevent this transmission of the heart's sounds, and therefore the mere absence of this sign is not valid evidence against the existence of tubercular consolidation.

A systolic pulmonary murmur is often present in these cases at the second left sterno-costal articulation, resulting from pressure on the artery caused by the tuberculous deposit, and so is a jerking or interrupted murmur in the subclavian artery on the left side, and in the innominate artery on the right. None of these signs, however, if viewed alone, affords trustworthy evidence of tubercle. In many phthisical patients the pulmonary murmur is absent, and in many anæmic, non-phthisical persons it is present, while the subclavian and the innominate murmur, if not audible during tranquil respiration, may be developed in many healthy individuals by a few successive deep and forcible respiratory efforts.[1]

It should, perhaps, be added that in the earlier stages of consumption more reliance can be placed on the altered quality and

[1] In order to ascertain the true significance of this subclavian sound, I carefully examined 100 healthy persons, who presented themselves for examination with a view to life assurance. In 3 the murmur was strongly marked in both sides of the chest, and in 9 others on the left side of the chest during tranquil respiration; in 19 it was temporarily induced on both sides, and in 46 others on the left side of the chest by the forcible effort of blowing the spirometer, whilst in the remaining 23 it was not heard either during tranquil or after forced respiration. The position of the left subclavian artery, which admits of its being compressed against the clavicle by the fully expanded lung, explains the frequency of the occurrence of the sound in the left as compared with the right side. Of course it would be more likely to be produced, and would prove a more persistent condition if the apex of the left lung were solidified by tubercle.

intensity of the respiration, and on the occurrence of jerking irregularity of rhythm, than on the existence of prolonged expiration or increased vocal resonance. Prolonged expiration, if unaccompanied by alteration in the quality of the respiration, may be a normal condition peculiar to the individual; and vocal resonance is subject to so many variations as to be almost valueless if taken by itself as the basis for an opinion as to the condition of the lung. At the right apex increased vocal resonance can hardly be regarded even as suggestive of tubercle, unless it be excessive in degree or has succeeded to naturally weak vocal resonance.

2d stage, or stage of softening.—Many of the signs already described as characteristic of the first stage of the disorder now exist in an advanced degree of development, and a series of new auscultatory signs, referable to the softened tubercular deposit, are met with for the first time.

Inspection.—The eye at once notes an abnormal frequency of the breathing, a decided flattening or falling in of the chest-walls on the affected side both above and below the clavicle, and great deficiency of local expansion, especially during forced inspiration.

Mensuration with the callipers shows a marked diminution in the size of the chest, both in its transverse and antero-posterior diameter, the decrease being referable in part to atrophy and interstitial contraction of the lung, and in part to retraction caused by firm pleural adhesions.

Percussion in most instances elicits evidence of a wider spread and more intense dulness.

Auscultation.—In addition to the phenomena discovered during the first stage of the disease, auscultation now announces the important fact of the presence of thin, irregular-sized bubbling râles, caused by the passage of the inspired air through the softened and liquefied tubercular deposit; it further proves that coarse and hollow-sounding respiration exists over a more extensive surface than heretofore, or possibly that the respiratory sounds have assumed a distinctly blowing character. Vocal resonance still remains extremely variable, and cannot be relied upon as a guide to the condition of the subjacent lung; and the existence or non-existence of vocal fremitus is an equally uncertain indication, inasmuch as it is dependent upon the amount of pulmonary consolidation, and not upon the occurrence of softening.

3d stage, or stage of excavation.—*Inspection* reveals increasing rapidity of the respiration and extraordinary prominence of the clavicles, referable to the increasing depression or falling in of the supra-clavicular and infra-clavicular regions, consequent on the advancing tubercular disease and decreasing pulmonary expansion. Further, it sometimes makes us aware of the existence of a fluctuating impulsive movement in the upper intercostal spaces, occasionally on the right, but more frequently on the left of the sternum, caused in most instances by the action of the pulmonary artery or of the base of the heart, which has been brought into unnatural contiguity with the anterior walls of the chest by the destruction of the intervening lung-tissue, but referable in some cases to the systole of the left auricle, as proved by the synchronism of the impulse with the diastole of the heart.

Palpation makes us aware of a further decrease in the expansion of the upper part of the chest; of marked vocal fremitus when the cavity is large, superficially seated, and in free communication with the upper air-passages; and sometimes of rhonchial and gurgling fremitus, or even of distinct fluctuation. Further, if, as sometimes happens, the disease exists principally in one lung, and that lung has shrunk considerably, and has dragged the heart along with it, palpation makes us aware of the fact that the heart is beating out of its proper place; it also informs us of any abnormal pulsation which may exist in consequence of the base of the heart or the left auricle or the pulmonary artery having been brought into contact with the anterior walls of the chest, as a result of an excavation in the intervening lung-tissue.

Mensuration shows progressive diminution in the diameters of the upper part of the chest.

Percussion varies greatly in its results, according to the varying conditions of the vomicæ, and of the tissues which lie between them and the chest-walls. If one or more small vomicæ exist, filled in great measure by purulent secretion, or surrounded by tubercular deposit, or by consolidated lung-tissue, the sound, on percussion, will be absolutely dull; whereas, if the same cavities be empty and seated superficially, if their walls be thin and the pleural membrane be not thickened by plastic exudation, the dulness will be comparatively slight, or the sound may be even resonant, though of a shallow, amphoric character. If, again, a tolerably thick layer of healthy permeable lung-tissue intervenes between the chest-walls and vomicæ partially filled with fluid or surrounded by tubercular deposit, gentle percussion may give rise to almost healthy resonance, while firm and forcible percussion will elicit the deepseated dulness referable to the vomicæ and the consolidated lung-tissue beneath. Large empty superficial cavities, with thin tense walls, yield an amphoric, and possibly a cracked-pot resonance; whereas, when the walls of the cavity are thick or flaccid, or when a denser thickened pleural membrane, or a portion of consolidated lung-tissue intervenes between the cavity and the chest-walls, the sound elicited may be absolutely dull, or it may be of a mixed dull and amphoric character. Except in rare instances of enormous superficial vomicæ with thin tense walls, it is almost impossible to judge of the size of a cavity by the results of percussion, inasmuch as the condition of the thoracic parietes, the distance of the vomicæ from the chest-walls, the variations in the relative proportion of air and fluid in the cavity, and the state of the intervening lung-tissue, offer insuperable obstacles to the formation of a sound judgment on the matter. When an empty cavity with resilient walls is seated superficially, and has a free communication with the upper air-passages, forcible percussion will often elicit a cracked-pot sound, especially when the mouth is open, and all obstacle to the egress of air is thus removed.

Auscultation.—Provided the cavity is tolerably empty, and in free communication with the upper air-passages, and that the sounds emanating from it are not masked by the intervention of a thick stratum of healthy permeable lung-tissue, the respiration will be heard of a hollow, blowing, or even an amphoric and metallic character; and when the fluid contents of the cavity have accumulated sufficiently to rise above the level of the permeable bronchi which lead to it, large irregular bubbling râles or distinct gurgling, often of a metallic quality, may also be heard. In some instances well-marked metallic tinkling is produced by the cough, the respiration, and the action of the heart in large cavities which possess smooth, tense, sound-reflecting walls. Vocal resonance varies greatly, according to the precise condition of the parts, and therefore cannot be relied upon as evidence of the existence or non-existence of a cavity. If a vomica be partially filled with secretion, and the bronchi leading to it be obstructed, no vocal resonance may be heard; whereas, after the air-tubes have been cleared by coughing, the resonance in one case may be nearly natural, in another simply feeble, and in yet another intensely loud, and of a bronchophonic, pectoriloquous, or amphoric and metallic character; these differences being due to the varying conditions of the walls of the cavity, and to the freedom or otherwise of its communication with the bronchi. Well-marked pectoriloquy resulting from a mere whisper is the form of vocal resonance most pathognomonic of tubercular exudation; but in many instances it is not met with, whilst, on the other hand, this and every other form of increased vocal resonance, though perhaps in a less intensely marked degree, may result from dilated bronchi, surrounded by hepatized, or otherwise consolidated lung-tissue.

Small cavities, partially filled with fluid, and deeply seated in the lung, seldom produce signs characteristic of a vomica, but simply give rise to imperfect hollow respiration, occasional irregular bubbling râles, and more or less modified vocal resonance —sounds which may originate in a large bronchus surrounded by consolidated lung. When, as often happens in cases of extensive pulmonary excavation, the base of the heart is brought into contact with the anterior surface of the chest, causing pulsation in one or more of the upper intercostal spaces, auscultation shows that the sounds of the heart are abnormally loud in that situation, and may be even louder, and apparently seated more superficially than at the fifth left interspace—the natural seat of the apex beat. Further, auscultation often announces the existence of intense systolic murmur at the second left sterno-costal articulation, and sometimes in the first and second intercostal spaces on either side, referable to the pressure of hardened lung-tissue or tubercular deposit on one or other of the larger arteries.

Succussion.—If a large cavity, possessing thin, tense, sound-reflecting walls, be seated superficially in the chest, and contain a moderate quantity of thin liquid secretion, a splashing noise may be elicited when the patient is forcibly and abruptly shaken. In a cavity of this description, placed under favorable conditions, even the action of the heart may sometimes give rise to this sound.

Consumption is a malady which assumes various forms, and is associated with a variety of pathological conditions and distinct anatomical changes. It is a malady, therefore, beyond all others, remarkable for the great diversity of its symptoms in different cases, and for extreme variation in the course which it runs; and as these differences are not associated with any particular stage of the tubercular deposit, but rather with peculiarities in the patient's constitution, and with the form which the malady assumes, it will be desirable to give a few sketches of the disease as it presents itself under different circumstances. In no other manner is it possible to convey an adequate idea of the complaint, or to furnish such a connected history of its symptoms as shall afford a clue to its diagnosis.

The heads under which I shall attempt to portray the disease are, 1st, the ordinary forms of chronic phthisis; 2dly, the acute forms of the complaint; and, 3dly, the latent or insidious forms of the disease. Its special diagnostic symptoms, and some of its more important complications, must be reserved for separate consideration.

To begin, then, with the ordinary forms of phthisis.

In many instances the earliest, and for some time the only symptoms of its approach are dyspepsia, with sick headaches, biliousness, loss of appetite, gradually increasing languor, and debility and depression of spirits. The patient feels unequal to his ordinary avocations, his nights are restless, and in the morning he rises weary and unrefreshed. After a time emaciation commences, the flesh becomes flabby, the countenance pale, the pupil of the eye dilated, and the conjunctiva of a pearl-like whiteness; the hair falls, and in many instances the fingers become clubbed at their extremities and the finger-nails incurvated and adunc. Occasionally the patient suffers from weakness and huskiness of voice, soreness of throat, and tightness across the upper part of the chest, with dull aching fugitive pains about the collar-bones, or under one or both shoulder-blades, and although he has no running at the eyes or nose, he fancies he must have "caught cold," for he feels chilly and uncomfortable, and has a short, dry, hacking cough. This occurs principally on rising in the morning and on going to bed at night. At first the cough may be so slight as not to cause him any annoyance, or even to excite the apprehension of his friends; it is regarded simply as a clearing of the throat, and appears to be occasioned by relaxation of the uvula, or by irritation of the pharynx, which in the early stage of phthisis is often rough, red, and covered with mucus. But after a time the cough increases in frequency and violence, recurs at intervals throughout the day, especially after exertion, and becomes attended by a scanty expectoration of ropy or glairy mucus, occasionally specked or tinged or streaked with blood. Little suspecting the cause of his ailment, he complains that he is short-breathed on going up stairs, and is soon exhausted by active exertion; and the physician finds that his breathing is quicker, and the temperature of his body higher than natural, usually ranging from 99.5° to 102°, even at midday, and that his pulse is accelerated, especially towards evening, and very deficient in force. His face flushes on the slightest excitement, and particularly after meals, whilst, in some instances, febrile paroxysms, marked by alternate chills and heats by night, and by perspirations towards morning, form a cause of gradually increasing weakness and of serious annoyance and complaint. The bowels may act regularly, but more commonly are confined; the tongue may be clean or more or less coated, and the pulse weak or irritable, varying in frequency from 60 to 140 in the minute. The urine is at one time clear and pale, at another high-colored, scanty, and turbid, but it simply varies with the state of the system, and does not throw any light on the condition of his chest. The menstrual discharge may be natural in quantity and quality, or it may be excessive in quantity, and may recur too frequently, but more commonly it is scanty or entirely suppressed, and replaced by profuse leucorrhœa. Not unfrequently suppression of the monthly periods is the event from which the patient dates her illness.

In another class of cases the symptoms do not differ greatly from those just described, except that they are unaccompanied by dyspepsia, and that in the earlier stages the febrile paroxysms are almost wholly absent. The appetite is fairly good, and nothing seems to disagree; the bowels act regularly, the pulse is not much accelerated, the urine is clear and natural in appearance, and the patient sleeps quietly, unless disturbed by the cough. He is weak and languid, and loses flesh; but even when pressed for a full confession of his feelings, he will assert that there is not much amiss, and that were it not for occasional slight cough, huskiness of voice, and a sense of weakness and undue exhaustion after exercise, he should have no cause for complaint.

In yet other cases, the patient, who is usually of a florid complexion, may consider himself in good health, when possibly, after some straining effort, he is suddenly attacked with profuse hemorrhage from the lungs, or after exposure to cold is seized with bronchitis, attended by blood-tinged or blood-streaked, frothy mucous expectoration. The only premonitory symptoms will probably have been occasional huskiness of voice, and an unusual desire to "clear his throat" on rising in the morning, more than ordinary exhaustion after taking active exercise, and a tendency to perspire on the slightest exertion. Even these symptoms will have failed to attract his attention, so that pointed and earnest inquiry has to be made before he will admit that he was at all out of health when the attack commenced. In these cases there is generally undue excitement of the circulation, with a quick and irritable pulse, feverish heat of skin, a chronic and undue elevation of the temperature of the body, which often reaches 103°, restlessness at night, and a frequent hard cough attended by pain in the chest. Sometimes, however, though the pulse be quick, the cough is unattended by pain, and there is entire absence of fever; the difference apparently being referable to the existence or non-existence of bronchitis, or of inflammation of the pulmonary tissue.

But whatever symptoms may have inaugurated the accession of the disease, its progress will depend in part on the amount of tubercle which has been deposited, in part on the nature of the patient's constitution and the precise form which the local changes assume, and in part on the judgment displayed in the treatment. Very commonly a large amount of tubercle is deposited in the first attack; the intervening portions of lung are congested, there is considerable bronchial irritation, and the function of respiration is seriously interfered with. If this condition of lung occurs in a person of a hitherto unimpaired constitution, the more active symptoms may usually be subdued by judicious treatment, and considerable improvement may take place; the cough may remain troublesome, and the patient will be short-breathed, and somewhat of an invalid; but nevertheless he may gradually mend, and the symptoms of rapidly progressive disease may for a time be stayed. But if he be of a weakly constitution or strong scrofulous tendency, the disease will run on from bad to worse, notwithstanding any treatment which may be adopted. The tubercle will soften and break down into cavities, and while the disintegrated tubercular matter is being got rid of by expectoration, fresh tubercles will be deposited, and the healthy lung-tissue will be still further encroached upon; the bronchitic or pneumonic symptoms may be subdued for a time, but only to reappear after a brief interval, as fresh local irritation is excited; the breathing becomes shorter and more oppressed, the cough more frequent and harassing, exciting sickness and rejection of food, and the expectoration profuse and of a purulent character; often streaked or mixed with blood. The complexion is usually pale, though the skin is hot, owing to the febrile excitement which prevails; the lips are of a dusky yet pallid hue, or else of a brilliant red color; the cheeks flush on the slightest occasion, and especially after food; the pulse is much accelerated, and distinct febrile paroxysms, which commence with alternate chills and heats, and terminate in profuse sweating and great exhaustion, occur towards noon and again at night. Meanwhile the patient wastes, and becomes rapidly weaker; the monthly periods cease, the nails become adunc, the fingers clubbed, the hair falls; pain or stitch in the side, resulting from partial or local pleurisy, is of frequent occurrence; hoarseness and partial loss of voice often supervene, arising from tubercular ulceration of the larynx, and diarrhœa sets in, occasioned by the same form of mischief in the intestines; the tongue becomes extremely furred, or else clean, smooth and red; aphthæ make their appearance in the mouth and throat, the appetite fails, and sleep is only to be wooed by full and repeated doses of sedatives: the integuments of the back become inflamed and sore, or even ulcerated, in consequence of the pressure to which they are subjected; the features become sharp and collapsed, the feet and ankles swollen and œdematous, and as death approaches, slight wandering or delirium often occurs. Sometimes coma ultimately supervenes, and the patient, unconscious of his approaching end, sinks quietly and without a struggle; in other instances the brain remains unaffected, and the mind is clear and collected; the cough almost ceases, and the patient passes away in perfect tranquillity, almost as if asleep; in others, again, the mind remains unaffected, but for many days, if not for a longer period before death, the patient suffers constantly from a distressing sense of sinking and exhaustion, and is tormented with pains in the chest and bowels, oppressive shortness of breath, and excessive difficulty of expectoration, leading to frequent fits of suffocative dyspnœa, which renders the death-struggle extremely painful.

The duration of the disease in the cases above described varies from a few weeks to six or eight months; and from the time when the active symptoms commence the patient is a confirmed invalid.

But in most instances the malady runs a much more protracted course, and presents marked intermissions in its progress. In such cases the area of tubercular deposit at first is small, the tubercles are confined to a portion of the upper lobe, and the remainder of the lung is perfectly healthy, so that the respiratory function is not seriously interfered with. Under these circumstances, if the general health can be improved, and the patient's system invigorated, the deposit of tubercles may be arrested; those already deposited may remain in a quiescent state for years, or even throughout a tolerably long life, the irritation they cause may pass off, the cough and other symptoms will then subside, and the patient may recover a certain degree of health. Sometimes he regains full bodily vigor, but more commonly he remains short-breathed, and liable to be affected by slight disturbing causes. Nevertheless, though somewhat of a valetudinarian, and prone "to catch cold," he is able to resume his ordinary avocations, and is regarded by his acquaintances as in tolerable health. Two or more such attacks as these may occur at intervals of as many months or years, before any softening of the tubercles takes place; the system rallies quickly after each attack, and the patient recovers more or less completely; though as each attack results in a fresh crop of tubercles, it leaves the lungs less equal to carry on their respiratory function, and the general health more seriously impaired. At length the deposit reaches the limit which is compatible with even feeble health; the general power of the system fails, the tubercles soften, cavities are formed, the patient expectorates a large quantity of puriform matter, suffers greatly from hectic fever, has frequent attacks of diarrhœa, sweats profusely at night, becomes rapidly emaciated, and sinks, as in the former case, after a few months of suffering.

Sometimes, again, the course of the disease is somewhat different. The tuberculous deposit is not extensive, and there is very little bronchitic complication, so that, as in the former case, sufficient healthy lung-tissue remains to carry on respiration in a tolerably efficient manner; but the tubercles are of the soft, yellow variety, and speedily soften and break up into vomicæ. The expectoration therefore becomes purulent, and is often streaked with blood; hectic fever supervenes, and the patient loses flesh and strength. But, notwithstanding the stethoscopic signs and general symptoms of far-advanced disease, there are indications which, to the practised observer, denote the possibility of the patient's rallying. The respiration is not so quick nor the pulse so much accelerated as in rapidly progressive phthisis. There is not the same burning heat of skin, or the same amount of perspiration, and the aspect of the patient is not expressive of much anxiety or of fatal constitutional disturbance. Accordingly, after a time, if the tone of the system be judiciously upheld, and no fresh deposit of tubercle take place, the tubercular matter already deposited may be expectorated and got rid of, or may cease to excite local irritation; the cough may subside, the night-sweats cease, the appetite return, the patient may regain flesh, and in great measure recover strength. In this case, as in the former, there may be intervals of months or even years, during which the patient may enjoy comparatively good health, and in some instances the patient may at length shake off his tendency to the disease, so that no further deposit of tubercle may take place, and practically the disorder may be regarded as cured. But in most cases the symptoms recur at a future period, and each attack, being accompanied by fresh tubercular deposit and destruction of lung-tissue, leaves the patient less able to recruit his health and strength. At length, if relapses continue to recur, the time must ultimately arrive when the amount of deposit in the lung, or the breaking down of its tissue, will be such as to render the due continuance of respiration impossible, and then the patient "will fall into a decline," or, in other words, the disease will run its fatal course in much the same manner as in the former instance. But it should be clearly understood that, notwithstanding the more formidable character of the physical signs and general symptoms, recovery may take place from the earlier attacks as perfectly as when the tubercles do not soften and break down; and that if the deposit of tubercle has not been extensive, the patient may regain his health, and may live many years, nay, may attain to a fairly advanced age, even after several vomicæ have been formed.

2. Under the title of *Acute Phthisis*, two distinct forms of disease are usually comprised—the one resembling ordinary phthisis in many of its symptoms, and remarkable chiefly for the severity of the constitutional disturbance, and the extreme rapidity of its course; the other more nearly allied in its general characters to a low type of irritative fever, accompanied by congestive pneumonia. They both run their course in from three weeks to three months from the time when the patient was supposed to have been in good health, and unless caution be exercised, they are both apt to be mistaken for some other form of malady.

The first variety, which is connected with extensive deposition of yellow tubercle, or else with tubercular infiltration of the lungs, is usually ushered in by severe rigors, followed by anxiety of countenance, heat of skin, the temperature often rising to 104° or 105°, a quick, irritable pulse, a short, hard cough, and considerable oppression of the breathing. There is pain and tightness across the chest, and a sense of heavy aching in the shoulder-blades. In the course of a few hours the cough is accompanied by frothy mucous expectoration often specked with blood, but before many days have elapsed the tubercles begin to soften, the character of the sputa is altered, and they become muco-purulent, or even purulent. At the same time the feverish heat of skin which marked the first invasion of the disease is replaced by alternate "chills and heats," and by the profuse perspirations which characterize hectic fever: the anxiety of the countenance increases; the skin is of a pallid yet dusky hue; the pulse becomes more rapid, the cough more frequent, the breathing more hurried and more oppressed, and the lips and face livid; the brain sympathizes with the general distress, and wandering or low muttering delirium ensues, especially at night. So matters go on from bad to worse, until, after the lapse of a few weeks, the oppression of the breathing becomes excessive, the sensation of depression and exhaustion extreme, the skin is constantly bedewed with a clammy perspiration, a vesicular eruption of sudamina makes its appearance, emaciation progresses rapidly, the pulse can hardly be counted, the tongue becomes dry, the teeth are covered with sordes; and after a few days more of extreme restlessness and of constantly impending suffocation, the patient usually lapses into a state of coma or of semi-consciousness, attended with muttering delirium, and so gradually sinks. In some instances yellow tubercles are found after death extensively disseminated throughout the lungs; there are numerous small cavities resulting from the softening of isolated tubercular masses, and much of the intervening lung-structure is extremely congested, and oftentimes consolidated by inflammatory exudation. In other cases large portions of the lung-structure are thoroughly infiltrated with tubercle, which in some parts is seen in a softened state, and in others has been expectorated, and has thus given rise to the formation of cavities, or so-called vomicæ.

The physical signs which are met with in these cases are by no means characteristic, and can be interpreted correctly only by constant reference to the general symptoms. There may be deficiency in the expansion of the chest, dulness on percussion, and increased vocal fremitus, but not necessarily at the apices of the lungs; the respiration may be weak or deficient in some spots, coarse and exaggerated, or hollow-sounding, and faintly metallic in others; whilst in others, again, the respiratory murmur may be overpowered or replaced by loud rhonchi, and irregular-sized, bubbling râles.

The true character of the disease which has produced these signs may generally be diagnosed if due caution be observed. The only malady with which it is likely to be confounded is pneumonia;[1] and between the two disorders there are many well-marked points of difference. In the tubercular disease the intensity of the constitutional disturbance is out of all proportion greater than the apparent amount of chest affection; the disease occurs equally on both sides of the chest, and is seldom accompanied by much pain or catching of the breathing; there is absence of marked and limited dulness on percussion, of true pneumonic crepitation, and of the intensely metallic tubular breathing and sniffling bronchophonic resonance of the voice which characterizes pneumonic consolidation, whilst, on the other hand, the depressed aspect of the patient, the profuse perspirations, the intense prostration, the gradually increasing but diffused dulness on percussion, and the occurrence of large irregular bubbling, and other signs of pulmonary excavation, point clearly to tubercular consolidation and excavation as the cause of all the mischief.

The second variety is connected with the deposit of innumerable gray miliary tubercles. Its symptoms resemble those which

[1] In making this statement I exclude the hypothesis of cancerous and other malignant deposits in the lungs.

characterize the first variety, except that there is often an absence of cough and of purulent or muco-purulent expectoration, and that the feverish symptoms thereby acquire a greater prominence. If cough is present, it is usually very slight, and hardly attracts attention; and the sputa, if any exist, are scanty, and consist of little more than frothy mucus. But the fever is considerable, the elevation of the temperature of the body persistent and very great, often ranging from 102° to 106°; the nervous depression extreme; the breathing is excessively short and oppressed, the lips and face are livid, and the whole surface of the body is of a dusky hue, and usually bathed in perspiration. In short, the general symptoms are those of low remittent fever attended by congestive pneumonia, rather than those which are usually regarded as characteristic of pulmonary consumption.

After death, the lungs are found extremely congested, and everywhere studded with innumerable gray semi-transparent miliary tubercles, sometimes so minute as to be scarcely visible; but there are no softened tubercles, and no vomicæ.

The physical examination of the chest gives little more than negative results. There is no observable difference in the size or expansion of the chest on the two sides, nor in the intensity of the vocal fremitus, nor in the sound elicited on percussion, nor in the vocal resonance, nor in the sounds produced by cough or respiration. There is great acceleration of the breathing, and there may be slight harshness and unevenness of the respiratory murmur, and occasionally a few rhonchi or a few small râles; but there are no large bubbling râles, there is no fine crepitation, no tubular breathing, no snuffling bronchophonic resonance of the voice, and no marked and limited dulness: indeed, the dulness on percussion, if any exists, is extremely slight, and is met with equally on both sides.

Thus the physical signs, which vary little from day to day, show the fallacy of the opinion that pneumonia is present, and, if duly weighed in connection with the persistent elevation of temperature[1] and the other general symptoms, can hardly fail to lead to a correct diagnosis. If the oppression of the breathing be not due to pleurisy or pneumonia, it must be referable to bronchitis or phthisis. But the intense heat of the skin, the comparative absence of râles and rhonchi, the almost entire

[1] The temperature in pneumonia often rises to 104° or 105°; but it attains its maximum elevation at from the third to the sixth day of the attack, after which it falls with tolerable rapidity and regularity, unless a fresh attack of pneumonia or pleurisy be set up—an event which may be discovered by a stethoscopic examination. On the other hand, the elevation of temperature in cases of acute phthisis is remarkably persistent.

absence of sputa, and the complete absence of muco-purulent expectoration as the malady progresses, conspire to prove that the malady is not bronchitis, and, therefore, that it must be of a tubercular character. In some of these cases of acute phthisis, the stomach is irritable, and vomiting occurs; in others, symptoms resembling those of meningitis arise, and in a few diarrhœa supervenes; but, if the disease runs a rapid course, these symptoms, with the exception of delirium, are seldom met with in a marked degree. Even emaciation is not a constant symptom. I have seen several patients die of acute phthisis in St. George's Hospital, partly suffocated by the enormous crop of tubercles in their lungs, and partly poisoned by the noxious quality of their own blood, whilst as yet no emaciation had taken place. Indeed, their bodies after death were not only plump, but had large quantities of fat stored up in the omentum and other parts.

3. The latent or insidious forms of phthisis present a remarkable contrast to those last described. They are characterized by persistent mental and physical depression, and by an absence of fever and of the more prominent symptoms of pulmonary disease. The patient feels low and weak and depressed, but is utterly at a loss to account for his illness. His complaint is of general lassitude and weariness, of inability to fix his attention, and of being unequal to much exertion; but he has little or no cough, no pain in the chest, no feverish heat of skin—no symptoms, in short, to direct attention to the real seat of mischief, and being more or less dyspeptic, he usually imagines that "it is all stomach." His friends, probably, consider him nervous and hypochondriacal, and rally him on his indolence and lowness of spirits; but the physician observes signs of loss of health, which he justly attributes to deepseated and serious mischief. He remarks that his patient's aspect is unhealthy; his pupils dilated, the conjunctivæ of a pearl-like color, the skin unusually pale, the muscles flabby, the extremities often cold, and the pulse feeble—not necessarily much accelerated; he learns by inquiry that the appetite is impaired, the digestion weak, his sleep disturbed and unrefreshing; and together with these signs of general functional derangement and impaired constitutional power, he notes a tendency to emaciation, acceleration of the breathing, and an occasional slight bark or short dry hacking cough after exertion. The cough probably is slight, and has escaped observation, so that the patient will deny its existence; and it often remains so slight throughout the whole course of the disease, as not to form a subject of complaint. Thus the patient will go on for months, suffering, as it is thought, from nervous debility, becoming gradually thinner, but otherwise without

material change in his symptoms. At length he finds himself short-breathed and unable to go up stairs without panting: he perspires on the slightest exertion; experiences chills down the spine, followed by burning heat of the palms of the hand, and not unfrequently by perspirations at night; his appetite becomes more than usually capricious, the liver disordered and the stomach irritable, so that sickness and vomiting occur; the bowels are disturbed; and, after a time, griping pain in the belly and obstinate diarrhœa set in. For weeks, or even months, this may persist, in spite of ordinary remedies; and even when it is controlled by appropriate treatment, it is apt to recur as soon as the medicines are discontinued. Meanwhile emaciation proceeds somewhat rapidly, the legs are apt to become œdematous, profuse exhausting perspirations alternate with the diarrhœa, pains occur from time to time in the chest, the breathing becomes more oppressed, aphthous ulceration occurs in the mouth and throat, the appetite fails, and the patient sinks. In many instances these symptoms are found associated with amyloid degeneration of the tissues.

In another class of cases the symptoms at first do not differ materially from those just described, but after a time severe abdominal pain occurs, accompanied by sickness and irregularity of the bowels; there is gradually increasing distension of the belly, caused partly by the effusion of lymph and serum into the abdominal cavity, and partly by flatulent distension of the bowels; and there are great tenderness on pressure and severe febrile excitement. Those symptoms which are due to ulceration of the glandular structure of the intestines, and a tuberculous condition of the peritoneum with chronic peritoneal inflammation consequent thereupon, may be brought more or less under the influence of treatment, and thus may run a longer or shorter course; but their occurrence is commonly the sign for the commencement of uncontrollable diarrhœa, rapid emaciation, hectic fever, and other symptoms which bring the patient rapidly to the grave.

In yet another class of cases the symptoms resemble those above described, except in the occurrence of fistula in ano. The cough is just as slight, the nervous and physical depression as great, and emaciation is also well-marked; but diarrhœa is a far less prominent feature, and appears to be replaced by the discharge from the fistula. Sometimes, indeed, a harassing cough or a troublesome diarrhœa may coexist with a fistula in ano; but in the cases of latent phthisis now under consideration, neither the cough nor the diarrhœa assume any importance unless the discharge from the fistula is checked. Nor need they do so even then; for if the patient's health and strength be improved, and the tendency to further mischief arrested, the discharge may be stopped and the fistula cured without any fear of untoward consequences: but if, on the contrary, the discharge be checked while the constitutional disorder is still progressing, severe cough or profuse sweating will be excited, or diarrhœa will set in and the malady will assume the more ordinary features of pulmonary consumption.

In some instances of latent phthisis there is almost entire absence of cough from first to last; in others there is slight cough throughout, but it never becomes a prominent symptom; and in others, again, it may be scarcely noticeable until the diarrhœa is checked, or the fistula in ano cured, when it often proves extremely troublesome, and the case assumes the ordinary characters of pulmonary consumption. In either case tubercles may be found in the lungs after death in every stage of development and degeneration, the absence of cough being in no degree attributable to the paucity of tubercular matter in the lung, or to any particular stage of its development, but rather to the absence of pulmonary irritation resulting from the local determination elsewhere set up, by the irritation of the fistula in ano, or by tubercular deposits in the glandular structure of the bowels and of the ulceration occurring there.

The general character of the symptoms of consumption will be obvious from the description already given of the malady, but their special features require careful consideration. Cough is one of the earliest symptoms, and in some instances is peculiar and characteristic; it is generally unpreceded and unattended by running at the eyes and nose, is rarely paroxysmal, and at first is dry, being due to sympathetic irritation of the larynx, and hence it is often supposed to be a nervous or stomach cough, or is referred to relaxation of the throat. At first it occurs chiefly if not exclusively in the early morning, but, as the malady proceeds, it is observed during the day, and especially after exertion, or on any change of temperature. After a time, it ceases to be dry, and is accompanied by more or less expectoration; it recurs at short intervals throughout the day, is especially troublesome at night and in the morning, is often very severe, and not unfrequently gives rise to vomiting, especially when it occurs soon after a meal; but, in exceptional cases, it is extremely slight throughout the whole course of the disease, and is not attended by expectoration; whilst in others in which the disease arises in sequel of bronchitis, it is attended by expectoration from the date of its invasion. The expectoration is at first scanty, thin, colorless, and transparent, somewhat resembling saliva or gum-water; or it may be of a grayish color, and more or less frothy. After a time, the thin, colorless sputa lose some of their transparency, and are seen to contain specks of opaque matter, which gradually subside and form a deposit resembling the sediment in barley-water, or else remain suspended by the more ropy portions of the secretion, and float as streaks in the transparent mucous fluid. If the expectoration be at first grayish colored and frothy, it gradually becomes less aerated, more glairy and more tenacious; loses its purely gray color, and is seen to be mixed with specks or streaks of an opaque white or buff color, and not unfrequently with specks or streaks of blood.

As the malady progresses, the character of the sputa changes again; they gradually become opaque, of a whitish or yellowish hue, and are coughed up in more distinct and more homogeneous masses; sometimes they form ragged pellets of a yellowish-white color, somewhat resembling boiled rice in appearance, which sink or else partially float in a colorless semi-transparent, ropy, non-aerated mucous fluid; and at others they are accompanied by very little of this mucous fluid, but form larger masses of a buff or yellowish-green color, flocculent in appearance, yet perfectly smooth in outline, which do not easily coalesce, but remain perfectly distinct, and separate from one another if expectorated into a vessel containing water. These sputa have been termed nummular, from their assuming the flat, circular form of a piece of money.

At a still later stage of the disease the sputa assume an ash-gray color, become perfectly purulent and homogeneous, and are coughed up in smooth, roundish pieces, which run together and form one mass if expectorated into a dry vessel, or sink, being perfectly free from admixture with air, if expectorated into water. None of these forms of sputa, however, are perfectly characteristic of phthisis. The boiled-rice sputa—the thin, semi-transparent sputa with a deposit resembling barley-water, and the opaque, circular, nummulated sputa are those which are most distinctive; but they all occur occasionally in chronic bronchitis, and therefore cannot be relied upon as evidence of consumption.

In certain instances, sometimes through a period of many years, pieces of calcareous matter are expectorated, varying in size from that of a pin's head to that of a pea or a small bean; and in hardness, from that of soft gritty matter up to that of a urinary phosphatic calculus. Sometimes these harder concretions are branched, being moulded in the minute subdivisions of the bronchi. They all indicate a retrocession of the tubercle at the spot where they are formed, but not necessarily a subsidence of the disease, which oftentimes is advancing steadily in other parts of the lung at the very time when these concretions are ejected. In other instances branched *fibrinous* concretions are expectorated, formed either by the coagulation of blood in the smaller bronchi, or else by exudation from their surface.[1] This, however, is exceedingly rare.

The quantity of expectoration varies greatly in different cases, and also at different periods of the disorder. Sometimes it is profuse throughout the greater part of the attack; in other instances it is scanty at one period, and profuse—suddenly profuse, as if from the evacuation of a vomica—at another; and in others, again, it is comparatively insignificant in amount throughout, although large cavities are found in the lungs after death. In exceptional cases it is altogether absent, the sputa being swallowed, as they constantly are by children.

The taste of phthisical sputa is at first saline and disagreeable, but at a later period it is often sweet, owing to the formation and presence of sugar; its odor is ordinarily faint and nauseous—not fetid, unless mortification of some portion of the respiratory apparatus has taken place, when it becomes nauseous in the extreme, and acquires what is known as the gangrenous odor.

The microscopical appearance of phthisical sputa varies according to the source whence they are derived. Before the tubercles have softened, or before the softened tubercular matter has found its way into the bronchi, the expectoration is derived from the mouth, and from the congested and inflamed mucous lining of the bronchi. It then consists, 1st, of salivary fluid with epithelium from the mouth; and, 2dly, of epithelium from the trachea, bronchi, and air-cells, blood corpuscles, mucous corpuscles, exudation cells, and pus globules. But when the softened tubercular matter has made its way into the air-passages, the products of disintegration of the lung and the tubercle are found mixed with the matters already mentioned. They consist, 1stly, of pus globules and exudation cells in large numbers, tubercle corpuscles, oil globules, saline crystals, amorphous earthy material, and pigmentary matter—the products more especially of disintegration of tubercle; and, 2dly, fragments of the curly yellow elastic tissue of the lung, capillary vessels, and blood corpuscles—the products of the disintegrated pulmonary tissue. These fragments of lung-tissue often afford the earliest obtainable evidence of the existence of tubercle, and in the absence of symptoms of pneumonic abscess may be regarded as almost a distinctive characteristic of phthisical sputa. In every

[1] See *ante*, pp. 324, 326.

instance, but especially in obscure cases and in the earlier stages of the disease, the discovery of the curly elastic fibre of the lung is most important, as it removes all doubt as to the nature of the disorder, and imparts a significance to symptoms which might otherwise be regarded as trivial, and not requiring active medical interference.[1]

Spitting of blood is often one of the earliest, and is certainly one of the most frequent symptoms of phthisis. It may vary from a mere speck or streak of blood, to an ounce, a pint, or even a larger quantity; and in some instances it proves rapidly, if not immediately fatal, producing instant suffocation, or killing indirectly after a few days by exhaustion. Seldom, indeed, does consumption run its course without spitting of blood to a greater or less extent; but fatal hemorrhage, as the result of tubercular ulceration, is an event of very rare occurrence. The cause of this is, that the larger vessels resist ulceration in a remarkable degree, and before complete erosion of their coats has taken place, clots usually form in their interior, and the circulation through them is arrested. Dr. Walshe names 81 per cent. as the proportion of cases in which spitting of blood to a greater or less extent occurs, and he reports having met with two instances of rapidly fatal hemorrhage amongst 131 analyzed cases. My own experience leads me to affirm, that, if the observation be confined to adults, and if cases of acute and rapidly fatal phthisis be excluded from the calculation, spitting of blood to a greater or less extent occurs in every instance; and it also warrants my asserting that fatal hemorrhage is far less frequent than is represented by Dr. Walshe's numbers. I have only met with seven such cases, though I have seen many hundreds of deaths from phthisis.[2] In children profuse hæmoptysis is even less frequent.

[1] Thanks to Dr. Fenwick, the discovery of minute fragments of pulmonary tissue by the microscope is no longer the tedious and uncertain process which it was formerly. The mode of proceeding which he has suggested is as follows: A portion of suspected sputa should be mixed with an equal quantity of a solution of caustic soda containing fifteen grains to an ounce of distilled water, and should then be boiled for two or three minutes with the view of dissolving all the particles capable of solution by the soda. The boiling being completed, three or four times its bulk of distilled water should be added gradually to the mixture, which should then be poured into a conical-shaped glass, and set aside to deposit any insoluble matter. If any lung-tissue exists, it will be found in the fluid which is in contact with the sides of the conical glass near its bottom.

[2] I have seen several patients, supposed to be phthisical, die suddenly of hemorrhage from the lungs; but post-mortem examination has revealed the existence of an unsuspected aortic aneurism, from which, and not from tuberculous ulceration, it was proved that the hemorrhage took place.

Spitting of blood very commonly takes place at the outset of the disease, and recurs at longer or shorter intervals throughout its progress. In most instances the blood is small in quantity, and only specks or streaks the sputa; but sometimes, especially in florid persons, profuse hemorrhage is the first or earliest noticed symptom. This has led to the erroneous impression that spitting of blood occasionally gives rise to consumption. The blood is usually of a florid, red color, not unfrequently of a brick-red hue, and sometimes, though rarely, except in the later stages of the complaint, of a venous tint. As the disease advances, it is ejected more frequently, and in rather larger quantity, forming small clots; and it is more commonly met with in the expectoration of males than in that of females,[1] and in that of adults than in that of children. When it is ejected in large quantity, the blood is frothy from admixture with air.

When the hemorrhage is very profuse, it tends to serious obstruction of the breathing, and possibly may thus accelerate the progress of the tubercular disease; but in smaller quantity it may act beneficially by relieving the local congestion, and it certainly does not hasten the fatal termination of the disorder. Even when very profuse, it does not necessarily shorten life. It may take place once, and may not recur, or it may take place frequently without producing permanent ill effects. A gentleman, whose mother was phthisical, and who died of consumption at the age of fifty-six, had coughed up almost yearly during the last thirty-three years, and sometimes two or three times a year, blood varying in quantity from a few ounces to more than a pint, and yet, until the last two years of his life, he had enjoyed good health during the intervals of the attacks. Such cases as these, however, are quite exceptional. In some instances the outpouring of blood into the air-passages, in connection with phthisis, gives rise to the production of pulmonary apoplexy; but this occurrence also is rare—so rare that it was met with only eight times amongst the 517 consumptive patients examined in the dead-house of St. George's Hospital during the ten years ending December 31st, 1850.[2]

From what has been already stated, it is obvious that the prognosis of spitting of blood in consumptive cases must be somewhat uncertain. Ordinarily, little fear need be entertained of a fatal issue, unless the hemorrhage be very profuse; and even then the chances are immensely in favor of temporary recovery.

[1] This is contrary to Louis' observation. He reports that spitting of blood is more frequent in females than in males, in the proportion of three to two. Dr. Walshe's experience coincides with mine in reference to this matter.

[2] See "Decennium Path.," p. 66.

The most unfavorable signs are persistent frequency and irritability of the pulse, and chronic elevation of the temperature of the body, which indicate considerable constitutional disturbance and undue excitement of the circulation. In such cases a very cautious opinion should be given; for when once a severe attack of bleeding has occurred, profuse hemorrhage is apt to take place at some future date. This at least is the result of my observation. In each of the seven rapidly fatal cases which I have met with, profuse bleeding had previously taken place from the lungs.[1]

Pain in the chest, especially acute pain, is rarely an attendant on the early stages of consumption, but a dull aching uneasiness under the clavicles or the shoulder-blades is one of its most constant symptoms. As the disease advances, pain—sharp pain, though not very acute or catching—is often experienced in the chest, caused sometimes by simple pleurodynia, or by the morbid changes going on in the lung, but in most instances by local attacks of dry pleurisy. They differ from the pains which accompany bronchitis, in being felt in the side, or in the back, or under the shoulder-blades, rather than under the sternum, and in being aggravated by inspiration almost as much as by coughing, which alone produces much pain in bronchitis.

Difficulty of breathing is by no means a frequent symptom. When it does occur in any marked degree, it is not referable to unmixed phthisis, but to some coexistent mischief, such as heart disease, pleurisy, pneumonia, bronchitis, or pulmonary emphysema. In one case which fell under my care at St. George's Hospital, dyspnœa occurred to a frightful extent, and was found to be referable to rapidly developed and extensive emphysema of the lung.[2]

But though difficulty of breathing is not observed except on active exertion, *acceleration* of the breathing is a constant and most valuable diagnostic symptom. It is one of the earliest indications of consumption, and in the absence of heart disease, and of any bronchitic, pneumonic, or pleuritic complication, it serves as a tolerable index to the extent of tubercular mischief. Even when the patient is at rest, and appears to be breathing tranquilly, the frequency of the respirations will prove to be above the normal standard, and on the slightest exertion may rise to a degree which is quite inconsistent with a healthy chest.

[1] For further particulars respecting hæmoptysis, see *ante*, chap. ii, pp. 269-276.

[2] This case, which is most interesting in a pathological point of view, is reported in the "Trans. Path. Soc. of London," vol. xi, p. 15.

Their frequency bears a marked relation to the state of the pulse, and consequently, in many instances, to the severity of the febrile symptoms; but when the lungs are extensively occupied by tubercles, the respiration is very frequent, although there be little febrile excitement.

Febrile symptoms are seldom observed in the very earliest stage of the disorder, unless it occurs in an acute form; but they begin to show themselves at a somewhat later period, and in the second and third stages assume the form which is recognized as hectic. Sometimes, however, in chronic cases they are not well-marked, even after cavities have been formed in the lungs. They usually commence with chilliness, followed by burning heat of skin, the temperature of which very commonly reaches 104° or may range as high as 106°, and then by profuse perspiration; and this train of symptoms, which constitute a perfect paroxysm, may occur at noon, and again towards evening. More commonly, however, they occur only at night, and the perspiration continues until early morning. The paroxysm is often imperfectly developed, the shivering being slight, though the perspiration is profuse, and not unfrequently the sweatings take place even when no distinct chilliness has been previously experienced. Indeed, at an advanced stage of the disorder perspirations break out whenever the patient falls asleep, and are extremely copious and enfeebling. No symptom is more remarkable or more distressing than these colliquative sweats in phthisis, and no source of drain on the system appears to be more exhausting. Thus, it usually happens, that when the perspirations are profuse, and persist in spite of treatment, the case runs on rapidly to a fatal termination.

Emaciation is one of the most striking and characteristic features of consumption, and one which is always present, unless the disease has run a rapid course, and has cut off the patient ere time enough has elapsed to admit of wasting taking place. Not unfrequently it is the symptom which first attracts notice, and excites the apprehension of the patient and his friends. Referable, without doubt, to the general cachexia and to the mal-nutrition consequent thereupon, it often commences while the cough is as yet trifling, and before loss of appetite, diarrhœa, or perspiration have set in. When these sources of waste are superadded, it progresses with increased rapidity. The fat which is stored up in the body disappears, the cellular tissue shrinks, the muscles waste, even the heart decreases in weight, and the whole body becomes rapidly thinner and lighter. The only structure which appears to withstand the general loss of volume is the tongue, which even up to the time of death does not materially diminish in size.

There are exceptions, however, to this law of waste. In acute and rapidly fatal phthisis the patient is sometimes suffocated before emaciation has taken place; and in certain chronic cases I have seen patients die whilst as yet plump, although cavities had long existed in the lungs. But in these latter cases the patients, although consumptive, have not succumbed to phthisis, but have been carried off by inflammation of the lungs, acute bronchitis, or some other disorder.

As a general rule, emaciation progresses least rapidly when the appetite remains good. It should be clearly understood, however, that the mere amount of food taken does not afford a trustworthy clue to the rapidity of its progress, for excessive wasting will often occur while nourishment is being taken in large quantity; and in other cases it will proceed slowly, even though food be eaten sparingly. So, again, it will sometimes go on more rapidly in persons who are free from the ordinary symptoms of dyspepsia than in those who are troubled with flatulence, acid eructations, and other signs of indigestion. The fact appears to be, that it is regulated more by the mode in which the whole process of assimilation is performed, than it is by the quantity of food taken, or by the occurrence or non-occurrence of mere stomach derangement. My own impression is, that it is closely connected with the extent of tuberculization of the mesenteric glands. Be this as it may, emaciation, though not necessarily proportioned to the extent of pulmonary disease, and not even a necessary accompaniment of it, is, when it exists, a very trustworthy guide. It is always an alarming symptom of disease; and when it occurs without any apparent cause, coincidentally with shortness of breath, acceleration of the pulse, and a slight, hacking cough, it may be regarded as almost conclusive evidence that tuberculization is making havoc in the lungs, and probably is implicating the mesenteric glands.

Sore throat is a symptom experienced by many consumptive patients long before pectoral mischief has declared itself; and as the tubercular disease progresses, it becomes a serious aggravation of the patient's sufferings. The affection now referred to is not the aphthous ulceration frequently met with in the later stages of consumption, but congestion of the mucous membrane lining the posterior fauces, the pharynx, and the larynx, accompanied by a diseased condition of the follicles, and a morbidly increased secretion of viscid mucus. It is an affection analogous to that known as the "clergyman's sore throat," and is especially prone to occur in persons between the ages of twenty and forty who are weak and cachectic, and whose digestive organs are out of order. It gives rise to a constant soreness or uneasy sensation in the throat, a frequent inclination to swallow, and repeated hawking to clear the throat, which, after a time, results in the ejection of a small quantity of viscid, opaque mucus. It is often confined to the posterior fauces or pharynx, and in such cases begins and ends as a "relaxed sore throat;" but when it extends to the epiglottis and larynx, it produces more or less huskiness or hoarseness. How far this affection is connected with the deposit of tubercle in the diseased follicles of the lining membrane is a point on which different opinions are entertained; and it is probable that the truth lies between the theory which denies that tubercle is ever secreted by those follicles, and that which refers the disease invariably to the deposit of tubercle in the follicles. 'Certain it is, on the one hand, that the affection is met with and permanently got rid of in persons who, though weakly and dyspeptic, are not consumptive; and, on the other, that in some instances enlarged follicles may be seen filled with a matter closely resembling tubercular matter. Further, it is certain that in persons who, in the earlier periods of their illness, have suffered from this form of sore throat, the more distinct tubercular affection of the larynx, hereafter to be described, is apt to arise at a later period.

The pulse affords an early, and therefore an important indication of consumption. In exceptional cases it may not range above 60 or 70 in a minute throughout the whole course of the disease; but more commonly it rises to 80 or 90 at an early period, and as the malady advances may number 120 or 140. A pulse constantly above 80 in a person who is short-breathed, and has a slight hacking cough, is always suggestive of tubercular disease, however stout and otherwise healthy he may appear, and this more especially if the temperature of the body usually exceeds the normal standard; and a confessedly phthisical patient, whose pulse is habitually very quick and weak, has little chance of amendment.

The digestive organs sometimes remain in tolerable order, and the appetite, though capricious, is not materially impaired; but more commonly the appetite fails, and the digestive organs are disturbed to a greater or less extent, even in the earlier stage of the disease; and at a more advanced period the disturbance is very serious. At first there are pain and tenderness at the epigastrium, increased after food, together with nausea and occasional vomiting; the bowels are costive or extremely irregular, and small aphthous ulcerations occur in the mouth. As the disease advances, tuberculization of the mesenteric glands and of the glandular follicles of the bowels occurs, the abdomen becomes tender on pressure, ulceration of the bowels takes place, and diarrhœa sets in, accompanied by extremely offensive evacuations, sometimes mixed with blood; nausea and vomiting after food become more frequent, and the mouth and fauces are apt to become sore and extensively aphthous, so that deglutition is rendered painful and difficult. This, at least, is the ordinary course of the disease; but experience has proved that extensive tubercular ulceration of the bowels may be going on although constipation exists, and although there be no pain in the abdomen, no nausea or vomiting, and no tenderness on pressure. Any considerable increase in the temperature of the body which is not accompanied by an accession of pectoral mischief, or by other obvious cause, gives fair ground for suspecting that abdominal mischief is going on. In some instances peritonitis is suddenly set up, and occasionally, though rarely, ulceration of the bowel terminates in perforation and the escape of fecal matter into the abdominal cavity. In this case excruciating pain occurs in the abdomen, collapse ensues, and the patient usually sinks in a few hours.[1]

Diarrhœa is of common occurrence, and is a symptom of grave significance in doubtful cases of consumption. Oftentimes, indeed, it serves as an index to the true nature of symptoms about which some doubt may have been previously entertained. The diarrhœa of phthisis is not an ordinary looseness of bowels, nor is it due to ordinary causes; it has a special origin, and is marked by several peculiarities. It is troublesome but not painful: it is very persistent, and is seldom amenable to ordinary remedies; it is apt to recur; and it is not usually accompanied by the furring of the tongue, and the sickness and vomiting which attend other forms of diarrhœa. Hence its character may be recognized after a few days' observation; and when, in a person suffering from cough, the bowels, which previously have been costive, become habitually relaxed, and the looseness assumes the characters above mentioned, there can be little doubt as to the true significance of the cough, or as to the nature of the coexisting constitutional malady. If any doubt exists it may be solved by observing the temperature of the body, which in these cases is persistently above the natural standard, and usually ranges from 99.5° to 104°.

[1] A patient in whom this complication arose was admitted under my care into St. George's Hospital on the 21st of December, 1859. For full particulars of the case, which was remarkable for the entire absence of diarrhœa and other abdominal symptoms during life, see "Trans. Path. Soc. of London," vol. xi, pp. 103–4.

The tongue is sometimes clean, at others coated; but when ulceration of the bowels is going on, it is oftentimes clean, red, and beef-steaky in appearance, and not unfrequently ulcerated or aphthous. Its condition, however, does not throw any light on the progress of the pectoral disease.

The urine in the earlier stages of the complaint is sometimes normal, but more commonly it is remarkable for the rapid variations in its character. At one time it is clear, pale, abundant, and of low specific gravity; in a few hours, or on the following day, scanty, high-colored, and possibly loaded with lithates. When the disease is more advanced, it generally contains an excess of lithic acid, and is often turbid. Albumen is not usually one of its constituents, though a minute quantity may be detected in it from time to time, referable to temporary congestion of the kidneys. In a certain proportion of cases, however, it is more constantly present, owing to the existence of tubercular or other disease of those organs.

The monthly periods may continue regular almost to the last, and the discharge natural both in quantity and quality. More commonly, however, the menses are very profuse or else very scanty at the commencement of the disease, and cease as soon as the symptoms of consumption are more fully developed. In some instances they cease at the very commencement of the malady, and their sudden cessation is referred to by the patient as the cause of all her subsequent illness: indeed, the apparent connection between their suppression and the setting up of pectoral disease is sometimes so remarkable, that many writers on consumption have referred to cases under the title of amenorrhœal phthisis. It is almost needless to add, that the term is an improper one, the phthisical symptoms being in no wise referable to the stoppage of the monthly discharge, but rather connected with the morbid condition of the system, which led to its cessation. In many of these cases leucorrhœa is a prominent symptom.

The external and superficial glands are especially liable to become the seat of tubercular deposit in children, and form knotty swellings, which greatly disfigure the patient. In adults they are rarely enlarged. When the enlarged cervical glands suppurate, as they often do, the pectoral symptoms are usually quiescent; and it is a rare occurrence to find active mischief in the chest coexisting with tubercular suppuration in the neck. Indeed, glandular enlargement, though symptomatic of tuberculosis, and sometimes coexisting with phthisis pulmonalis, can hardly be regarded as a symptom of that disease. On the contrary, though indicative of a scrofulous disposition, it seems to act as a derivative, and to be in some measure antagonistic to tubercular deposition or suppuration in the lungs.

The fingers and nails undergo a change which requires special notice; the more so as diagnostic importance has been attached to it. Sometimes, in the earlier stage of the disease, but more commonly when emaciation has proceeded to some extent, the last joint of the fingers enlarges, their tips shrink and diminish in size, and the nails become remarkably convex and adunc. The ends of the fingers thus acquire a peculiarly rounded or clubbed appearance. There cannot be a doubt that this change of form is very remarkable, and is observed to a greater or less degree in most cases of phthisis; but, nevertheless it cannot be regarded as distinctive. In several healthy, middle-aged, and elderly persons, I have seen this condition of the fingers strongly developed, and in some consumptive patients it does not occur even up to the last hour of their existence.

So, also, in regard to falling of the hair, which is a symptom of great frequency in consumptive cases. It is sometimes observed at the very outset of the disease, and is a common cause of complaint long before the patient or her friends suspect the nature of the malady which is impending; but in certain instances, and more especially when the disease runs a chronic course, unaccompanied by much sweating, the hair is retained until the last.

Dropsical swelling of the extremities is not of frequent occurrence; indeed, its absence is rather remarkable, when we consider the impoverished condition of the blood, and the serious impediment offered to its circulation through the lungs. Nevertheless, the feet and ankles are apt to become puffy and swollen towards evening, especially at the close of the disease, so that the occurrence of such swelling may be regarded as of evil augury. But, except in the later stage of the complaint, and to the extent just described, I believe that œdema does not occur simply as the result of the conditions above referred to. In certain instances the legs become anasarcous, as a result of coexisting disease of the kidneys: in others, in consequence of feebleness of the heart, dependent on fatty degeneration of its tissue; whilst, in certain cases, one or both of the lower extremities become excessively swollen and anasarcous, in consequence of embolism or of the coagulation of blood in the veins. In this latter event, if phlebitis be present, the swelling is attended with more or less pain in the course of the veins.

Sometimes a red line makes its appearance on the gums, and is regarded by many persons as very distinctive of the tubercular cachexia.[1] My own observation, however, inclines me to believe that it is seldom developed in the earlier stages of the disease, when alone its presence would be of practical value in a diagnostic point of view, and that it sometimes accompanies derangement of the stomach when no tubercular mischief exists. Under these circumstances its occurrence, though suggestive of consumption, cannot be viewed as declaratory of the disease.

The functions of the brain are seldom interfered with to any great extent. Slight wandering at night, or on waking out of sleep, is often observed towards the close of the disease; but, even in such cases, the mental powers remain unimpaired during the day. Indeed, the perceptive faculties oftentimes acquire unwonted vigor, and the imagination is unusually active. Sometimes, however, violent delirium sets in, or the patient passes into a state of muttering delirium, and then becomes comatose. In these cases, if the irritation results from the deposit of tubercle in the brain, or in its investing membranes, a rapidly fatal issue may be anticipated. If, on the other hand, there is reason to believe that the delirium has been caused, as sometimes happens, by injudicious and depressing treatment, the prognosis would be less unfavorable, inasmuch as the mischief may then be overcome by appropriate remedies.

A strange waywardness of temper usually marks the disease throughout. The patient is at one time irritable and depressed; at another placid and singularly hopeful, maintaining up to the last a delusive hope that the malady will be subdued. So unaccountable is the apparent hopefulness in many instances, that it might almost be supposed referable to an anxiety on the part of the patient to deceive herself and her friends as to the actual state of the case. But the absence of this feature in other forms of complaint which are equally lingering and equally fatal, proves that some different explanation must be sought, and stamps it as a characteristic of consumption.

The complications of phthisis pulmonalis have been already alluded to. Pleurisy, pneumothorax, pneumonia, bronchitis, and spitting of blood, which are some of the more important, have been discussed under their respective heads, and ulceration of the bowels has been briefly described in the present chapter. But there are yet others which require further consideration. Amongst these may be mentioned chronic tubercular inflammation of the epiglottis, larynx, and trachea, tuberculization of the bronchial glands, and irritation and inflammation of the brain or its membranes consequent on the deposit of tubercle.

[1] See "Lancet," Oct. 8th, 1859.

It has been already stated, that hoarseness and partial loss of voice are apt to supervene in the course of pulmonary consumption. They are due to congestion and inflammation of the epiglottis and larynx, arising in consequence of local tubercular mischief, and when once established, they are apt to continue in a chronic form, and not unfrequently increase gradually in intensity. Sometimes this alteration of the voice is slight, and does not occur until the pectoral disease is far advanced; at others, it is observed very early in the attack, and becomes a prominent symptom soon after the commencement of the catarrhal symptoms, by which the invasion of the disease is marked. Many writers have alluded to cases of this kind under the title of "Laryngeal Phthisis," and have implied, if they have not actually affirmed, that tubercular mischief may occur in the larynx, giving rise to symptoms of the so-called laryngeal phthisis, without the existence of any tubercular disease in the lungs. This, however, is not the fact. The tubercular deposit in the lungs may be more or less extensive, and the pectoral symptoms may be developed to a greater and less degree, but the so-called laryngeal phthisis never runs its course without the coexistence of tubercular disease of the lungs. In short, there is no such disease as laryngeal phthisis. The chronic inflammation or tubercular ulceration in the larynx often met with is merely an accessory to or an accidental complication of phthisis pulmonalis.

The laryngeal symptoms, however, are somewhat peculiar, and impart an entirely new complexion to the complaint. They are characterized by an alteration in the force and quality of the voice and cough, and by pain, tenderness, and a sense of dryness, pricking, and constriction over the larynx. At first there is simply occasional huskiness and "cracking" of the voice, which disappears after a forcible "hemming" or "clearing of the throat," but gradually more permanent hoarseness and partial loss of voice supervene, with a muffled and cracked sound on coughing—the cough being frequent and accompanied by a scanty mucous expectoration. As the disease proceeds, and the vocal cords are destroyed by ulceration, the voice degenerates into a hoarse or a squeaking whisper; the breathing in chronic cases becomes markedly stridulous; and the cough, which is constant, and attended by the expectoration of a large quantity of thin, frothy mucous or muco-purulent fluid, sometimes tinged or streaked with blood, produces a strangely mixed hoarse and stridulous sound. Yet with all this evidence of local mischief, there is seldom much febrile excitement, and at first, not unfrequently, very little pain or difficulty of breathing.

If the epiglottis remains free from mischief, there is no dysphagia, but when, as often happens in the latter stage of the disease, it becomes inflamed or ulcerated, each attempt at swallowing may give rise to a paroxysm of suffocative dyspnœa and spasmodic cough, the liquid which the patient was in the act of swallowing being rejected through the nose as well as through the mouth. This, however, is not always the case, as post-mortem investigations have clearly proved that the epiglottis may be ulcerated, and, nevertheless, there may be no difficulty of swallowing.

Sometimes, though rarely, portions of the cricoid and arytænoid cartilages become detached during the process of ulceration, or cretaceous deposits, which have formed in the larynx, become loose, giving rise to excessive cough and a sense of impending suffocation. In most instances these hardened masses are ejected with expectoration during a violent fit of coughing, but occasionally they pass into the trachea, and occasion great local irritation and an aggravated attack of dyspnœa. Death commonly takes place, as in uncomplicated cases of pulmonary consumption; but occasionally it results from paralysis of the glottis, induced by the extension of ulceration, or from spasm excited by the passage of foreign matters into the diseased larynx. After death, the congested mucous membrane of the larynx often exhibits a granular appearance; the edges and laryngeal surface of the epiglottis are found ulcerated, or ulcers are discovered in the ventricles of the larynx, or between the epiglottis and the chordæ vocales; the laryngeal muscles are wasted and often softened, and portions of the cartilages are sometimes ulcerated and necrosed. In many instances ulcerations exist in the upper part of the trachea, but seldom give rise to any marked symptom; in others false membrane is found effused into the larynx and upper part of the trachea,[1] and occasionally ulceration perforates the larynx and extends into the œsophagus, giving rise to excessive difficulty of swallowing. So common is ulceration in some part of the upper air-passages, that Louis reports having met with it in the epiglottis and larynx in one out of every five consumptive patients whom he examined after death, and in the trachea in no less than one out of every three patients.

[1] A case in which this occurred, and in which tracheotomy had to be performed, is at present (Nov. 7th, 1861) in St. George's Hospital, under the care of my colleague, Dr. Bence Jones. The patient, a man named Kent, in Fuller Ward, ejected through the opening pieces of false membrane an inch and an inch and a half in length, and about the breadth and thickness of a quill pen.

When the larynx is affected it is often difficult to diagnose the mischief in the chest, inasmuch as the mischief in the upper part of the air-passages obscures the sounds obtainable through the stethoscope. The general symptoms, however, are sufficiently characteristic; and when, in addition to hoarseness and partial loss of voice, there are present the peculiar laryngeal cough, with thin frothy muco-purulent expectoration, semi-stridulous respiration, night sweats, emaciation, and other of the ordinary symptoms of consumption, there can be little doubt as to the true nature of the malady, however masked the chest symptoms may be. If, however, any doubt remains as to whether the stridor and other symptoms of laryngeal affection may not be referable to aneurismal pressure or to malignant or other disease of the larynx, recourse may be had to the laryngoscope. By its aid the physical condition of the larynx can be readily ascertained, and all questions as to the existence of laryngeal disease set at rest.

The prognosis of the disease is always unfavorable, from the fact already mentioned, that the laryngeal mischief is accompanied by tubercular deposit in the lungs. Generally, when hoarseness and aphonia have set in, the progress of the case is rapid, and the patient does not survive beyond a few months. Nevertheless, as with other forms of tubercular disease, its course may be almost indefinitely protracted. A consumptive patient at present under my care has suffered from hoarseness, partial loss of voice, and occasional pain and constriction in the larynx for a period of seven years, and at the present time is again rallying from an attack which for some weeks threatened to terminate fatally.

Tuberculization of the bronchial glands sometimes occurs when the lungs are free from tubercular deposit, but more commonly, it is associated with pulmonary disease, being merely a complication of that disorder. Like other scrofulous affections of the lymphatic system, it is specially prone to arise in children, and may be regarded as an exceptional occurrence in adults. Insignificant in many instances in clinical importance, it is sometimes so extensive as to give rise to symptoms far outweighing those produced by the coexistent pulmonary disease. Thus a congeries of tuberculized bronchial glands may form a large, hard, and irregularly lobulated mass in the chest,[1] which will compress, and may ultimately produce ulceration and perfora-

[1] In the case of Henry Truton, aged twenty-five, who was admitted under my care into the King's Ward of St. George's Hospital, on November 13th, 1861, a mass of tuberculous bronchial glands was found, which weighed more than two ounces.

tion of the adjacent structures, giving rise to the various local and physical signs of mediastinal tumor, and sometimes to symptoms which may be mistaken for croup. Twice I have been called in consultation to decide upon the propriety of performing tracheotomy in cases of supposed croup, when the symptoms during life, which were subsequently corroborated by post-mortem examination, convinced me that the mischief was due simply to a mass of enlarged bronchial glands pressing upon the trachea just above the bifurcation of the bronchi. A similar case, the particulars of which will be found in the post-mortem and case-book in the museum of St. George's Hospital, came under my observation in the year 1861. In each instance I was enabled to decide against the existence of croup, by observing that the stridor accompanied expiration rather than inspiration, and to diagnose the true nature of the malady by finding this peculiar respiration accompanied by dulness on percussion between the scapulæ, and by symptoms indicative of a scrofulous habit of body, and by an absence of all evidence of any other form of disease capable of producing the stridulous expiration.

Tubercle, when deposited in the bronchial glands, is prone to undergo degeneration and softening, just as it does in other parts of the body,[1] and sometimes it becomes the seat of cretaceous transformation. When the enlarged glands suppurate, the softened tubercular matter may make its way by ulceration into a bronchus, or into the trachea, and may then be expectorated, or it may even open into the œsophagus, and occasion more or less difficulty of swallowing. In either case the excavated gland may give rise to the physical signs of a vomica.

It is needless to detail the local symptoms of mechanical origin produced by enlarged bronchial glands, as they are identical with those occasioned by other intrathoracic tumors.[2] Suffice it to say that swelling of the superficial veins of the neck or chest occurring in a child, when accompanied by lividity of the lips, puffiness of the face, œdema limited to one side, or affecting both sides of the neck and chest, constitute a train of symp-

[1] A case in which this softening occurred was admitted under my care in St. George's Hospital, on May 6th, 1861. The child Harriet Bennett, aged four, was sent into the hospital as a case of croup with a view to operation, but as the stertor was almost confined to expiration, and there was abundant evidence of scrofulous disease, I did not consider it a case of croup, but regarded the symptoms as attributable to the pressure of a mass of bronchial glands on the trachea. After death this was found to be the case, and it was discovered further that this scrofulous mass had softened and formed an abscess which had burst into the trachea (See "Hospital museum Post-mortem and Case Book" for 1861.)

[2] They are fully discussed in the section relating to that subject.

toms which are extremely suggestive of the disease under discussion, and that when to these symptoms are superadded dysphagia and ringing, paroxysmal cough, stridulous breathing, hectic, and emaciation, there can be no reasonable doubt as to the nature of the malady. It may be added, that whereas aneurismal and other intrathoracic tumors are of extreme rarity or almost unknown in children, bronchial gland tumors are of somewhat common occurrence. Therefore, if pneumonia or bronchitis be not present, the presence of labored respiration in children, attended by greatly prolonged expiration, and dulness of percussion between the scapulæ, affords sufficient grounds for suspecting the existence of enlarged bronchial glands.

The general symptoms vary greatly, according to the degree to which the enlarged glands press on the surrounding structures, and to the amount and stage of the tubercular mischief in the lungs. Ordinarily the patient's face is pale and anxious, the breathing hurried, often labored, and sometimes stridulous; there is frequent cough, which is harsh, dry, ringing, and paroxysmal, when referable to pressure on the trachea or bronchi, or to irritation of the vagus or recurrent nerves, but hoarse, loose and non-paroxysmal, when due principally to tubercular mischief in the lungs. If the patient be a child, there will be little or no expectoration throughout, inasmuch as the softened tubercular matter will be swallowed; whereas if the patient be an adult, the sputa will be catarrhal, muco-purulent, or purulent, according to the stage at which the lung disease has arrived. Sometimes cretaceous matter from an enlarged bronchial gland may be discovered in the expectoration. If the tuberculous mass causes pressure on the trachea or bronchi, or irritation of the recurrent nerve, not only will the breathing be stridulous, but violent paroxysms of suffocative dyspnœa may arise, accompanied by lividity of the surface, extreme anxiety of countenance, and cold, clammy perspiration. If there be pressure on and subsequently ulceration of the œsophagus, there will be gradually increasing difficulty of swallowing, and ultimately extreme dysphagia, with a sense of soreness or pain on attempting to swallow. The ordinary symptoms of pulmonary consumption will be gradually superadded to these, according as mischief in the lungs is more and more developed.

Thus, then, it will be seen that in the earlier stage of the disease there is no positive evidence of enlargement of the bronchial glands, and that inasmuch as the signs to which this malady gives rise are simply those of pulmonary irritation and local pressure—signs which may be produced by any intrathoracic tumor—the diagnosis of enlarged bronchial glands is extremely difficult; indeed, it can only be a matter of inference from the absence of symptoms indicating the existence of other forms of mischief, which are equally liable to be accompanied by the development of tumors causing pressure on the structures within the chest. In adults, therefore, in whom intrathoracic tumors of all sorts are apt to arise, the diagnosis must necessarily be very uncertain. But when, as is commonly the case, this form of disease is met with in children, the diagnosis is less difficult and less uncertain, for the reason that in them no other form of mischief is liable to occur capable of exciting many of the symptoms ordinarily attendant on this affection, and that percussion between the shoulders elicits dulness referable to the enlarged glands so much more readily than it does in adults.

The last complication of pulmonary consumption which requires special notice is inflammation of the brain or its investing membranes, due to the local irritation of tubercular deposit. Tubercles may occur in the nervous centres, independently of tubercular mischief in the lungs; and by reference to Dr. Chambers's analysis of the post-mortem examinations in St. George's Hospital it will be found that they did so in five cases which proved fatal during the ten years ending December, 1850. But more commonly this form of mischief is associated with tubercular disease in the lungs, and forms one of the many complications which are apt to arise during its progress. It is especially prone to occur in childhood. The first symptoms which attract attention are unusual peevishness and irritability of temper, restlessness at night, and grinding of the teeth during sleep. Shortly the patient complains of headache and occasional giddiness, the scalp is hot, and there is aversion to light and noise; febrile excitement ensues, with constipation and obstinate vomiting, and the countenance becomes distressed and anxious. After the lapse of a few hours, or possibly a few days, the headache increases in severity, the pulse becomes irregular, delirium is apt to occur, especially in adults, and twitchings of the face or actual convulsions take place, followed in many instances by squinting, and by more or less hemiplegia or general paralysis. The pupils, which at first were contracted, eventually dilate, the delirium ceases, and extreme drowsiness or actual insensibility supervenes, and this may endure for periods varying from a few hours to ten days, or even longer, varied only by occasional convulsions. Sometimes the delirium is violent and boisterous, more commonly of a quiet character, and occasionally the cerebral disturbance does not result in true delirium, but rather in a peculiar loss of mental power, which renders the patient speechless, and gives a dull, meaningless expression to his countenance. He seems to understand what is said to him: but after gazing at the speaker for a few moments, deliberately turns his head away, without attemping to speak or give an answer to any question which may be put to him.

The prognosis in these cases is extremely unfavorable. Not only is there in most instances coexistent disease in the lungs sufficient of itself to prove eventually fatal to the patient, but even temporary recovery from these attacks of delirium is rare. Cases, indeed, are met with occasionally in which the symptoms seem to betoken tubercular mischief in the brain or its membranes, and in which nevertheless the cerebral disturbance gradually subsides, and the patient regains the full possession of his faculties. But more commonly, notwithstanding occasional remissions, the extreme drowsiness or stupor, or actual coma, continue to a greater or less extent, and the patient sinks after a few days, or, at all events, within a fortnight or three weeks.

It should, perhaps, be stated, that during these attacks the chest symptoms sometimes remain in abeyance to a great extent, and, on the other hand, that constipation, which so frequently is an attendant on cerebral disorders, is seldom met with when head symptoms occur during the advanced stage of phthisis, in which ulceration of the bowels so commonly prevails. These facts are both important, as, if not borne in mind, they might be apt to mislead.

It will be obvious, from what has been already stated, that the diagnosis of pulmonary consumption must be a matter of extreme uncertainty to a person unpractised in conducting a physical examination of the chest. Not only may the general symptoms of phthisis be obscure, but bronchitis, pneumonia, or pleurisy may be present, and mask the true nature of the disorder. On the other hand, it may not be without a certain degree of difficulty even to the most experienced auscultator; for in the earlier stages of the disease there is no one sign by which the existence of tubercle is clearly indicated, and the general symptoms may each one separately accompany bronchitis, pneumonia, or pleurisy; nay more, when referable to the presence of tubercle, they may be complicated, and masked by the existence of either of those forms of disease. And thus it happens, that if great care be not exercised in conducting the examination of the chest, and much caution observed in drawing inferences from the signs observed, an incorrect conclusion will be arrived at. In short, the physical signs alone are oftentimes as insufficient as the general symptoms to form trustworthy data for an opinion; and the existence of consumption is only to be established by a careful inquiry into the history of the patient, by a jealous investigation into the general symptoms, and by a comparison of those symptoms with the physical signs which are revealed by an examination of the chest.

Bronchitis is the disease with which phthisis is most liable to be confounded, and it may be well, therefore, to place in a tabular form the salient diagnostic features of the two diseases.

Incipient Phthisis.	*Bronchitis.*
The cough commences gradually without fever, and without the running at the eyes and nose which characterize an ordinary "cold"	*The cough* commences suddenly, and is usually ushered in by feverishness and running at the eyes and nose.
The cough is generally dry and hacking for some time after its commencement.	*The cough* is accompanied by expectoration almost from the first.
The expectoration is often specked or streaked with blood.	*The expectoration* is not tinged or mixed with blood.
Pain.—There is seldom much pain in the chest, and rarely any fixed pain. When pain does occur, it is usually of a dull, aching character; is met with in the sides or under the collar-bones or the shoulder-blades; and is not usually aggravated by coughing.	*Pain.*—There is almost invariably pain or tightness or uneasiness under the breast-bone; but rarely any pain in the sides, except immediately after, and as a result of coughing.
The morbid sounds are usually confined to the upper lobe of the lung, and are often confined to one side of the chest; they are very persistent, and even, if met with on both sides at first, are apt to subside partially or wholly on one side, whilst they continue, or even increase on the other.	*The morbid sounds* usually predominate in the lower lobes, and exist equally in both sides of the chest; they are of temporary duration, and subside gradually and equally on both sides of the chest.

For the sake of those who are commencing the investigation of diseases of the chest, it may be desirable to point out certain combinations of symptoms which are always suggestive of tubercular disease of the lungs, and should therefore lead to more than usually careful inquiry. Foremost amongst them is obstinate cough, commencing gradually, without any running of the eyes or nose, or pain in the chest—a cough which at first was dry and hacking or attended by scanty expectoration, and has persisted in spite of ordinary remedies. If a person who gives such an account as this of his cough admits that one or more of his relatives have been delicate, asthmatic, or consumptive; that he has experienced dull, aching pain about the collar-bones and upper part of the chest; that the cough and expectoration have gradually increased, and are most troublesome at night or towards morning; that on one or more occasions the sputa have been mixed or streaked or specked with blood; that he has lost strength and flesh, and has become short-breathed on going up stairs—the general evidence of early phthisis will be complete, though it will require confirmation or refutation by means of a physical examination of the chest. So, too, if emaciation has occurred without obvious cause, and especially if it is accompanied by any of the general symptoms just mentioned, or by night-sweats, or by hoarseness which is not connected with a syphilitic taint, and does not speedily subside under treatment, the gravest suspicion of consumption may be entertained, and a careful examination of the chest is necessary. So, again, if obstinate diarrhœa occurs, and in spite of ordinary remedies and careful dieting, persists for months with scarcely any intermission, and especially if it is not accompanied by much coating of the tongue, or by sickness or other signs of biliary derangement, but on the contrary is attended by a red, glazed, or beef-steaky tongue, and by more or less abdominal tenderness on deep pressure, the case is calculated to excite a suspicion of consumption—a suspicion which would amount almost to a certainty if the diarrhœa were attended by even the slightest cough, or by loss of flesh or shortness of breath. In like manner, chronic peritonitis occurring in an adult is almost always connected with the presence of tubercle, and renders imperative a close investigation of the state of the chest. Under any of these circumstances, a slight flattening of the chest-walls or deficiency of expansion at the summit of one or both lungs, a decrease in the vital capacity of the chest, the occurrence of more or less dulness on percussion in the clavicular, infra-clavicular, or supra-scapular regions, with a harshness or feebleness of the respiratory murmur, an irregularity or jerkiness in its rhythm, the existence of prolonged expiration, or of increased vocal resonance, especially at the left apex, and the occurrence of occasional dry clickings, in addition to any other râles or rhonchi which may be present,—would one and all warrant a more serious interpretation than that which they would justify if unaccompanied by the aforementioned general symptoms. Râles, of whatever character, if persistent at and confined to the apices of the lungs, and still more so if persistent at one apex only, and accompanied by spitting of blood, emaciation, night-sweats, and other of the general symptoms of consumption, may be regarded as almost certainly indicative of tubercular disease; but a cautious practitioner will never express a confident opinion in a doubtful case from the result of a single examination. Time is an important auxiliary in the diagnosis of obscure cases. Râles, which are due to pneumonia, bronchitis, or the presence of blood effused in an attack of hæmoptysis, and dulness which is referable to pleuritic effusion, or to œdema of the lung, or to pneumonic consolidation, will gradually disappear, and cease to complicate and mask the signs which are referable to more permanent and organic changes in the lungs. On the other hand, signs which are due to the presence of tubercle in course of development will gradually pass from one phase to another, and thus establish the true character of the disease. In all cases the aid of the microscope should be appealed to if any uncertainty exists as to the nature of the symptoms. The discovery of a portion of the curly elastic tissue of the lungs in the expectoration would at once serve to dissipate any doubt on the subject.

The prognosis of pulmonary consumption is necessarily unfavorable. In acute phthisis the downward course of the patient is steadily and rapidly progressive; but in certain instances of chronic phthisis the disease is of very long duration; whilst in others, and certainly the majority of cases, it runs a less protracted course, but is nevertheless marked by distinct remissions. The question therefore arises whether it is possible to prognosticate which course it will pursue, by reference to the character of the symptoms which accompany it? In some instances, undoubtedly, it is possible to do so, but in others there are not sufficient data for our guidance. It has been already stated that tubercles when once deposited may remain for years in *statu quo*, or may speedily undergo softening or disintegration, and that these peculiarities are referable in great measure to the intensity of the consumption or constitutional disorder under the influence of which the tubercle is deposited. The same conditions regulate the entire course of the disease, and modify the effects of treatment. Thus, in any given case, if the intensity of the constitutional disturbance can be ascertained, it will not be difficult to predicate with some degree of certainty, whether the course of the pulmonary disease will be slow or rapid—whether temporary remission or suspension of the symptoms may be expected, or whether they will run on from bad to worse, unchecked by remedies. Now it happens that phthisis, or that condition of constitution which results in the deposition of tubercle, is marked by tolerably characteristic symptoms, and when the tendency to tubercular deposition and disintegration is strongly developed, these symptoms are most prominent. Amongst them may be mentioned extreme rapidity and softness of the pulse, an unnatural softness and dampness of the skin, the early occurrence of emaciation, and of hectic with profuse perspiration, sleeplessness, with entire loss of appetite, derangement of the bowels, and diarrhœa. When these symptoms are observed coincidently with great prostration, hurried breathing, and dyspnœa on the slightest exertion, the probability is, not only that there is extensive tubercular deposit in the lungs, but that the constitutional derangement is deep-seated and excessive, and that the retrograde metamorphosis of tissue will continue unchecked, and will bring the patient rapidly to his grave. On the other hand, when the pulse remains slow or is but little accelerated, when the skin maintains its normal elasticity, temperature, and moisture, when there is little quickening of the respiration, little tendency to hectic, little perspiration, and no diarrhœa, when the appetite remains good and emaciation takes place slowly—the system obviously is not overwhelmed by the disease, but possesses considerable power of resistance; the progress of the malady, therefore, will be slow; and not improbably, if appropriate treatment be adopted, remissions may occur, or the onward course of the disorder may be arrested. In cases such as these, if our patient is placed under favorable hygienic circumstances, the prognosis will be comparatively favorable, and partial or practically complete recovery may be anticipated.

It has been attempted to prognosticate the issue of the complaint solely by reference to the condition of the tubercular deposit; and for this purpose it has been assumed that signs of pulmonary softening and excavation are also signs that the disorder is tending rapidly to a fatal issue.' But nothing can be more at variance with the truth. A single mass of tubercle may soften and form a cavity, and nevertheless the patient may rally, and may remain for years in tolerable health, whilst on the other hand, tubercular infiltration of the lungs may take place, and though little or no breaking down of the lung-tissue may occur, the patient may sink rapidly, worn out by the hectic induced by the vitiated condition of his blood, or suffocated as the result of the local interference with his respiration. In short, the amount of the tubercular deposit, viewed in connection with the general symptoms already described, forms a far more trustworthy guide to a prognosis than any evidence, however satisfactory, of the tubercle being in an advanced stage of degeneration and softening.

Of late years it has been asserted not only that remissions of the disease may occur, but that the malady may be permanently arrested. In former days such an event was regarded as simply impossible, the very fact of recovery being admitted as conclusive evidence that the diagnosis was at fault, and that the disease under which the patient was suffering could not have been phthisis. Even at the present day there are not wanting those who maintain, and act as if they believed, that consumption is incurable. But there are no valid grounds for such an opinion. Assuming, for the sake of argument, that the deposit of tubercle has not been so extensive as to interfere materially with the function of respiration, there is no reason why the constitutional derangement on which the formation of tubercle depends should not be checked, and the tubercle already formed got rid of. Experience fully justifies the statement that consumption does not ordinarily pursue a steadily progressive course, but like gout consists of a series of attacks, which often occur at an interval of years, and each one of which results in a deposit of tubercle which forms an additional impediment to the due performance of the function of the lungs. Fortunately, the extent of the respiratory apparatus is greater than is needed for our ordinary requirements, and therefore the presence of a certain amount of tubercle is not incompatible with the due oxygenation of the blood and with the maintenance of tolerable health. Consequently, if the amount of tubercle already deposited, be not excessive, and the general health can be improved, and the tubercular cachexia got rid of during the intervals between the attacks, the tendency to the further deposit of tubercle will be prevented, and virtually the patient will be cured. The tubercle already deposited may be absorbed, or it may remain unchanged and imbedded in the lung, or it may be transformed into a cretaceous mass, or it may soften and be got rid of by expectoration; but in either case, if no further deposit takes place, the functions of life will not be materially interfered with, and the patient may attain to longevity. It is only after several attacks have occurred, and successive crops of tubercle have been deposited, or after a single crop has occurred sufficiently extensive to interfere materially with the function of respiration, and to impede the processes necessary to the oxygenation of the blood and the maintenance of health, that it is out of the power of medical treatment to restore the tone of the system, and arrest the further development of mischief. Clinical observation and pathological research have left little to be desired in the way of proof, not only that recovery from consumption is possible, but that it does take place in a certain proportion of cases. They have shown that the disease is not essentially local, but consists of constitutional derangement productive of impaired nutrition and deterioration of the blood; and as other constitutional disorders, which equally with phthisis result in an alteration of the quality of the blood, admit of removal, there is no *à priori* reason for doubting that under favorable circumstances the faulty condition of the system may in this instance also be rectified, and that then the local mischief would cease.

In numberless cases encysted tubercles have been discovered in the lungs of persons who for years before their death had not been subject to cough, and had not exhibited any symptoms of consumption, and thus it has been proved, that even after the tubercular cachexia has been fully developed, and has resulted in the deposition of tubercle, complete and permanent recovery may take place. It has been shown that tubercle is an exudation from the blood; that it is of low vitality, and not prone to become organized; and inasmuch as all other exudations from the liquor sanguinis, so long as they are unorganized, admit of being absorbed, we are constrained to conclude that the same applies to tubercle. Further, it is a matter of experience, that when tubercle is deposited in the external glands it may remain for years in a quiescent state without producing serious inconvenience; that more commonly it gradually undergoes absorption; and that even when it softens and is eliminated by suppuration no further deposit of tubercle may take place, but the patient may live to an advanced age without the slightest recurrence of the disease. Thus we are constrained to believe that the same holds good in respect to the lungs, and that, whether tubercular deposits in these organs remain quiescent or undergo absorption or calcareous transformation, or be got rid of by suppuration and expectoration, the patient may recover and attain to longevity, provided only that his general health can be improved, and the condition of his blood altered, so that no fresh deposit of tubercle shall occur.

In corroboration of this view, cavities or empty vomicæ in the lungs have been discovered after death in process of contracting and healing, and in many instances they have been found entirely healed and cicatrized, causing deep indentations on the surface of the lung. But observation carries us even a point beyond this; we have not been left to infer, simply as a matter of analogy, that absorption of tubercle may take place in the lung; for, as if to prove that where the reparative process is at work, tubercular matter may be absorbed, tubercle is seldom found in large quantity in the vicinity of pulmonary cicatrices—nay, rather, is often altogether absent, or has left as its only representatives a few pieces of cretaceous or calcareous matter, the whole of the animal constituents of the tubercle having disappeared; whilst in other instances a few small, isolated masses of tubercle—some of which are in their original crude condition, some in a state of partial transformation into cretaceous matter, some completely cretafied or calcified, and some partially or wholly encysted—alone remain to establish the nature of the disease which led to the excavation and cicatrization of the pulmonary tissue.

But there is evidence which goes even beyond this. Cases have been met with in which the general symptoms and the results of repeated examinations of the chest have mutually borne witness to the existence of consumption, and in which, nevertheless, the symptoms have gradually subsided, the physical signs have disappeared, and the patient has regained health, and for years has remained free from chest affection. Several such cases have come under my own observation, in which the subsequent career of the patient, and the facts elicited by post-mortem investigation, have furnished me with ocular and demonstrative proof that complete recovery may take place under favorable hygienic conditions—recovery so complete at all events as to enable the patient to attain to an advanced age, and to live for the remainder of his life in tolerably robust health and free from any fresh accessions of pulmonary symptoms.

Without entering largely into the detail of cases of arrested phthisis, many of which I have had the opportunity of noticing, I will briefly give an outline of three cases in which post-mortem examination confirmed, and thus gave additional interest to the result of careful observation during life.[1] The first was that of a young lady aged nineteen, whose parents were healthy and whose family were not consumptive. She fell into ill health in consequence of the distress and worry incident to a disappointment respecting marriage, and speedily became short-breathed, had spitting of blood, and evinced unmistakable symptoms of consumption. The late Dr. Chambers and Dr. C. J. B. Williams were consulted, and both pronounced her phthisical, and asserted that she had cavities at the apices of the lungs. After the lapse of a twelvemonth, she began to rally, the cough in measure ceased, and the shortness of breath subsided. She lived for nine years after this, and died eventually of dropsy resulting from disease of the heart, the signs of pulmonary excavation existing at either apex up to the last. My friend and colleague, Mr. G. D. Pollock, performed the post-mortem examination. In the upper lobe of both lungs were two or three encysted masses of crude yellow tubercle, about the size of a hazel-nut; at the right apex was a large firm cicatrix, resulting from an old vomica, and a small cavity about the size of a large hazel-nut still open and lined by a dense, smooth, and highly vascular false membrane; at the left apex was a similar sized cavity also lined by a dense and vascular false membrane; the other portions of the

[1] For full details of many cases of arrested phthisis see Dr. Hughes Bennett, on "Tuberculosis;" also, one or two cases briefly referred to by Dr. Walshe.

lungs were healthy and free from tubercular deposit. From the whole history of the case, there can be little doubt, firstly, that tubercle must originally have existed in far larger quantity than was manifest on the post-mortem examination, and probably had been removed by absorption; secondly, that the tubercle which still remained was in a perfectly quiescent state, and was not seriously prejudicial to her health; thirdly, that no deposit of tubercle had taken place for some years before death, and that in fact consumption, or the tubercular cachexia, had ceased to exist; fourthly, that for all practical purposes the patient would have recovered thoroughly, and would have got rid of her cough, had it not been for the accidental formation of the dense membrane which lined the two open cavities in the lungs, and prevented their collapse and cicatrization.

The second case was that of a gentleman who died at the age of fifty-four. Born of consumptive parents, and himself of slender frame and delicate appearance, he had profuse spitting of blood at the age of twenty, and was pronounced consumptive by the late Drs. Chambers and Nevinson. For more than twenty years, he had rarely passed a twelvemonth without severe cough and somewhat copious hæmoptysis, and when I saw him, twelve years ago, his symptoms both general and physical were those of advanced phthisis. He was thin, short-breathed, and unhealthy in appearance; there was dulness on percussion and flattening under both clavicles; auscultatory signs of cavities existed at the summit of each lung, and there was great deficiency of breathing in other parts of the chest. Knowing it to be impossible that he could continue at work, I recommended him to take a holiday for a twelvemonth, and to travel in the East in search of health. He determined to do so, but his departure was unavoidably delayed for six months. He left London, however, and constantly shifted his quarters in England; and before he started on his trip his constitution seemed to have undergone an entire change; his complexion had lost its unhealthy hue, he had gained flesh, his cough had greatly diminished, and he stated that he was becoming less short-breathed than formerly. After an excursion of eleven months up the Nile, and through Palestine and the northern parts of Syria, he returned home apparently quite well. His cough had ceased, and he had become tolerably stout; the infra-clavicular regions had lost their flattened appearance, the dulness on percussion had in great measure disappeared, and all stethoscopic signs of pulmonary excavation had ceased. Indeed, the only indications which remained of an abnormal condition of the chest were, imperfect expansion of the infra-clavicular regions on full inspiration, and weakness or deficiency of breathing in some parts of the upper lobes of the lungs. From the time of his return home until the day of his death, which occurred eleven years afterwards, he had very little cough, and never had the slightest spitting of blood. He died of ascites, resulting from disease of the liver. After death, large cicatrices were found in the upper part of both lungs, and calcareous matter existed in the cicatrices; several small masses of chalky matter about the size of a pea were distributed through the upper lobes of both lungs, and there were also a few small, isolated, and encysted masses of crude tubercle. In this case, the family history, the spitting of blood, and other general symptoms observed during life, together with repeated physical examinations of the chest, left no doubt as to the nature of the case; and the fact that he had been consumptive was further proved by inspection of the body after death; yet for nine years at least, he had not exhibited any symptoms of consumption; and inspection after death confirmed the opinion which had been previously entertained, that he no longer suffered from the tuberculous cachexia, and that the tubercle which had been deposited or formed in the lungs had been in great measure removed by absorption and expectoration.

The last case was that of Mary Liddon, aged fifty, who died in St. George's Hospital in the month of June, 1858, with whose earlier history I happen to be acquainted. Her mother died young, of consumption, and she herself had frequently spat blood, and had been "asthmatic" for many years; otherwise she had enjoyed fair average health, and had been enabled to keep at work. When she first applied to me for relief in the year 1849 she presented most of the general symptoms of phthisis. She had lost flesh, though she was not much emaciated; she was cachectic in appearance, had adunc nails, was short-breathed, and was expectorating a large quantity of thin homogeneous pus. At that time there was considerable deficiency of expansion during inspiration in the right infra-clavicular region, with marked flattening and dulness on percussion in that situation; and the stethoscope informed me that in the same situation there existed gurgling, and intensely increased vocal resonance; at the left apex there was some flattening of the chest-walls, with dulness on percussion, deficiency of the respiratory sounds, and prolonged expiration. A tonic plan of treatment was pursued, and after a time the general symptoms of consumption began to subside, the physical signs of pulmonary excavation to disappear, and she rallied in a surprising manner. Before the expiration of eighteen months her general aspect was that of tolerably good health; the "asthma" had almost wholly ceased, and the most careful examination of the chest could only detect slight deficiency of breathing at the summit of the right lung, and feebleness of respiration, with prolongation of the expiratory sound at the upper part of the left. I then lost sight of her for some years, but in the year 1865 she again applied to me. At that time she was suffering from dyspepsia, but she assured me that she "had quite got rid of her asthma." Her chest remained in precisely the same state as at the date of my last examination, except that the respiration in the upper part of the left lung had become almost normal. On the 2d of June, 1858, she was admitted into St. George's Hospital under my care, suffering from diffuse cellular inflammation of the leg and hand, of which she died in a few days.[1] After death a large old cicatrix was found in the right apex, and there was a small mass of old crude tubercle, of putty-like consistence, immediately beneath it. In other respects the lungs were sound. There was no trace of tubercle, except in the one spot immediately adjacent to the old cicatrix, and there was no emphysema. In this case the family history of the patient, no less than the general symptoms and physical signs of disease, concurred in pointing to tubercular mischief, as the cause of all her trouble, and the post-mortem examination confirmed the opinion which had been formed, by revealing the existence of an old cicatrized cavity at the spot where, during life, the stethoscopic signs of a vomica had been observed. After careful consideration of the history of the case and of the fact that no trace of tubercle was discovered in the left lung, and only one small mass in the right, the legitimate inference appears to be, that as her general health improved, her constitution underwent an entire change, the tendency to phthisis was overcome and the tubercle already deposited was almost wholly absorbed. Her lungs were restored very nearly to their normal condition, and, practically speaking her consumption was cured.

A similar case, in which our knowledge of the patient's history extended back only about three years, occurred under the care of my colleague, Dr. Pitman, in the year 1861; and another, in which the tubercular disease had made rapid progress towards recovery, was admitted in the year 1860 under the care of Dr. Page, the full particulars of which will be found at p. 35 of the "Hospital Post-mortem and Case-Book" for that year. Our museum records state that the man, who was twenty-two years of age, and died of renal dropsy, was in good bodily condition at the time of his death. His consumptive symp-

1 See "Hospital Post-mortem and Case-book" for 1858.

toms, including spitting of blood, dated back nearly two years. After death "the upper lobe of the left lung was found *occupied in its greater part by a mass of cicatrices,*" and there was also "a small, smooth, lined cavity." Several masses also of crude tubercle, not softened, existed at the upper part of the right lung, indicating clearly the nature of the mischief which had resulted in the cicatrization and healing of the lung.

Thus, then, it must be admitted that recovery from pulmonary consumption is possible, nay more, that recovery does take place in a considerable proportion of cases. It may therefore be desirable to inquire, whether there are any general or physical signs indicative of the retrocession of the disease, and of the probability of permanent recovery. The peculiarities of the general symptoms, which denote a slow progression of the disease, have been already described, and it only remains to be stated, that when the disease is arrested there is not only no further evidence of declining health, but the patient regains a healthy appearance, recovers flesh and strength, and gradually loses his cough. Such an occurrence, however, would only indicate a temporary lull in the progress of the disease; it need not necessarily denote its permanent arrest. And if reference be made to a physical examination of the chest, it will be found that little assistance towards a correct prognosis is to be derived from that quarter. The physical indications of active tubercular mischief may subside, and the only remaining morbid signs may be slight flattening of the chest-walls, with dulness on percussion, increased vocal resonance, and feebleness or harshness of the respiration—signs indicative simply of pulmonary contraction and consolidation. But the same phenomena are observed during a temporary suspension of the disease, and are in no degree characteristic of permanent recovery. In short, it must be admitted that there are no marks diagnostic of a permanent arrest of the disease—a fact which can only be judged of after the event by observing not only that the various physical signs which indicate pulmonary excavation and consolidation have disappeared, but that the general symptoms of active disease have subsided, and that there is progressive improvement in the health.

But, although there are no general or physical signs indicative of permanent recovery, the symptoms above referred to as denoting the probability of a slow advance of the disease, and as characterizing its temporary arrest, will go far towards justifying a hope of permanent recovery, provided the malady be not hereditary. For when a tendency to the tubercular cachexia is not thoroughly ingrained in a man's constitution, but has only been developed as a result of temporarily depressing cause, it is reasonable to hope that if these causes can be got rid of, and a fresh impetus given to the function of nutrition, the further progress of the disease may be stayed. Theory points to this conclusion, and practice largely confirms it. Therefore, as the evidence of gradually improving health may be regarded as indicative of an effort of nature—and to some extent an effectual effort—to rectify the derangement which has occurred, it is fair to infer that if that improvement can be sustained, the tendency to the deposit of tubercle will be permanently overcome. In most hereditary cases the disposition to tuberculosis is too strong to admit of removal, and therefore, except under peculiar circumstances, our utmost efforts will not avail in those cases to do more than postpone the fatal event. But experience is decisive as to the possibility of recovery, even in these cases, and still more so in non-hereditary cases; and when the patient has once begun to rally, there is no reason why the improvement should not continue, provided only that he be placed under conditions favorable to his general health. It is indispensable, however, to his permanent recovery, that he be subjected to these conditions for a lengthened period—not for a few weeks or months only, but for several years. If an opposite course is pursued, and at the expiration of a few months, when the symptoms have in some measure subsided, the patient ceases to consider himself an invalid, and subjects himself to the various depressing influences which originally upset his health and induced his illness, the phthisical tendency will probably recur with increased intensity, a fresh deposit of tubercle will take place, and as his lungs are already partly occupied by the deposit, he will "fall into a galloping consumption." It is the neglect of rational means to sustain the general health for a sufficiently long time after improvement has commenced, or in other words until the system has entirely shaken off its tendency to the disease, which leads to the fearful mortality from phthisis.

The average duration of the disease is a point on which it is difficult to adduce any trustworthy evidence, inasmuch as it is well-nigh impossible, in most cases, to ascertain when tubercular deposition commenced. It is not improbable that tubercles may be slowly deposited in the lungs, and may long exist there in a quiescent state without giving rise to any indication of their presence. They are constantly discovered after death in the bodies of persons who, during life, were not only not suspected of being consumptive, but had not been subject to cough, and had not exhibited any symptoms of phthisis; whilst, on the other hand, persons who have spat blood, and from time to time have been subject to cough, and have manifested well-marked symptoms of consumption, and who eventually fall victims to that disease, are often observed to pass months, or even years, during a temporary suspension of the malady, in entire freedom from cough or other notable symptom of pectoral disease. This being the case, it will be admitted that, even when the prior history of the patient can be ascertained, great uncertainty must attach to any opinion concerning the date at which tubercular deposition commenced, and that when, as in the case of inmates of hospitals, it is impossible to obtain any reliable account of the patient's ailments for some years previously, any opinion as to the duration of the malady must be a matter of the merest conjecture. All that can be safely stated is, that the disease may kill in three weeks or a month from the date of its accession, or may run a chronic course of twenty or thirty years. The average duration of the complaint is ordinarily, I believe, very much understated, from the fact that the inferences respecting its duration are drawn from the statements of hospital patients, who pay little heed to the earlier, and, as they imagine, unimportant symptoms of disease, and pertinaciously date their malady from the occasion on which they first experienced pain in the chest, or were frightened by the occurrence of spitting of blood, or found themselves unequal to their daily work. The statistics collected at the Hospital for Consumption, and others recorded by Louis and Bayle, accord generally with those which I have collected at St. George's Hospital;[1] but they differ completely

[1] The subjoined table exhibits the duration of pulmonary consumption, so far as the fact could be ascertained, in 803 cases:

Duration of disease.	Physicians' cases at Hospital for Consumption.	M. Louis' cases.	M. Bayle's cases.	Cases investigated by Dr. Fuller, at St. George's Hospital.	Cases in private practice investigated by Dr. Fuller.
Less than 1 month, . . .		1	1	—	—
In 1 month, ...		3	1	—	—
From 1 to 3 months, . .	1	11	14	1	—
" 3 — 6 " . .	22	52	44	9	1
" 6 — 12 " . .	66	62	64	21	3
" 12 — 18 " . .	34	24	30	25	8
" 18 — 24 " . .	22	17	18	27	9
" 2 — 3 years, . . .	29	—	—	8	9
" 3 — 4 " . . .	13	—	—	9	11
" 4 — 5 " . . .	—	—	—	3	14
" 5 — 6 " . . .	—	—	—	1	6
" 6 — 7 " . . .	—	—	—	—	5
" 7 — 8 " . . .	—	—	—	3	2
" 8 — 20 " . . .	—	—	10	5	2
" 4 years and upwards,	14	—	—	—	—
" 2 to 8 years, . . .	—	23	18	—	—
Doubtful, ...	14	—	—	6	7
Total, ...	215	193	200	118	77

from those obtained by the careful investigation of cases in private practice, where every little symptom is noted, and a much more trustworthy account can be obtained of the patient's health for years prior to the final outburst of pulmonary disease. In the former cases, by far the larger number of patients appear to die in from three to eighteen months; whereas in the latter, the usual duration of the complaint varies from about eighteen months to seven years, a discrepancy which cannot be explained by the different social positions of the sufferers. Doubtless, the advantages enjoyed by the upper ranks of society in respect to medical treatment, change of air, and proper regimen, may go far towards accounting for a certain amount of the variations observed; but the difference is too great to admit of being wholly explained away in this manner, and can only be accounted for by the greater jealousy with which the upper and more educated classes are wont to watch their health, and note the earlier inroads of disease.

It is often asserted that pregnancy retards the progress of consumption, and many instances have come within my own knowledge in which marriage has been most improperly recommended under this idea. My experience amongst pregnant women is too limited to enable me to speak with confidence as to the effect which may be produced in this way; but theoretical considerations and the issue of several instances which I have met with, induce me to believe that pregnancy operates like other agencies which influence the general health. When the tubercular disease is not far advanced, and the excitement incident to marriage and pregnancy serves to rouse the system, and improve the patient's general health, it may also operate indirectly in impartially checking the progress of the disease, or in leading to its suspension. But, even in this case, the debility and exhaustion consequent on childbearing will often occasion a fresh outbreak of the disease soon after parturition, so that the patient will go just as speedily to her grave as if pregnancy had never existed. In cases in which a patient becomes pregnant when far advanced in consumption, I am satisfied that the disease is not only not suspended, but is urged more rapidly along its downward course. Some of the general symptoms may for a time become less prominent, so that the patient or her friends may flatter themselves that the malady is stayed; but the stethoscope will reveal the fallacy of their expectations, and the case will tend speedily to a fatal issue. Such at least has been the course of events in the cases which have come under my immediate notice; and the inquiries I have made amongst those of my friends who are specially engaged in the practice of midwifery confirm the view I have been led to entertain. In a moral point of view, these marriages of consumptive invalids cannot be deprecated too strongly; and socially, everybody is interested in preventing them; for the offspring of such marriages can scarcely fail to be weakly and scrofulous, and even if they attain to manhood, will only perpetuate disease, and swell the list of those who, unequal to the fatigues and duties of life, sink prematurely to the grave.

There is yet one question which demands consideration, before we enter on the subject of treatment; I mean, as to whether consumption is contagious or infectious? In Italy an impression is very prevalent that the malady is communicable from one person to another; and not only is the consumptive invalid avoided, as if he were plague-stricken, but after his death, his clothes, bedding, and everything he has made use of, are frequently burned or destroyed.[1] As an example of this feeling I may cite a statement of the celebrated Morgagni, which is contained in one of his letters. He says, "phthisicorum cadavera fugi adolescens, fugio etiam senex." Even in our own time and in our own country, instances are often cited in proof of its contagious character. A wife, hitherto healthy, and apparently free from phthisical taint, nurses her consumptive husband, and falls a victim to the same form of pulmonary disease; a sister tends her gradually failing phthisical brother, and soon afterwards shows symptoms of decline. The sequence is remarkable; and as it naturally places the two facts before us in the relation of cause and effect, it is apt to excite unnecessary apprehension; for there really is no ground for supposing that, in most instances, they hold that relationship to each other. The sister, probably, inherited the taint in common with her brother and other members of her family; and the wife, if free from an inherited predisposition, was not unlikely to become consumptive, worn out as she must have been by her weary, anxious watching, the sleepless nights and long exposure to the confined atmosphere, and other depressing influences of the sick-room—circumstances beyond all others powerful in fostering or exciting a tendency to phthisis. The marvel rather is, that notwithstanding the strain of long-continued anxiety, and the depression inseparable from the protracted nursing of a dearly loved relative, consumption should not be developed in the survivor

[1] For a full exposition of the arguments by which the contagiousness of phthisis is advocated, see a lecture by H. Guenneau de Mussy, in "l'Union Médicale," No. 138, 1859; also the experiments of M. Villemin, detailed in the "Gazette Hebdomadaire," 1865-6.

more frequently than observation has proved it to be. The fact, however, serves to confirm the theoretical impression that a diathesis, or peculiar state of constitution, cannot be transmitted from one person to another by the agency of infection.[1]

But, though the non-infectious character of phthisis be admitted, it behooves the physician to warn the patient's friends of the dangers incident to long-continued attendance on him, especially if the disease be in an advanced stage. It would be the height of imprudence for a healthy person, especially if young, and of a scrofulous diathesis, to sleep in the same bed, or even in the same apartment with a consumptive patient; for, although the malady might not be communicated directly from one to the other, unless possibly under the condition of some tubercular matter being accidentally introduced into his air-passages or into some other part of his system, the close, relaxing atmosphere, and faint, nauseating odor of the sick-room, and the painful scenes he would necessarily witness, would be eminently calculated to depress him, and predispose him to disease.

There are three distinct questions which require consideration in respect to the treatment of pulmonary consumption, viz.,—first, how to prevent the occurrence of the disease in persons who seem predisposed to its invasion; secondly, how to arrest its progress whilst as yet the pulmonary mischief is limited in extent; thirdly, how to alleviate the patient's suffering and smooth his passage to the grave, when it is beyond our power to avert a fatal issue.

The first two questions involve several points on which medical men are frequently consulted, and on which much difference of opinion exists. Modern research has shown that, by invigorating the general powers of the system, and improving the functions of assimilation and nutrition, the tubercular cachexia may be removed, the formation of tubercle prevented, and its further development arrested. The question therefore arises—How is this desirable result to be brought about? Many persons imagine that warmth is the panacea which will effect our object, if the patient be placed under otherwise favorable hy-

[1] Some recent experiments by M. Villemin ("Gazette Hebdomadaire," 1865-6), on the inoculation of tubercle, would seem to show that if tubercle cells were introduced into the blood, they might possibly reproduce themselves like cancer cells, and lead, not only to local mischief, but to the contamination of the whole system. Experiments of this kind, however, are open to so many and such grave objections that they ought to be repeated on a large scale before any inference is founded on them. Moreover, if their full force be admitted, they serve only to prove that the disease may be communicated by inoculation; they certainly do not show that it is transmissible by infection.

gienic circumstances. "The first and most effectual prophylactic," says one of our most esteemed authorities, "is residence in a warm climate." Now, here at once I find myself at issue with the opinions of many excellent and learned physicians, and no less so with popular prejudice. From our cradle to our grave we are taught to regard cold as an unmitigated evil; and as warmth is intimately mixed up in our minds with the notion of comfort, and that again with everything which is conducive to our bodily well-being, the recommendation to seek a warmer and more genial climate is quite in keeping with our inbred feelings, and is regarded as consistent with sound sense and the exercise of common prudence. But, before adopting this view, let us look more closely than is ordinarily done into certain facts of every-day life, and into others which have been elicited by recent statistical inquiry. Nothing is more certain than that many persons are conscious of enjoying better health and greater vigor in winter than in summer; the cold "seems to brace them and do them good," they breathe in it more freely and are equal to more exertion, their digestive organs are more active, and they are enabled to take many articles of diet which disagree with them and render them "bilious" in the warmer and more enervating weather of summer. There are others, and possibly they form a majority, in whom cold produces a precisely opposite effect; it "pinches and prostrates" them, lowers the whole tone of their system, interferes with their digestion, and disturbs the regular course of their other functions. The absurdity of subjecting these two classes of persons to the same regulations in respect to temperature and climate is too obvious to be insisted on, and yet that practice is almost universal!

To take another class of facts. The result of careful observation has been to show that, although many consumptive patients are better in summer than in winter, yet that others—and amongst them many even of those whose necessities oblige them to go regularly out of doors in search of their daily bread—lose flesh less rapidly, and preserve their strength better in winter than in summer. The experience of physicians at all large hospitals, no less than amongst the upper classes in private practice, proves that to a large proportion of consumptive invalids the warm weather of spring and summer is most fatal; they complain of it as enervating their whole system, they lose whatever remains they have of appetite, feel unequal to the least exertion, and fade away apparently without the power to fight against the general depression which overwhelms them.

And yet again, the results of medical investigation into the geography of phthisis prove that cold is, if anything, preventive of the disease. Of all countries in the world there are none so exempt from the ravages of consumption as those which are included within the isothermal lines of 30° and 40° mean annual temperature. In St. Petersburg and Moscow, with a mean annual temperature of about 38° Fahr.; in Canada and the northern districts of North America; in Iceland and in the Faroe Islands, and in the northern parts of Norway, Sweden, and Lapland, the disease is comparatively rare,[1] whilst in France, Italy, and along the northern shores of the Mediterranean, in Malta, Madeira, and other localities to which consumptive invalids are usually consigned, the ratio of mortality among the natives from phthisis equals, and in many instances exceeds, that which obtains in many parts of our own country.[2] And these results of careful inquiry and statistical research are corroborated by facts which from time to time are forced on our attention in the course of our professional career. There are few physicians who have not met with instances of health restored and vigor regained by removal to a colder climate. Some of the most remarkable recoveries from consumption which have come within my own cognizance have been in the persons of those whose occupations or necessities have driven them to the cold regions of the north.

The climates which next to those of the cold northern regions appear to impart the greatest immunity from consumption are those of places situated at high altitudes above the sea-level—climates remarkable for their coolness and uniformity of temperature, accompanied by a low barometric pressure. The residents in many of the towns in tropical America, placed at an elevation of 6000 feet and upwards above the level of the sea, enjoy almost complete immunity from phthisis, though the disease is very prevalent amongst inhabitants of towns situated at the lower and warmer levels in the same region; and the same has been observed in the Himalayas and other mountain districts.[3]

[1] Dr. Thorstenson, who practised many years in Iceland, Professor Retzius, Drs. Roberts, Schleisner, and other physicians, have testified to this fact. See Schleisner, "Brit. and For. Med.-Chir Rev," April, 1850.

[2] Andral, Lugol, Burgess, Gourbay, and many other physicians, have borne witness to this important fact, and their statements are corroborated by the valuable reports of Colonel Tulloch relative to the influence of those climates on our soldiers.

[3] In illustration of this fact I would refer to an interesting paper by Dr. Archibald Smith, in the "Dublin Quart. Journ." for February, 1866, entitled "Climate of the Swiss Alps and Peruvian Andes compared." The author states, p. 350, "The equable climate of the strath of Huanuco, with a night and day range of temperature from 66° to 72° Fahr. all the year over, is not favorable in similar cases," *i.e.* in cases of incipient consumption and spitting of blood; but (p. 349) the disease is "almost certainly curable if

It must be admitted, then, not only that cold is not productive of consumption, but that in many instances it invigorates the animal economy, and thus proves antagonistic to the accession and subsequent progress of the disease. And if this be so, it follows that in certain instances, at all events, warmth must not only enervate the patient, and thus expose him to an attack of phthisis, but when the disorder is once developed must have a prejudicial effect on its progress.

I do not wish to imply by the above statement, nor to adduce the foregoing facts to prove, that warmth is always or even generally prejudicial to persons of a consumptive tendency. That would be an error precisely similar to the one against which it is my wish to protest. Nothing is more certain than that a warm atmosphere is of all things that by which many persons suffering from the tubercular cachexia are most likely to benefit. Their organization is delicate, their circulation weak, their extremities are often cold, they are pinched and prostrated by a low temperature, and they are very susceptible of damp. To such persons a warm atmosphere is invigorating in the highest degree, and without its aid medicine and the most carefully regulated diet are of little avail. My wish is rather to point out that, whereas many persons are benefited by warmth, others are equally benefited by cold, and that the opposite opinion, which is commonly and rigorously enforced in practice, is constantly leading to lamentable results. The rapid progress which consumption sometimes makes in patients who have gone abroad for their health, and which is often attributed to their having delayed their journey too long, is referable in most instances to the fact of the individuals in question being persons who are naturally benefited by a cool, bracing atmosphere, and who sink at once when sent to the warm and enervating climate of the south.

It may be asked, what facts are calculated to assist us in determining the sort of climate by which a patient is likely to be benefited? The question is somewhat difficult to answer, inasmuch as a change of climate implies not only a change of temperature, but a change in the humidity and electrical condition of the atmosphere, in the degree of its barometric pressure, in the force and direction of the wind, and probably in many unknown but potent telluric influences productive of marked effects on the animal economy.[1] Nevertheless, though much uncer-

taken in time by removing the coast patient so attacked to the open inland valley of Janja (where the temperature ranges from 50° to 59° or 60° Fahr.), which runs from ten to eleven thousand feet above the sea-level."

[1] On what other supposition is it possible to account for the vast difference

tainty must necessarily exist, there are certain points which will serve in most instances to guide us to a correct opinion. The most important of these are—first, the sort of climate and the degree of temperature which formerly suited the patient's constitution, or, in other words, agreed best with him when he was in health; and, secondly, the state of the patient's bronchial mucous membrane at the time when his removal to another climate comes under consideration. It is obvious that if the bronchial mucous membrane is irritable the invalid cannot bear the effect of a very dry and stimulating atmosphere, however warm the locality may be. His symptoms require a soft atmosphere, and its temperature and the precise degree of humidity which is necessary must be determined by reference to his constitutional peculiarities. Thus, if he formerly enjoyed better health in summer than in winter, and felt in greatest vigor in very warm weather, and in an atmosphere devoid of markedly stimulating or relaxing qualities, the probability is that the climate of Syria, Persia, Rhodes, Egypt, Algeria, and other parts of Northern Africa, would exercise an influence on his system the good effect of which could hardly be over-estimated.

If, again, though usually better in summer than in winter, he was formerly oppressed by excessively dry heat, but enjoyed a warm and humid atmosphere, such as that of South Devon or Cornwall, the probability is that the climate of Cauterets, Eaux-Bonnes, or the Righi, during the summer, and of Torquay, Dawlish, Penzance, or Jersey, of Pau, Rome, the Azores, Madeira, Santa Cruz, the Mauritius, or Ceylon, during the winter, according to the degree of temperature required, would be found to suit his general health, and assist in subduing the irritability of the air-passages.

And yet again, if he is constitutionally disposed to general languor, and has not only felt depressed and enervated by heat, but pinched and prostrated by cold, then, notwithstanding the irritability of his bronchial mucous membrane, a medium climate must be sought—a climate such as is to be found in Queenstown

existing between the so-called "bracing air" of Brighton and the "relaxing" atmosphere of Hastings? The two places are within a few miles of each other, and both face the south; both are washed by the same water, shone upon by the same sun, and do not differ more than one or two degrees in average thermometric range. It is useless to appeal to the more sheltered position of Hastings; for the same difference in their effect on the system is felt when the wind is sweeping over the sea from the south or southwest, and when, therefore, Hastings is not more sheltered than Brighton. The fact, I believe, is rather attributable to the difference in the geological formation of the two localities, which plays a far more important part than is commonly supposed in determining the effects of "climate" on the system.

and other parts of the coast of Ireland, on the western coast of Scotland, at Cheltenham, St. Leonards, Ventnor, and Bournemouth; or, if a higher range of temperature is necessary, at Hyères, Pisa, Montreaux, or Malaga, or in New Zealand, Brisbane, or the Cape of Good Hope.

But a large class of consumptive patients exist in whom there is little or no irritability of the mucous membrane. In these a more bracing air will generally prove of the greatest benefit; but nevertheless, as in the former cases, the selection of a locality in each particular instance must be regulated by the constitutional peculiarities of the invalid. If his circulation is languid, and he has usually felt more vigorous in summer than in winter, the invalid must repair to a warm locality; and in such a case the climate of Mentone, St. Remo, Cannes, Malta, Nubia, Algeria, Upper Egypt, the northern districts of Syria, and New South Wales, are likely to prove extremely beneficial. In some such cases the air of the villages high up the Peak of Teneriffe, the Himalayas, the more elevated parts of the Andes, and other hill districts, has been found remarkably serviceable. Indeed, there is abundant evidence to prove that on the elevated slopes of the Andes and in the higher regions of the Peruvian Cordillera phthisis is almost unknown, and consumptive invalids who take up their abode there, in most cases, make good recoveries.

If, again, the patient has an active circulation, and has usually enjoyed better health in winter than in summer, feeling braced and invigorated by cold, he will probably derive benefit from a residence at Brighton, Margate, Aldborough, Cromer, Harrowgate, or Malvern; if a warmer air is needed, at Nice or Florence; or if a cooler and still keener air is required, in Montreal, or other places in Canada, or in certain dry localities in Norway, Russia, or other northern countries. In these cases, too, the stimulating effect of the air at high ranges above the sea-level is oftentimes productive of extraordinary beneficial results. Some of the most remarkable recoveries from consumption which have come within my own cognizance, have occurred under the bracing influence of a northern clime and the diminished barometric pressure which obtains at considerable altitudes above the sea-level.

Thus, then, to revert to the point at which we started in this discussion, respecting warmth and climate, it must be clearly understood that, whether in relation to the prevention, or alleviation, or cure of phthisis, the influence of warmth and of change of climate will prove beneficial or otherwise, in proportion as the use made of these agencies is suited to the idiosyncrasy of the patient, and the existing condition of his bronchial mucous membrane. If the consumptive invalid be not surrounded by atmospheric and climatic influences adapted to the requirements of his organism, the skill of the physician and the utmost care and attention of the patient will generally prove unavailing to modify the morbid action which is going on, or to stay the development of the tubercular cachexia; whereas, if those influences be favorable to his health and care be taken to regulate his diet, and sustain the tone of his system, I am satisfied that more may be done for his relief—nay, even for his cure—than is possible in most other constitutional disorders of equal severity.

Inasmuch, then, as it is of the utmost importance to place the patient under favorable hygienic influences, it may be desirable to consider whether change of climate is always needed, and if not, what precautions should be taken to modify the character of the existing atmosphere.

There are certain cases in which change of air is indispensable. Persons of a very "chilly nature," if resident in a cold and bracing locality, must shift their quarters ere their health will improve; so too, must those who require "bracing," whose lot is cast in a warm and humid climate. These and all others who happen to be placed under circumstances unfavorable to their general health, will do wisely to migrate at the earliest opportunity. But their migration need not necessarily be to foreign climes. There are many spots in our own country, nay, even within a few miles of London, which afford to the average of consumptive patients all the changes which their organism demands, and it is only the comparatively few who require a warmer climate than this country can offer to whom a residence abroad is really necessary. All persons will do well to select as their residence a spot which has ordinarily agreed well with them, and if their own home comes within this category there is no reason why they should leave it for other quarters. Temporary change of air is useful now and then, by imparting a stimulus which is not to be obtained in any other way, and to persons to whom travelling is a pleasurable excitement, and who, so long as they remain at home, are unable to shake off the cares and anxieties of business, a residence abroad may be almost indispensable. But, the majority of consumptive invalids will fare better in their own country than even in the most favored regions abroad, especially now that the railways afford facilities for frequent change, and the Crystal Palace at Sydenham offers, to Londoners at least, an opportunity of combining mental occupation and rational enjoyment with daily exercise in a warm and agreeable atmosphere. Mere absence from home, with its comforts and associations, is to many invalids a constant source of annoyance and regret, while the fatigue of a long journey and the privations incident thereto, go far towards neutralizing any good effect which otherwise might be produced. Therefore with this as with other remedies, its efficacy will depend upon the judgment with which it is employed. As a preventive of the disease, or as a corrective of that derangement of the health which marks its earlier inroads, a residence abroad, or foreign travel, with the pleasurable excitement incident thereto, may, for many persons, though not for all, be regarded as one of our most potent and most valuable remedies. But in an advanced stage of consumption, when the patient is weak, emaciated and exhausted, the recommendation which is often given to go abroad for the winter is cruel and unwarrantable. Under these circumstances no difference of climate can compensate for the loss of home comforts, or make up for the immediate increase of suffering which the journey entails; and an early death in a lodging-house abroad, which frequently ensues in the cases alluded to, is a painful commentary on the eagerness with which men vainly seek to prolong their brief existence here on earth, and on the thoughtless inconsistency of physicians who recommend the journey.

When it is determined to remain at home, every care should be taken to insure good ventilation, and to keep the temperature of the house at a point which is congenial to the feelings of the invalid. The objects to be attained are, that the patient shall not be chilled, nor on the other hand enervated by warmth, and that he shall have a free supply of pure air which he can breathe without exciting irritation of the bronchial mucous membrane. The temperature which ordinarily effects these purposes varies in different cases from 60° to 65° Fahr., and perhaps the best means of maintaining this heat and imparting the necessary moisture to the air is by heating the hall and staircase by the aid of hot-water pipes or of an Arnott's stove, on the top of which is placed a dish containing water. Yet, whilst counselling attention to the temperature of the house, I would guard against having it supposed that a close atmosphere is beneficial, or that confinement to the house should be recommended. Nothing can be more prejudicial to the consumptive invalid than the want of active exercise in the open air. During the summer he should spend almost the entire day out of doors, walking, riding on horseback, driving, travelling, yachting, or simply sitting in the open air; and even in the winter, provided the weather be fine, he should avail himself of the warmer hours in the middle of the day to take whatever out-door exercise his strength will permit.

If the cold air irritates the bronchial membrane, and excites cough, a Jeffries' respirator should be worn, with a view to heat the inspired air, and obviate this source of trouble. But ordinarily this will not be needed; and if care be taken to guard against chills by means of appropriate clothing, it will be found that although, during sedentary occupations in the house the invalid may require the temperature to be artificially raised, he will not often suffer from breathing the cool fresh air of heaven, during active exercise out of doors. It is essential, however, that his clothing be regulated to meet his necessities under the various circumstances in which he may be placed. Flannel should always be worn next to the skin; but in a warm room he should be clad less heavily than in the cold corridors or passages, and in these, again, less warmly than in the still colder air out of doors. The principle being to insulate the body and preserve its natural temperature, the amount of extra clothing should be strictly proportioned to the temperature of the atmosphere to which for the time he is exposed; in no other way can he hope to avoid chill, or to maintain his vital power. Excessive clothing in a warm room would enervate and weaken him, but deficient clothing in a cold room or out of doors would prove equally or even more prejudicial, by depressing his vital energy. It is a fact which cannot be too often impressed on the invalid, that warm clothing does not render the body more obnoxious to cold, but rather imparts power to resist it; and it is a fundamental rule which cannot safely be neglected, that a person whose circulation is feeble should be thoroughly warmed before he leaves the house, and should return immediately he finds himself unequal, even with the aid of exercise, to bear up against the existing cold without losing his warmth and becoming chilled.

This brings us to the consideration of other circumstances besides warmth and change of climate which exert an influence for good or for evil on a patient suffering from the tubercular cachexia. Of these none are more important than diet and exercise.

It has been already stated that out-door exercise is an important agent in the treatment of consumption; and so efficacious has it proved in certain cases, that Dr. Rush and other consumptive invalids, who have made trial of it in their own persons, have vaunted it as a specific and certain cure. But little observation is needed to prove that it does not deserve such unbounded praise. It is preventive of phthisis, and curative in its action, precisely in the ratio in which it is proportioned to the strength of the invalid. Its primary object is to impart tone and vigor to the system; and inasmuch as the vital force of the consumptive invalid is extremely low, it behooves us to be careful, lest by ordering an amount of exercise in excess of our patient's strength, we exhaust the little power that remains to him, and so impair the activity of the digestive functions on which the renewal of his strength depends. Active but moderate exercise excites the action of the heart and arteries, increases the appetite, and stimulates the process of assimilation; whereas, exercise carried beyond the patient's strength, exhausts his vital force, impairs the appetite, diminishes the power of assimilation, and so increases the general debility. Therefore, in the treatment of phthisis, it is essential that the physician should define the amount and sort of exercise which should be taken. Walking exercise is too fatiguing for many consumptive invalids, and when this is the case, horse exercise may be employed with advantage. Driving in a carriage, especially an open carriage, may be had recourse to when walking and riding prove too fatiguing; or walking may be alternated with riding, driving, or yachting. Violent exercise, even of a temporary duration, may exhaust a patient's vital power to a degree from which he may never recover, and, therefore, if he be weak, it is to be deprecated most earnestly. But moderate and frequent exercise is sure to prove beneficial, and may be carried so far as even to produce fatigue, provided the fatigue does not occasion persistent exhaustion, but is soon relieved by rest.

In addition to general bodily exercise, certain partial exercises have been found of service in the treatment of the earlier stages of consumption. Gymnastic movements of all sorts; the club exercise, fencing, the daily use of the dumb-bells; the chest expander and the skipping rope; even reading aloud, and singing, and the practice twice or three times a day, for a quarter of an hour at a time, of deep and forcible inspiration, are each and all important aids to treatment. They increase the action of the respiratory apparatus, stimulate the circulation through the lungs, promote the development of animal heat, and gradually increase the capacity of the chest. But in the employment of these partial exercises, the same rule must be observed as to the non-production of exhaustion as was laid down in regard to general bodily exercise. And, whereas these partial exercises excite the pulmonary circulation to a far greater degree than general bodily exercise, it is further necessary, whilst employing them, to watch for any indication of disturbance in the flow of blood through the lungs. Pain, or a weight or a tightness in the chest, or even excessive shortness of breath, should be a signal, either to suspend the exertion, or to modify its character; for they indicate a want of freedom in the return of blood through the pulmonary veins; and if, in disregard of their warning, the exercise be continued, the congested vessels may relieve themselves by hemorrhage.

Diet is a subject of equal importance to those we have hitherto discussed. Fresh air, exercise, change of climate, and judicious clothing, are all subsidiary to the one object of re-establishing healthy nutrition; and no one in the present day would venture to deny, that in the treatment of phthisis, our principal efforts should be directed to the assimilating and digestive organs, rather than to the organs of respiration. It happens that dyspepsia is one of the commonest of the early symptoms of consumption, and that during its existence, the appetite is very capricious, and fatty matters are generally disliked, and therefore avoided. Lehmann, however, has suggested that a certain amount of fat is essential to the due performance of the digestive process, and that oily matter plays an important part in influencing the healthy metamorphosis of the albuminous constituents of the blood, and promoting cell development. The fact itself is intelligible enough; and no one who is acquainted with the valuable researches of Dr. Hughes Bennett, and has watched the effect of cod-liver oil, and other oleaginous and fatty matters on the system, can doubt its importance in a therapeutical point of view. If, then, it be admitted that the assimilation of animal and fatty matters is important, our main efforts must be directed on the one hand to the avoidance of those articles of diet which are likely to prove indigestible, and to create or increase undue acidity of the stomach, whereby the assimilation of fatty matter is rendered difficult, and on the other, to supply food containing albuminous and oleaginous principles in a form in which the weakly and disordered stomach is able to assimilate and duly prepare them for the nutrition of the body. The most carefully regulated hygienic and therapeutical treatment will fail if the diet is insufficient in quantity or faulty in quality. The prevalent error has always been a too exclusive reliance on one particular sort of diet. Some physicians have extolled the virtues of fish, and have confined their patients to that sort of food; others have maintained the efficacy of farinaceous and vegetable food alone; others of milk, eggs, or more solid animal food; while not a few, especially in these later days, have insisted on the necessity of enforcing a diet in which fat or oil holds a conspicuous place. But it only needs the exercise of a little common sense to establish the fallacy of each of these doctrines. As one stomach, even in a state of health, differs greatly from another in its power of dealing with certain articles of diet, so, as might have been expected, in a state of disease these differences become even more apparent; and, without refining over much, it may be stated, that the diet suitable for consumptive patients must necessarily vary according to their individual peculiarities. "One man's meat is another man's poison," quite irrespective of the stage at which the disease has arrived, or of the symptoms by which it is accompanied. And not only so, the food which is suitable at one period of the disorder, and which the stomach proves itself equal to digest, may be quite unsuitable at another time, when that organ has acquired increased tone, or else has lost whatever little vigor it previously possessed. Thus, as it is necessary to consult our patient's feelings and constitutional peculiarities, in respect to warmth and climate, so also in respect to diet, individual idiosyncrasies must be carefully sought out and studied, if we wish to prescribe food which is likely to be properly digested. Remembering always the fundamental principle, that our aim must be to support our patient, and with that view to induce him to avoid indigestible matters, and to take nourishment in the form best adapted to the capacity of his stomach, we must endeavor to compass our end, not by adhering to any particular kind of food, however advantageous it may appear to be theoretically, but by prescribing a diet as nourishing and supporting as the stomach will bear, by carefully watching its effects, and varying it according to the result of our observation, and, at the same time, by administering such remedial agents as experience has proved likely to assist the weakened and disordered organ in the performance of its duties.

If the disease be not accompanied by symptoms of bronchitis or of active local congestion, and if the stomach be not materially deranged, a nutritious diet should be prescribed, comprising caviare, marrow, fat bacon, and fat meat, twice or three times daily. If there be much exhaustion and depression, and the pulse be slow, the skin cool and the urine clear, wine and malt liquor may be added with advantage, and the patient should be ordered to take some rum and milk, or half a pint of cream with a teaspoonful of brandy, on first waking in the morning, and some bread and milk, mock turtle soup, or other light, nourishing food, with a little brandy, if necessary, on retiring to rest at night. Indeed, stimulants in full doses are indispensable in many cases.

If the liver be disordered or the stomach loathes ordinary cooked butcher's meat, raw meat mixed with brandy may be tried, or venison or game may be given, or a light fish diet may for a time be substituted for the meat. Oysters are often digested very readily; and cod and other light white fish, including their livers, are especially to be recommended, and so is turtle soup. In many of these cases saccharine matters prove easy of digestion, and the greatest benefit is often derived from the administration of "eau sucré" in considerable quantities. Indeed, when cod-liver oil and pancreatic emulsion cannot be tolerated, and the patient evinces a repugnance to butcher's meat, it is my practice to recommend the free use of sugar. When properly digested it is essentially "fattening," and several patients have gained much in weight and have improved in health when taking little in the shape of food beyond bread, milk, and "eau sucré"—the sugar so consumed amounting to a pound and upwards daily. Another article which has proved useful as a nutritious beverage is milk caused to ferment by the admixture of yeast or other material. In some instances, if taken in considerable quantity,[1] it is said to be productive of great benefit.

If, again, the case be characterized by heat of skin, quickness of pulse, oppression and constriction across the chest, and a hard, irritable cough, the diet must be limited for a time to articles such as milk or cream, or farinaceous food. Linseed tea, the decoction of marshmallow, gum arabic, the mucilaginous material of the Fucus crispus, or Carragheen moss, and the bitter demulcent mucilage of the Iceland moss, are serviceable, not only in allaying the cough, but as bland, unirritating articles of food; and in many of these instances milk in which mutton suet has been boiled, proves extremely useful. Beverages such as whey and buttermilk, and soda, lime, or Carrara water and milk, are also to be recommended, as being at once grateful to the patient and useful adjuncts to the diet. Chocolate may be advantageously substituted for tea, and oranges, lemons, and grapes may be allowed if the stomach is not irritable and the bowels are not relaxed. It need only be added, that the amount, no less than the quality of the food, must always be proportioned to the tempera-

[1] The immunity from phthisis enjoyed by the nomadic tribes who wander about the steppes of Eastern Russia is popularly attributed to their great consumption of milk in a state of fermentation, and Dr. Schnepp, of Eaux-Bonnes, has lately been making trial of it as a remedy. The "galactozyme," as the Doctor calls it, or the "kumos," as it is more commonly termed, possesses a pleasant mild acidity, and a slightly vinous flavor, and is said to appease thirst and excite the appetite. The urine becomes clearer, more limpid, and larger in quantity; the pulse slower, softer, and larger; and when a considerable quantity of the galactozyme is taken, a peculiar kind of calm inebriety is produced, followed by somnolence or sleep and a strong inclination to quietude of mind and body. Three or four bottles may be taken daily, and a condition of embonpoint is attained in a few weeks. (See "Med. Times and Gazette," June 24, 1865.) I am informed that it is commonly used as a beverage in phthisis in the Shetland Isles, in Norway, and other countries besides Central Asia.

ment and idiosyncrasy of the patient, the temperature of the air in which he resides, and the amount of exercise he is taking.

This brings me to the consideration of cod-liver oil, which may be regarded partly as an article of food, and partly as a medicinal agent. To whichever of its constituents its efficacy may be due, there cannot be a doubt that it is beyond all others the remedy on which most reliance can be placed. In the north of Europe it has long been a popular and favorite remedy, and was introduced into this country many years ago, but practically we are indebted to Dr. Hughes Bennett for our earliest knowledge of its remarkable powers in the treatment of tuberculous disorders.[1] Chemistry has hitherto failed to explain its action, but clinical experience leaves not the slightest doubt as to its beneficial influence. Of course its virtues are most strikingly displayed in the earlier stages of the disorder, and in cases where there is not a strong inherited predisposition to tubercular disease, but this fact amounts to nothing more than that where there is little mischief to be overcome, the oil has less difficulty in subduing it than it has when the mischief is deeper seated or more advanced. And speaking generally, it may be stated that the oil proves serviceable at all ages, and at all stages of the disorder. The symptoms and physical signs ameliorate under its administration with a rapidity which is sometimes quite surprising; and in favorable cases the weight of the body increases, the cough and expectoration decrease, the appetite improves, and the night sweats and other unfavorable symptoms gradually disappear.

Three facts, which have been elicited by careful observation, appear to afford some clue to its mode of action. The first is, that unless the patients gain weight whilst using it, the oil seldom or never proves remedial; the second, that weight and flesh may be gained during its administration although the pulmonary disease be steadily progressing; the third, that when it does act remedially, the weight gained is far beyond what would result from the oil as a mere aliment. Hence, it would appear that its operation consists in aiding digestion by supplying some principle which is essential to the assimilation of food and the establishment of healthy nutrition. This view has been uniformly maintained by Dr. Hughes Bennett, who has stated, further, that its efficacy depends on its supplying the fatty matter which is essential to the formation of healthy chyle, and which the

[1] I cannot allow this opportunity to pass without recording my sense of the large debt of gratitude due to Dr. Hughes Bennett by the profession of this country for his persevering advocacy of cod-liver oil at a time when its value was not recognized.

digestive organs have become unequal to manufacture or separate from the food. My own impression is, that its influence is not so simple as is here suggested by Dr. Bennett; for, if his views were correct, it ought to act beneficially whenever the stomach is capable of retaining and digesting it, whereas cases not unfrequently occur in which, though the oil does not disagree, it fails to exercise any influence on the disease, which, nevertheless, may yield to other remedies. Further, the improvement which results from cod-liver oil is not found to follow the administration of neat's-foot and other oils which equally supply an oily matter; and as in administering cod-liver oil we are conveying into the system iodine, bromine, phosphorus, and other matters which are known to exercise a powerful influence on the animal economy, I am inclined to believe that the peculiar efficacy of cod-liver oil depends partly on the supply of oil which it affords, but partly also on its containing elements which are wanting in most cases of consumption, and which in such cases it satisfactorily supplies; and that when it fails to operate remedially, it does so in some instances, because it is not tolerated by the stomach, and in others, because it does not contain, and therefore does not supply the elements which in those particular instances are required to promote healthy assimilation. Be this as it may, its extraordinary virtues cannot be doubted, and it ought to be administered in all cases of consumption in which it does not derange the stomach. It rarely purges or otherwise disturbs the bowels, and it does not induce congestion of the lungs, or fatty enlargement of the liver or kidneys; on the contrary, it improves nutrition generally throughout the body. Not only does it produce increase of weight, but the patients whilst taking it gain strength and color. As a mere aliment it is extremely useful, and as a remedial agent in appropriate cases its value is inestimable.

It has been urged by Dr. Walshe,[1] that the oil operates less beneficially in proportion as age advances, an assertion which is quite opposed to my own experience. Had he stated that it less frequently agrees with persons advanced in years than with young persons, and therefore less commonly produces satisfactory results, I should have coincided entirely in his views; but some of the most remarkable instances of rapid improvement under its use, which have occurred in my own practice, have been in the persons of patients far beyond the middle period of life. Not only has flesh been gained in the cases referred to, but the gen-

[1] Loc. cit., p. 532.

eral symptoms and physical signs have improved to a corresponding degree.

Further, it has been stated, that "intrathoracic inflammation and hæmoptysis" are contraindications to its use. Experience at the bedside, no less than theoretical considerations, lead me to demur to this statement. In pulmonary consumption, the thoracic inflammations are attributable to the unhealthy condition of the nutrient fluid and the local irritation of the tubercular deposit, and the spitting of blood is due to congestion, induced by the same agencies. If then the oil effects what its advocates profess, viz., an alteration or improvement in the condition of the blood, and a consequent improvement in the process of nutrition, it is difficult to see why its use should be abandoned at the very moment when its aid is most urgently required. Moreover, experience does not afford any warrant for suspending its administration under the circumstances alluded to. My constant practice has been to continue its exhibition in combination with the remedies best calculated to relieve the inflammation or the hemorrhage, and I have never seen the slightest inconvenience produced thereby, in cases in which it otherwise agreed.

Sometimes, however, it thoroughly disagrees with the stomach, and even if it does not produce actual vomiting, it gives rise to nausea and acid eructations, together with feverish derangement of the system. In these instances it must either be omitted altogether, or administered in small doses, and at all events its use must be discontinued until the stomach is brought into better order, and has acquired the power to digest and assimilate it. The question of treatment in cases such as these resolves itself into that of imparting tone to the digestive organs, and regulating the action of the liver. There is generally undue acidity of the stomach, with deficiency or irregularity in the action of the liver and bowels; and the method to be adopted in overcoming the difficulty must depend upon the degree of biliary derangement. If the stomach appears to be the organ principally at fault, the administration of the mineral acids in combination with vegetable bitters and taraxacum will often serve to give it increased tone, and thus rectify the error; whilst if the acids are not well borne, the same result may be produced by the exhibition of an occasional alterative at night, and of the light vegetable bitters, with liq. potassæ, or the carbonates of the alkalies, or the hypophosphites of soda or lime, or the nitrate, or bicarbonate of bismuth, during the day. If, on the other hand, the stomach is not the principal offender, but the bowels are disordered, and the motions knotty, clay-colored, or of a dark green, or almost a black color, no good will be effected without the aid of an emetic, or the careful administration for some days of slight mercurials, with ipecacuanha, taraxacum, ox-gall, podophyllin, and other remedies which in some way influence the secretion of bile and the action of the bowels. Not unfrequently, after careful attention to the digestive organs, and equal care in respect to hygienic treatment, the stomach will acquire the power of digesting the oil, and the animal economy will then be improved by the administration of the very remedy which had previously upset it. In all such cases, however, it is advisable to recommence its administration in small doses. Under ordinary circumstances, half an ounce of the oil may be given twice, three times, or even four times daily, and the dose may be cautiously increased to one or two ounces. But, whenever an intolerance of the oil has been once manifested, it is prudent on recommencing its administration to give it in doses not exceeding a teaspoonful twice a day; and, if that quantity be digested, the dose may be gradually increased, and repeated more frequently. It should be taken shortly before or after a meal, in order that it may be presented to the stomach at a time when digestion is in full activity, and it may be taken floating on a glass of wine, or on a dose of tonic medicine containing the mineral acids or the alkalies. Not unfrequently, when the stomach rebels at the oil in its natural condition, it receives it readily when made into an emulsion with liquor potassæ, or some other alkali, and it is always advisable to try it in this form before abandoning its use. The kind of oil employed, whether the pale, light brown, or dark brown oil, is of little importance in a therapeutical point of view; but the paler kinds are purer and more palatable, and are sometimes retained when the coarser brown oil is rejected by the stomach.

Sometimes, however, do what we may, the stomach is unable to retain the oil, or to digest or assimilate any fatty food. Under these circumstances an attempt should be made to introduce the oil into the system through some other channel, and this may be effected either by means of the skin or the rectum. The practice of inunction is not an innovation of modern times, nor is it confined to cases of disease. Among the Romans it was employed as a hygienic luxury; and it is in constant use in the present day amongst the natives of Western Africa, who suffer little from the ravages of phthisis. Even in our own country, accidental inunction takes place to a limited extent among certain classes of the community. The young persons who are employed in wool factories where large quantities of oil are employed,[1] tallow chandlers, butchers, and others, are necessarily in contact with fatty

[1] See "Edin. Monthly Journal," April, 1853.

matter throughout the day, and it has been observed not only that they are unusually well nourished, but that they are less liable than others to suffer from consumption. But more than this; experience has established the fact that the endermic method of introducing the oil is often very serviceable in phthisis. Cases have been published by Dr. Simpson of Edinburgh, and by other observers, which are quite sufficient to prove its efficacy. In my own practice several instances have occurred in which its operation has been signally beneficial. One gentleman whom I saw in consultation with the late Dr. R. Bright, and subsequently with Dr. C. J. Williams, gained a stone and a half in weight in less than three months by the persevering use of it, and the physical signs of pulmonary disease subsided greatly at the same time. In most instances, however, its virtues are not displayed so strikingly as when it is administered by the mouth; and the smell of the oil when applied externally, is so extremely disagreeable, that many patients cannot be persuaded to use it. Under these circumstances, lard or the inodorous vegetable oils may be substituted for it, and undoubtedly will serve to check waste, and support the strength; but the patient's weight will seldom increase materially under their use, and I am doubtful whether, in any instance which has come under my notice, the progress of the local disease has been arrested.

Another method of introducing the oil is by the rectum, in the form of injection. Some physicians have reported favorably of its action when exhibited in this manner; but in the few cases in which I have seen it tried, it has not answered my expectations, and the strong objection which is commonly entertained to the constant use of enemata must render its administration in this way an expedient to be resorted to only in exceptional cases. When the stomach will not tolerate the oil, and the fatigue incident to its inunction, or the dislike of the patient to its employment in that manner forms a bar to its administration, then enemata containing the oil may be tried; but its exhibition in any other way than by the mouth must be regarded as a mere expedient to serve the purpose of supporting the patient temporarily, while the stomach—the proper receptacle for food—is being prepared for its reception, and fitted to digest it.

One word of caution may be added respecting the period of the disease at which the administration of the oil should be commenced, and the length of time for which its exhibition should be continued. It has been already stated that its effects are displayed most strikingly when it is administered early in the disease; and it need only be added, that it cannot be given too early. Pathological research has long since proved that pulmonary consumption, in its incipient stage, is a disease which admits of cure in a considerable proportion of cases—of cure so readily effected that it often takes place spontaneously, and so permanent that there is often no recurrence of the disease, even during a life protracted to an advanced age. My own experience has fully corroborated the facts thus gleaned from the field of pathology; and it has also confirmed another fact, drawn from the field of clinical observation, viz., that such cures are rarely effected if the pulmonary disease has made much progress. Therefore it becomes of vital importance to recognize the earliest inroads of the disease, in order that a proper regimen may be enforced, and the oil and other appropriate remedies administered whilst as yet our patient retains his strength, and the system its reparative power. It is not sufficient to persist in the administration of the oil only until some sensible improvement has taken place; the very fact of improvement should rather be regarded as a stimulus to further perseverance. The deposit of tubercle is so clearly connected with constitutional derangement, and any material alteration in the constitution is notoriously effected so very slowly, requiring not weeks or months but rather years for its completion, that if we would consult our patient's safety, we must urge him to take the oil for many months after he considers his health re-established. If he refuses to do so, the probability will be that the improvement will be only temporary, and after a short interval of comparative tranquillity, the machine will again get out of order, mischief will recommence in an active form, and our power may not avail to arrest it; whereas, if the patient can be persuaded to continue taking the oil for a year or two, omitting it only three or four times in the twelvemonth, for three weeks or a month at a time, while he is enjoying change of air, and is otherwise under peculiarly favorable hygienic conditions, my experience leads me to believe that, in a considerable proportion of cases, the tendency to the disease will not only be arrested, but the improvement which has occurred will be maintained, even after the oil has been discontinued.

The late Dr. Theophilus Thompson[1] and other physicians have advocated the use of ozonized cod-liver oil, and have stated, as the result of their experience, that the force and frequency of the circulation are subdued through its agency, and inflammatory action therefore averted. The statement at present rests, I think, on insufficient evidence, and my own experience of its action is too limited to enable me to pronounce decisively on

[1] See "Med.-Chir. Trans.," vol. xvii, p. 349.

the subject. In the few cases in which I have employed the ozonized oil it has failed to exercise the controlling influence over the pulse attributed to it by Dr. Thompson, and my impression is that the results he obtained were quite exceptional. But if, on more extended observation, it should prove to exert the power claimed for it by its advocates, it would be a valuable adjunct to the treatment of cases characterized by acceleration of the pulse, and especially so when hæmoptysis is a prominent feature.

It has been suggested by Dr. Dobell that the constitutional disturbance so remarkable in consumption is due to the imperfect assimiliation of fatty matters consequent on defective action of the pancreas, and he has proposed to remedy this deficiency by the administration of pancreatine, either alone or in combination with fatty matters, in the form of emulsion. He affirms that many persons who cannot digest fat, and are unable to assimilate, and therefore to profit by cod-liver oil, can take fat or oil when formed into an emulsion by means of pancreatine, and that under its use the strength and weight of the patient improve, and, in many instances, the progress of the disease is checked. Nay, more, he has published reports in the journals, wherein he endeavors to show by statistical reports that its action is almost uniformly beneficial. I wish it were in my power to indorse his statement or confirm his favorable report. Experience, however, leads me to demur to his theory, and to question the value of the practice founded on it in most cases of consumption. The pancreas is not the only organ concerned in the emulsification of fat, and no proof has been adduced that the pancreas is diseased in those cases in which fat is not easily digested. Further, it oftentimes happens in consumption that there is no disinclination to take fatty matters, and no apparent difficulty in digesting them; on the contrary, fat is often eaten greedily and digested readily by this class of invalids. In these instances, therefore, it can scarcely be pretended that a deficiency of pancreatine is the cause of the complaint, and the administration of pancreatine, whether alone or combined with fat in the form of emulsion, is found not to afford the slightest benefit. I have watched its effect under these circumstances too often to be mistaken in my conclusions. Nevertheless there are cases in which the utility of pancreatine does not admit of doubt. The repugnance to fatty and oleaginous matters exhibited by some consumptive patients is so great that they cannot be induced to take them, and even if they do swallow them they either eject them by vomiting or find their digestive organs completely upset. In these cases the pancreatic emulsion is useful in supplying an important element of food in a form in which the digestive organs can deal with it. Several patients under my care, who have evinced an utter indisposition to take fatty matters, have derived much benefit from the pancreatic emulsion. But it must not be assumed that this is the only form in which fatty matters can be taken under the conditions just referred to. Even when the stomach rejects fat at dinner, and is unable to digest oil in its natural state, it will often receive either the one or the other when made into an emulsion by means of liquor potassæ or some other alkali, in which case I have usually found that the cod-liver oil proves more beneficial than the pancreatic emulsion. But in cases characterized by such an intolerance of oil and fat that the addition of alkalies does not suffice to insure their favorable reception by the stomach, pancreatic emulsion may be fairly tried as a remedy from which much benefit may be expected. In some such cases it can be taken and digested readily, and under these circumstances it is of considerable value.

The remedies which will best assist in imparting tone to the digestive organs and improving the general health, must necessarily vary with the condition of the patient. Alkalies and their salts, especially the hypophosphites of soda and lime, phosphoric acid, or any of the mineral acids in combination with taraxacum and light bitter infusions, dulcamara in full doses, the iodide, the phosphate, the sesquichloride of iron and other chalybeates, cinchona, quinine, strychnine, and sarsaparilla, are amongst the internal remedies which I have found most useful; whilst in some instances warm baths, hot air baths, such as the Turkish bath, and cold shower baths, judiciously employed, are also extremely serviceable. The same may be said of the mineral waters and mineral baths, especially those of Ems, Vichy, Cauteret, Eaux Bonnes, and Spa abroad, and of Bath, Tunbridge Wells, and Harrowgate in this country; but, like other agents, they prove remedial only when employed in appropriate cases, and in others are useless or even mischievous. The Cimicifuga racemosa and the wild cherry bark are remedies which have obtained a high reputation in America, and an endless variety of other substances, which experience has proved to possess no special virtues, have been brought from time to time before the profession. Amongst these, the hypophosphites of lime and soda, naphtha, iodine and its compounds, hydrocyanic acid, digitalis, and other agents have been vaunted as specifics; but experience has amply proved that they are serviceable only when administered according to the requirements of the patient. In no case do they deserve the title which their advocates have arrogated for them, even when they operate beneficially. Like all other remedies, they are useless in some instances, and worse than useless in others. And the reason is not difficult to fathom. Not only does the condition of the stomach and other organs differ in different cases, and in the same case at different periods of the disease, but the general symptoms present every possible variety, according to the precise nature of the actions going on in the system.

Everybody must have observed how strangely variable the course of consumption ordinarily proves, and how frequent are the accessions of febrile disturbance and of increased cough and pain in the chest—symptoms which are commonly, though erroneously attributed to "catching cold," and are, in fact, referable to the irritation produced by fresh deposits of tubercle, or to the supervention of bronchitis, pneumonia, or pleurisy, excited by the unhealthy and irritating quality of the blood. It needs not much experience to understand that medicines which prove useful when the disease is pursuing its ordinary slow, uncomplicated course, must be of little avail, or positively injurious, when the whole conditions of the case are altered by the occurrence of one of these active complications. In short, to revert to my former statement, medicines to be of the slightest use in this or any other disorder, must be administered according to the exigencies of the case, and in strict relation to existing symptoms.

Oftentimes the secretions prove valuable guides to treatment, and their condition can never be safely disregarded. If the motions are pale-colored or else dark and offensive, a few alterative doses of mercury will be needed before tonics or other remedies will be of any avail, and so if the urine is persistently loaded with lithates, alkalies and the neutral salts will be required to fit the stomach for the reception of cod-liver oil, iron, and other supporting agents.[1] In this disease as in all others,

[1] This was strikingly exemplified in the case of Wm. Wilby, æt. 27, who was admitted under my care into the King's Ward of St. George's Hospital, on October 5th, 1864. This man had large vomicæ in both lungs, suffered frequently from spitting of blood, sweated profusely, and was reduced to such a state of emaciation that nobody imagined he could live many weeks. His condition seemed to render the administration of tonics, stimulants, and cod-liver oil a matter of necessity, and they were given accordingly in every variety of form and combination. Day by day, however, he was becoming thinner and weaker, when my attention having been specially directed to the pale color of his motions, and the dark color and loaded condition of his urine, I determined to omit all tonics, and to administer alterative doses of mercury and effervescing saline medicines with excess of alkali. From that time he began to improve, and in the course of a few weeks was able to resume the use of cod-liver oil. He left the hospital on the 15th of February,

the value of a remedy depends less upon its intrinsic properties than upon its adaptation to the requirements of the patient at the moment of its administration.

There are other expedients for giving relief which must not be passed by without special notice. I refer to bloodletting, baths, the external application of absorbents and rubefacients, and the use of persistent derivatives, such as issues, setons, blisters, and the production of pustular eruptions and discharges by means of tartar emetic, croton oil, or biniodide of mercury ointments.

Small and repeated venesections, and the frequent application of a few leeches, have been recommended by authors as calculated to relieve the congestion of the lungs which so constantly accompanies phthisis, and so indirectly to cure the disease. Nothing, however, can be more opposed to sound pathology, and to the results of modern experience. A moderate abstraction of blood may possibly be requisite, if perchance active pleurisy or pneumonia should intervene; but even in that case the symptoms may usually be subdued by appropriate treatment, without loss of blood; and the idea of relieving congestion by repeated bloodletting, whether local or general, is simply absurd and mischievous. The congestion is due to the unhealthy and irritating condition of the circulating fluid, and though the abstraction of blood may momentarily empty the vessels and relieve the pain, the impoverishment of the vital fluid resulting therefrom must necessarily tend to an increase of the congestive tendency. Efficient constitutional treatment, aided by dry cupping and counter-irritation—not bloodletting—is the appropriate remedy for this form of congestion.

Some authorities maintain that considerable benefit may be derived from the repeated application of rubefacients and absorbents to the chest-walls. My own observation induces me to concur in this opinion; but I question, nevertheless, whether the mode in which these agents are usually employed is well calculated to effect our object. No one can doubt that the application of mustard poultices and turpentine stupes are oftentimes productive of great and immediate relief; but in these instances the curative influence of the remedies is aided by the effect of warmth, and their irritant effect is not carried to the extent of damaging the cuticle. It is otherwise in regard to many rubefacients and absorbents which are commonly employed.

1865, plump and in good condition, having lost his cough for more than six weeks, with no more physical indications of the attacks he had undergone than was afforded by slight deficiency of respiration at either apex. The vomicæ had cicatrized, and it was obvious that very little tubercle remained in his lung. I saw him five months after he left the hospital, and he was then in good health.

The acetum or the linimentum cantharidis painted on the chest after the manner usually recommended, produce little immediate effect as rubefacients, but speedily destroy the cuticle, and cause continued annoyance to the patient. I have repeatedly watched the application of these remedies, and have failed to satisfy myself of their curative action. So also in regard to the tincture and the liniment of iodine—the favorite and most commonly employed absorbents. Applied to the chest-walls in its concentrated form, iodine may prove useful as a mild counter-irritant, but it certainly must fail in its action as an absorbent, for it dries in the course of a few minutes, and after one or two applications induces death and separation of the cuticle—a result which manifestly precludes the possibility of our obtaining the absorbent action of the remedy. If any good is to be effected by means of the absorbent properties of iodine, it should be so applied as to produce a moderate determination to the surface, but not so as to interfere with the function of the skin, and prevent its absorbent action continuing during a lengthened period. Therefore when I have recourse to the assistance of iodine, I make use of it as a weak, watery solution in combination with glycerine, after the manner in which I have always employed it for reducing chronic enlargement of the joints. A lotion is employed composed of six drachms or an ounce of tincture of iodine, an ounce and a half of glycerine, and three ounces and a half of water; and if a piece of flannel wetted with this be constantly applied to the surface of the chest, and covered with oiled silk, it will usually keep up a redness of the surface, without unduly irritating the skin or interfering with its absorbent power. And, without attempting to define the amount of good or the precise character of the changes which may be thus effected, I may fairly state that more relief, if not more permanent benefit, results from the employment of iodine in this manner than from the application of any other external remedy. If, as sometimes happens, the smell of the iodine proves offensive to the patient, or if an objection be raised to the constant application of a lotion to the chest, or to the stain which iodine leaves on the skin, a liniment, of which the formula is subjoined, may be diligently rubbed in night and morning. This lotion is nearly colorless, and operates powerfully in inducing the characteristic effects of iodine, of which it contains a large proportion.[1]

[1] Iodi., gr. xxx.
Liq. Ammoniæ Fort., ℥j;
Linimenti Saponis, ℥j;
Linimenti Camphoræ Comp., ℥ij. M. ft. linimentum.

The question respecting the use of derivatives is not easily disposed of. From the earliest periods of medical history they have been used and lauded as extremely beneficial; and the curious facts deduced from the result of clinical observation respecting the effects of apparently insignificant discharges from ulcerated surfaces, point to the necessity of caution in arriving at an opinion adverse to their employment. My own opinion is favorable to their use in aid of constitutional treatment in the early stage of the disease, provided they are so regulated as not to exhaust the patient's strength by the profuseness of the discharge induced, and not to depress him, morally and physically, by causing constant pain and annoyance, and preventing sleep. Their applicability, therefore, must depend on the mental and constitutional peculiarities of the patient. The means most commonly adopted in former days for the purpose of inducing discharge were the use of issues, either between the scapulæ or in the arm, or over the margin of the false ribs; but a less formidable and more efficient method is the application to the chest of one of the many ointments or liniments which induce pustular eruptions. None, in my experience, answers better than an ointment for which the formula is given below.[1] In the later stages of the disease derivatives of this kind almost invariably depress the patient, and therefore should not be employed.

The use of baths, whether warm, tepid, or cold, vapor or hot air, is too much neglected in the treatment of consumption. This has arisen, I believe, partly from the difficulty which, until the last few years, has been experienced in obtaining baths in this country, and partly from the mischief which has frequently resulted from their improper or injudicious employment. It is not to be doubted that agents like baths must exert a powerful influence on the animal economy. Theory suggests the fact, and experience confirms it. But their agency will be for good or for evil, according as they are judiciously or injudiciously made use of. The cold shower bath will stimulate and brace one patient, but will chill and depress another; the warm bath will soothe and tranquillize one person, but enervate and render another miserable; the vapor or the hot-air bath will refresh the man whose skin is dry and inactive, and whose nervous system is oppressed or rendered irritable by the presence in the blood of materials which ought to have been thrown off by perspiration, whilst it would exhaust and reduce to an unwarrantable degree the patient whose skin is already relaxed and acting immod-

[1] Hydrargyri Perchloridi, gr. viii; Iodi., ʒss; Spiritus Rectificati, ʒiss; Adipis, ℥j, ft. ung.

erately. Thus, it may be affirmed of baths as of other agents, that they become remedies only when they are adapted to the requirements of the case. But it may be added, that when so employed they are valuable adjuncts to other treatment, and ought never to be neglected. I have seen more striking benefit result from the cold shower bath and the dripping sheet, followed by active friction, or from the hot-air bath, followed by a cold douche or a cold shower bath, than from any other remedy, except cod-liver oil. The patient, if cold and chilly, is warmed by the hot-air bath to a degree which cannot be produced by any other agency, while the cold water braces the nervous system, and renders the skin less susceptible of draughts and of sudden variations of temperature. Sponging the body with cold or tepid vinegar and water is sometimes very serviceable.

There are certain symptoms in connection with phthisis which require consideration apart from the general treatment of the disease. Sometimes, for instance, dyspepsia is such a prominent feature, and so urgent in its character, as to demand special and almost exclusive treatment. Constant pain and weight at the stomach, increased after food, and oftentimes accompanied by a distressing sense of nausea, induce the patient to reject food almost entirely. These symptoms must be met by treatment specially addressed to their relief; and although it is not my purpose at present to discuss at any length the treatment of dyspepsia, it may be stated that, among the remedies which I have found most serviceable in this particular form of the complaint are bismuth, liquor potassæ, the alkaline carbonates, hydrocyanic acid, the mineral acids, gallic acid, strychnia, and opium, preceded, if necessary, by an alterative dose of calomel or blue-pill, and aided by counter-irritation at the pit of the stomach.

Cough is another symptom which requires special treatment—not when it is loose or slight in amount, but whenever it is hard, frequent, and harassing, and produces exhaustion during the day and sleeplessness at night. Amongst the remedies most serviceable in allaying the irritation are opium and its salts, belladonna, conium, lactucarium, digitalis, and hydrocyanic acid, and their action may be aided by ipecacuanha, or squills, or by tolu, marshmallow, liquorice, Iceland moss, linseed tea, and other demulcent materials, whether in the form of a beverage, cough-drop, or lozenge.

If the cough be hard and accompanied by symptoms of febrile excitement, with pain and constriction in the chest, and a stethoscopic examination reveals commencing bronchitis or pneumonia, salines with tartar emetic will be the appropriate remedies, and their action will be assisted by occasional dry cupping between the shoulders, by stimulating embrocations to the chest, or by mustard poultices, turpentine fomentations, or blisters, according to the urgency of the case. If the febrile exacerbation be referable to pleurisy, a few leeches to the seat of pain may possibly be needed, and the administration of opium, or of calomel and opium may even be required, whilst blisters or fomentations are employed externally. But it must never be forgotten that all inflammatory complications of phthisis are referable to the irritation of tubercle in the lung, or to the morbid condition of the blood in which the deposit of tubercle originates, and that although the remedies just named may be needed for a time to mitigate or subdue existing inflammation, yet that they must be replaced as soon as possible by cod-liver oil, and other supporting or restorative agents, which alone can prove effective in combating the disease.

If the pulse be very quick, and the cough, though extremely irritable, is not accompanied by pain or constriction across the chest, or by physical signs of acute pneumonia or pleurisy, I know of no remedies of equal value with aconite, digitalis, the prunus virginiana, and the veratrum viride, and they may be given to the exclusion of tartar emetic, or in combination with it. In some instances, especially in the early periods of the disease, the effect of these remedies in affording relief, after ordinary sedatives and expectorants have failed, is very remarkable.

When there is entire absence of febrile excitement, and the cough is unaccompanied by pain or constriction across the chest, and the distress which attends it is referable to difficulty in expectorating, the decoction of senega, aided by carbonate of ammonia, the balsam of Peru, the compound tincture of benzoin, tolu, squills, sulphuric or nitric ether, and other stimulating expectorants, appears to modify the character of the secretion, and thus, by facilitating its rejection from the bronchi, to relieve the strain of coughing. In some of these cases the action of emetics is extremely serviceable, not only by leading to the immediate ejection of the mucus and muco-purulent matter, which is loading the air-tubes and oppressing the patient, but by modifying the character of the bronchial secretion, inducing a free action of the skin, unloading the whole of the internal organs, and thus promoting increased activity of digestion. Indeed, where there is much nausea or repugnance to food with evident disturbance of the liver, I know of nothing more certainly productive of relief than the deobstruent action of an emetic.

If the cough is of a spasmodic character and the breathing

is accompanied by much wheezing, without any evidence of local inflammation, a cough-drop, containing stramonium, lobelia, hydrocyanic acid, and ether, answers better than a mere narcotic and demulcent mixture, and in some such cases, the inhalation of a few drops of ether or chloroform from off a handkerchief, or the inhalation of atomized fluids holding hydrocyanic acid and morphia, or other sedatives in solution, will subdue symptoms which have resisted the same remedies taken in a liquid form. Sometimes, when the mucous membrane of the air-passages is very sensitive, and is easily irritated in a dry atmosphere, the mere fact of moistening the air with steam will prove useful in alleviating the cough, whilst in others the inhalation of dilute medicated vapors will be found more serviceable. For this purpose aqueous vapor, charged with the essential properties of conium, hyoscyamus, stramonium, or other substances, may be diffused through the atmosphere of the apartment, and breathed through one of the many forms of inhalers, whilst in other cases, in which there is little tendency to inflammation, more stimulating substances, such as iodine, camphor, benzoin, pyroxylic spirit (naphtha), creasote, carbolic acid, and common tar, may be beneficially inhaled in the same manner. Not unfrequently, however, the inhalation of air highly charged with moisture—whether medicated or not is of little importance—is oppressive and disagreeable to the patient, and under such circumstances relief may sometimes be obtained by fumigating the apartments in which he resides by gently heating benzoin, myrrh, and various balsams and gum resins, or creasote, tar, camphor, iodine, and other substances, and it is stated—though I cannot vouch for the fact—by the inhalation of oxygen and compressed air.[1]

Sometimes the irritability of the cough is due to relaxation of the throat and elongation of the uvula, and in these cases the practice of snipping off a small portion of the uvula and of daily swabbing the throat with a strong solution of nitrate of silver, or some other stimulant or astringent material, will be found of the greatest benefit. A concentrated solution of tannin in glycerine has proved in my hands extremely serviceable, and so has powered alum, applied by means of a large camel-hair brush, which has been wetted with glycerine and then covered with the alum. In certain instances, however, in which the pharynx is seen to be dry and devoid of secretion, I have found greater

[1] See "Emploi médical de l'Air comprimé," par M. Pravaz; also "The Compressed Air Bath: a Therapeutic Agent in various Affections of the Respiratory Organs and other Diseases," by R. B. Grindrod, M.D.

relief produced by the application of a mixture of two parts of the tinctura pyrethri and one part of glycerine than by any more stimulating or astringent solution. The effect of the pyrethrum in inducing increased secretion from the mucous surface is quite remarkable.

When the larynx is ulcerated, the application of a solution of nitrate of silver is often serviceable, but experience has led me to doubt the possibility of introducing either a sponge or a brush, except in instances in which the pharynx, the epiglottis, and the larynx are more than usually devoid of sensibility. Even with the aid of the laryngoscope the operation in most cases is well-nigh impossible. But by means of a syringe the solution may be injected into the larynx without difficulty, and I have known it to give relief when thus employed. But inasmuch as its application to the pharynx has appeared to be almost equally efficacious, and is certainly less distressing than its introduction into the larynx, I prefer using the brush or the probang instead of the syringe. If it is desired to introduce the solution directly into the larynx, the better plan is to invoke the assistance of one of the instruments now commonly employed for pulverizing fluids, as the application of the caustic can thus be regulated without the slightest difficulty, and its use can be suspended the moment any undue irritation is produced. In some instances I have known considerable benefit produced by blowing a few grains of calomel down the pharynx whilst the patient is taking a deep inspiration,[1] and in others relief may be obtained by local blistering or by counter-irritation induced by means of turpentine, or of a lotion composed of biniodide of mercury,[2] applied on a rag and covered with oiled silk, followed by the application of a bread-and-water or linseed poultice. The inhalation of medicated vapors is also in many instances attended with relief, and so is the inhalation of atomized fluids containing iodine, and other remedial agents or solution. Gargles are almost useless. In certain instances, in which suffocation is imminent,

[1] I have devised an instrument for introducing calomel or any other powder into the larynx, which answers its purpose admirably. It consists of a curved canula, the extremity of which is finely perforated like a pepper-box, and unscrews to admit the introduction of the powder. This is passed back into the throat and held above the opening into the larynx. When it is in this position the patient is directed to inspire, and the calomel or other powder is then ejected by a current of air set in motion by means of an India-rubber ball attached to the other extremity of the canula. The instrument may be obtained from Whicker and Blaise, in St. James's Street.

[2] The following is a formula I often employ: Hydrargyri Perchloridi, gr. vj; Tr. Iodi., ℥j; Glycerini, ℥ss; M. ft. lotio.

trachcotomy may be had recourse to with the effect of prolonging life.[1]

Pain in the chest is to be met by counter-irritation. Some persons recommend that blisters should be applied, and kept open by dressing the blistered surface with the ung. sabinæ or some other stimulating application; but I much prefer the practice of letting the blistered surface heal, and then repeating the blister if necessary. The counter-irritant effect produced is greater and occasions less distress to the patient. When the pain is not so acute as to require a blister for its relief, stimulating embrocations, or mustard cataplasms, or turpentine fomentations, may be used, or the compound tincture of iodine or the acetum cantharides may be applied to the chest daily by means of a brush, a fresh portion of the chest being painted with the solution each day. In most instances, however, it answers better to keep constantly applied to the chest a piece of lint saturated with a weak solution of iodine combined with glycerine, and covered with oiled silk. This gives little inconvenience to the invalid, and oftentimes suffices to relieve the dull, aching pain in the infra-clavicular regions with which comsumptive patients are apt to be tormented.

Perspiration, which, when profuse, proves extremely weakening and distressing to the patient, must be combated, like all the other symptoms, by judicious constitutional treatment. But there are certain remedies and plans of treatment which appear to exercise a special control over this disagreeable symptom, and therefore should be tried when the sweating is profuse. Amongst internal remedies may be mentioned gallic acid in full doses, which often answers well when given at bed-time, the mineral acids, bark, the muriated tincture of iron and opium; whilst, as external applications, the cold shower-bath or the dripping sheet and sponging the body with vinegar and water, often prove serviceable. The symptom itself is referable to debility, and, as it gives rise to considerable exhaustion, active measures should be taken to uphold the patient and prevent its continuance. In no instance ought the administration of food, together with some alcoholic stimulant on going to bed, to be neglected.

Nausea and vomiting, if attended with acidity of the stomach and not dependent on derangement of the liver, may usually be subdued by lime-water, or by the use of effervescing draughts with hydrocyanic acid, a drachm of Schacht's solution of bismuth and a slight excess of soda. But, occasionally, when the tongue is clean and the stomach irritable without any apparent

[1] See case in point referred to at p. 410 of this treatise.

biliary derangement, I have found greater benefit from the use of strychnia, nux vomica, or gallic acid, and sometimes from chloroform, creasote, or opium. Brandy and soda-water is often serviceable in these cases, and mustard cataplasms or blisters to the epigastrium are almost invariably productive of comfort, even if they do not arrest the symptoms. If the tongue and mouth are covered with aphthæ, nothing will give such speedy relief as the repeated administration of sulphite of soda, which may also be employed in saturated solution as a wash for the mouth and throat.

Diarrhœa is another symptom the treatment of which requires the exercise of considerable judgment. If the tongue be furred and the alvine secretions pale or otherwise unhealthy, and if, moreover, there be not any evidence of inflammatory mischief in the abdomen, a dose of calomel and opium, followed by a warm rhubarb draught or half an ounce of castor oil, will probably relieve the symptoms; whereas if the tongue be nearly clean and very red, giving evidence of long-standing intestinal irritation, the probability is that ulceration of the bowels is going on, and that calomel and rhubarb would only aggravate the mischief. In such a case chalk mixture is not of much service, and sulphate of copper and opium, acetate of lead and opium, gallic acid, ipecacuanha in full doses, the nitrate of bismuth, kino, rhatany, tormentilla, and hæmotoxylum, are the more appropriate remedies, and they may be aided by starch and laudanum enemata. If the diarrhœa is attended by pain in the abdomen, poppy fomentations, poultices sprinkled with laudanum, or sinapisms, should be employed in aid of the internal remedies.

Fistula in ano is another symptom which must not be lightly dealt with. As long as the discharge is insignificant in amount and the patient's mind is not seriously disturbed by its continuance, so long is it advisable to confine our efforts to the treatment of the constitutional malady, and not to disturb the fistula. Any attempt at curing it by operation under these circumstances would probably be followed by an immediate increase of the cough and other pectoral symptoms, and, therefore, would be highly injudicious. But I do not hold with those who maintain that a fistula in ano occurring in the course of phthisis ought never to be interfered with. In some instances the discharge is profuse, and constitutes an important source of waste; and the patient is so distressed and alarmed at its continuance, that no treatment can be of avail until his nervous apprehensions are overcome. He can neither eat nor sleep for thinking of it, and his whole system is depressed in consequence.

In such cases I have known the greatest benefit result from an operation, combined with the formation of an issue in the arm, the use of a proper diet, and the administration of cod-liver oil, quinine, iron, and other appropriate remedies. Not only has the fistula healed, but the general health has improved, the patient has gained flesh, and the physical signs of pulmonary disease have materially decreased.

Chronic peritonitis, which sometimes occurs in connection with the deposit of tubercle in the peritoneum, is little under the control of medicine. Calomel, which is often serviceable in ordinary peritonitis, very generally proves mischievous in this form of disease; and perhaps opium, iodide of potassium, and cod-liver oil are the internal remedies which are not only most likely to afford relief to the local irritation, but also best calculated to subdue the constitutional irritation. Externally, applications of a remedial nature can be made, which, if they are unequal to arrest the disease, are yet efficacious in mitigating its symptoms and affording comfort to the patient. Amongst these may be mentioned bran or mustard poultices, or poppy or turpentine fomentations, or, better still, a lotion containing laudanum, belladonna, iodine, and glycerine, kept constantly applied to the abdomen on a piece of lint covered with flannel.

Except in connection with the deposit of tubercle in the brain or its membranes, delirium is of rare occurrence in phthisis, and does not require any special treatment; but when meningitis occurs it must be treated on general principles by calomel and active purgation, aided by a blister to the nape of the neck, the application of cold to the head, and, if necessary, of a few leeches to the temples—the extent to which the treatment is carried being modified by the tubercular character of the disease and the constitutional power of the patient. As soon as the first fury of the attack is overpast, iodide of potassium, with iodide of iron and cod-liver oil, are the appropriate remedies.

A few words may be added respecting the treatment of acute phthisis. Theoretically, there is no reason why the symptoms should not be combated on general principles, and practically, I believe the adoption of those principles will afford the patient the best chance of obtaining relief. This form of disease is characterized by so much asthenia that the abstraction of blood cannot be right, however great the pulmonary congestion may appear to be, and dry cuppings and counter-irritation are the expedients which locally offer most chances of benefit. The action of these external remedies should be assisted by the internal administration of salines, digitalis, aconite, the veratrum viride, the prunus virginiana, and sedatives. If the depression be not very great, and the existence of fine crepitation gives evidence of inflammatory mischief in the lungs, some antimonial wine may be cautiously added; and if the attack be complicated by pleurisy, it may be necessary to give opium and small doses of calomel. In most instances, however, the use of fomentations and judicious counter-irritation will supersede the necessity for more active treatment. The immersion of the feet and legs in hot water for the space of an hour is useful in inducing revulsive action and promoting perspiration if the skin is hot and dry. When the more acute symptoms have subsided, and the state of the tongue indicates an improved condition of the digestive organs, cod-liver oil should be prescribed, and the other remedies given which prove serviceable in chronic forms of the malady.

CHAPTER V.

INTRA-THORACIC TUMORS.

In the sections devoted to percussion and auscultation frequent reference was made to the signs produced by intra-thoracic tumors, and the subject was partially mooted again when tuberculous enlargement of the bronchial glands was under discussion. It may be desirable, however, to point out more distinctly what can and what cannot be done towards their diagnosis and treatment.

It should be premised that tumors in the chest are either aneurismal or produced by morbid growths or deposits, usually of a malignant or scrofulous character. Malignant tumors are sometimes attached to the parietes of the chest, but more commonly, like aneurism, they are developed in the mediastina, and ordinarily take their origin in the glandular structures. The nature of the growth has little influence on the symptoms produced, for the phenomena are not peculiar to any special form of disease, but are referable to the interference with the respiration and circulation produced by the pressure of the tumor on the surrounding parts. Thus the physical signs and general symptoms vary not only with the seat and origin of the tumor, but with the size which it has attained, the direction in which it enlarges, and the nature of the organs on which consequently it exerts pressure.

The physical signs are of two kinds, viz., those which indicate the presence of consolidation and those which bespeak pressure on the adjacent structures. When the tumor is attached to the parietes or occupies the anterior portion of the chest, there will be intense dulness on percussion in the region of the tumor and strongly marked parietal resistance; weakness, or absence, or some modification of the respiratory sounds; and, according as the tumor is large or small, and of an inelastic or of a vibratile nature, there will be increase or absence of vocal resonance and vocal fremitus, and increased transmission or otherwise of the sounds of the heart. But the signs of pressure will be even more important. The intercostal spaces may bulge under the influence of the pressure from within, and even the bony structures, such as the ribs, the sternum, and the clavicles, may gradually wear away, until perforation takes place, and the tumor presents itself externally. Indeed, circumscribed pulsation is often felt, and with it a thrill transmitted from the heart or large vessels. The heart may be pushed out of its place, or the aorta may be compressed, occasioning systolic murmur; or the innominate and subclavian arteries may be pressed upon, with the effect of producing innominate and subclavian systolic murmurs, and weakness of the corresponding carotid and radial pulses; or the superior vena cava may be compressed or surrounded by the tumor, in which cases the venous circulation is interfered with, so that distension of the jugular, subclavian, axillary, and other veins occurs, tortuous and swollen veins show themselves superficially on the chest and abdomen, œdema commences in the face, neck and arm, and upper part of the body, and ultimately the patient becomes drowsy from congestion of the brain and stagnation of the blood in the venous sinuses of the dura mater; or the pulmonary artery may be subjected to pressure, which may induce a systolic pulmonary murmur, dyspnœa consequent on a deficiency in the supply of blood to the lungs, and intense congestion in the whole venous system; or the pulmonary veins may be more especially subjected to pressure, whereby dyspnœa, hæmoptysis, œdema of the lungs, and hydrothorax, will be produced; or the inferior cava may be pressed upon, and congestion of all the abdominal organs produced, together with distension of the veins of the lower extremities, and ultimately of ascites and œdema of the legs and thighs; or the bronchi may be pressed upon and irritated, so that increased mucous secretion is induced, accompanied by râles, rhonchi, and prolonged expiration; yet further, the bronchi may even be perforated by the pressure, and hemorrhage may take place, with frequently recurring hæmoptysis, and a violent, spasmodic, ringing cough;

or the lung itself may be seriously encroached upon and deprived of air by the long continuance of pressure, until respiration ceases to be heard over the part, and percussion elicits intense dulness; or the tumor may give rise to excessive local irritation, and pleurisy, bronchitis, or pneumonia, may be set up, or the vessels supplying the lung may be pressed upon, and its nutrition so far interfered with that gangrene of the part may be the result.[1] When the tumor occupies the central and posterior portions of the chest, the dulness on percussion is seldom so marked as in the former instances, and may not be perceptible on gentle percussion, though it is brought out by a forcible stroke.

The auscultatory signs are very variable, being modified by the size and physical nature of the tumor, and by its precise relation to the surrounding organs and the walls of the chest. Sometimes the breathing is weak or almost absent, either locally or over an entire side, and vocal resonance and vocal fremitus are also absent; but not unfrequently, when the trachea or the larger bronchi are pressed upon, the percussion-note is of a tubular or amphoric character: the breathing, instead of being weak or absent, may be intensely loud and tubular, the expiratory sound prolonged, the vocal fremitus and vocal resonance increased, and the cardiac sounds transmitted far beyond their usual limit. If the phrenic and the vagi nerves or the pulmonary plexus are pressed upon by the tumor, disturbance occurs in the respiration; whilst, if the recurrent nerve is implicated, aphonia or some other modification of the voice is induced. If, as often happens, the œsophagus is pressed upon, dysphagia is the inevitable result, and if it is perforated hæmatemesis may occur; if the thoracic duct were compressed, emaciation would proceed with extreme rapidity; if the trachea is subjected to pressure, stridulous breathing and weakness of voice are induced, the cough, which is paroxysmal, is either very weak or else loud, shrill, and clanging, and the expectoration is often streaked with blood. The trachea, the bronchi, the pulmonary artery, the œsophagus, and other structures, are liable to be perforated by the tumor.

The general symptoms of intra-thoracic tumor possess little of a distinctive character. A tumor may long exist in the chest, without giving rise to any notable symptoms, and the patient

[1] For the details of a remarkable instance of pulmonary gangrene, which occurred under my care in St. George's Hospital, as the result of pressure by an aneurismal tumor, see "Hospital Post-mortem Register and Casebook," for February 3d, 1860, under the head of Henry Barnes; also "Trans. Path. Soc.," vol. xi, p. 62.

may not be aware of its presence until its development has led to pressure on the surrounding structures. Then occur pain, which is variable in amount, and is sometimes altogether absent, but which, when constant, as it often is in the spine at the junction of the neck and back, is a symptom of grave significance; dyspnœa, frequently influenced by posture; wheezing and stridor, more or less distinctly marked, and peculiarly so with the expiration; cough of a paroxysmal character and peculiarly weak, or else loud, ringing, and of a metallic quality; expectoration, which for a long time is simply catarrhal, or else consists of clear, gelatiniform mucus, which may subsequently become blood-tinged or more or less largely mixed with blood; palpitation; frequent attacks of giddiness and fainting; excessive restlessness during the day and want of sleep and slight wandering at night. Oftentimes the patient is able to obtain repose in one position only—the posture which happens, by force of gravity, to relieve the suffering organ from the pressure of the tumor. Thus, he sometimes sits erect in bed; or remains day and night in an arm-chair, afraid to go to bed; or, sitting erect, he may lean towards one side or the other; or he may lie down in bed comfortably on one side, though the least attempt to move on to the other will induce violent spasmodic cough and a sense of impending suffocation; or, again, he may lean forward, with his elbows resting on his knees, and his head supported by his hands. In addition to these symptoms, there may be inequality or entire absence of the pulse at one or both wrists, inequality of the pupils arising from pressure on the sympathetic nerve, ptosis of the eyelid, difficulty in swallowing, hæmatemesis, aphonia, distension of the superficial veins, and œdema of the face, trunk or limbs, according as the tumor presses on one organ or another.

There are four circumstances which will sometimes enable us to arrive approximately at a decision as to whether a tumor in the chest is aneurismal, malignant, or scrofulous. The first is the existence of aneurism, or malignant or scrofulous disease elsewhere in the body. The second is that when the tumor is aneurismal it is seldom accompanied by local inflammation, and that a murmur may be sometimes heard in a distant part of the chest, although none exists in the region of the heart; or a loud murmur may exist in both situations, but may be lost or almost inaudible at intermediate points. The third is that the symptoms induced by the pressure of an aneurismal tumor are apt to be extremely variable. A diseased mass, whether malignant, scrofulous, or fibrous, having broad and firm adhesions in the chest—arising, that is, from a broad, extended surface, as such masses do almost invariably—would not be influenced by change of posture sufficiently to effect any material diminution in the amount of pressure, whether on the spine, the œsophagus, the bronchi or other parts; whereas an aneurismal tumor, arising, as it often does, from a narrow base, and attached only to a vessel which admits of considerable motion, may have its position altered considerably by gravity alone, so that the suffering organ may be relieved by change of posture, and the symptoms referable to posture may be mitigated. The fourth is one to which attention was first directed by Dr. George Budd,[1] viz., that when the tumor is cancerous, it is more apt than in the other cases to be accompanied by local symptoms of an inflammatory nature, a circumstance which is due to the fact that malignant growths often involve the root of the lungs, and implicate or destroy the greater part of the nerves which supply those organs. The result is that the nutrition of the lungs is interfered with, and inflammation is set up, just as inflammatory destruction of the eyeball ensues after division of the fifth nerve within the cranium.

It is not difficult to distinguish intra-thoracic tumors from the effect of phthisis, pneumonia, and pleurisy, though, in the absence of a careful physical examination of the chest, mistakes of this kind have been made. Thus intra-thoracic tumors may produce modifications of the percussion-sound, the respiratory sounds, and the vocal resonance, analogous to those which accompany phthisis; and may also give rise to râles and rhonchi, which are not distinguishable from those which are met with in that disease. But the probability is that in the former case the abnormal phenomena would be confined to one side of the chest, whereas in phthisis they almost always exist in both; that the course which the disease has run would not tally with that which phthisis ordinarily pursues; and that the appearance of the patient and the character and intensity of the general symptoms would not correspond with those which would be expected at the particular stage of the tubercular disorder which the physical signs would appear to indicate. Further, in the case of malignant disease the cervical glands, and especially those immediately above the clavicle, are very prone to be affected. The aphonia, which results from pressure on the recurrent nerve, is apt sometimes to complicate the case, and cause it to simulate laryngeal phthisis; but the voice, though weak, or even whispering, will not be materially altered in tone, whereas, in laryngeal phthisis, it would not only be weak, but thoroughly hoarse, harsh, and altered in character. The only case in which

[1] "Med.-Chir. Trans," vol. xlii.

any doubt ought to arise is when a mass of enlarged bronchial glands excites symptoms of pressure coincidentally with the symptoms of phthisis, and even then the evidence of intra-thoracic pressure, interpreted through the agency of the tubercular disease in the lungs, ought to indicate the existence and the nature of the intra-thoracic tumor.

With regard to pneumonia, its whole course is so unlike that of an intra-thoracic tumor that there is no possibility of the two diseases being confounded, unless the patient be not seen until all active symptoms have subsided. Even then the absence of the widespread tubular breathing and increased vocal fremitus and vocal resonance which characterize pneumonia, and, on the other hand, the presence of intense dulness on percussion, and of hæmoptysis and red gelatiniform expectoration, together with enlargement of the side, and the existence of signs resulting from the pressure and ulceration which usually accompany intra-thoracic tumors, ought to leave little doubt as to the true nature of the disorder.

The symptoms produced by chronic pleurisy, especially when it is circumscribed, are somewhat more perplexing, but ought to be unravelled without much difficulty. The intensity of the dulness on percussion may not vary greatly in the two cases, neither may the limits of the dulness be altered materially by change of posture, and in both cases the vocal fremitus and vocal resonance may be annihilated; but in chronic pleurisy the enlargement of the side will be uniform, and the intercostal spaces will often be convex, fluctuation will be perceptible in them, and there will not be enlargement of the superficial veins of the chest or other signs referable to circumscribed pressure; whereas, in the case of intra-thoracic tumor, the side, if altered in form, will be irregularly enlarged, there will be no intercostal fluctuation, and signs of circumscribed pressure will almost certainly present themselves; there will usually be found some spots within the area of dulness, where the percussion-sound is not completely gone or where the hollow breathing-sound is still to be heard, which could not well be if liquid were the cause of dulness, and there will often be hæmoptysis and red or colorless gelatiniform mucous expectoration.

Of course, if intra-thoracic tumor coexists with chronic pleurisy, and the pleuritic effusion be on the same side as the tumor, the diagnosis is more difficult, the more so when the early history of the disease is wanting. But if the signs produced by pressure on the intra-thoracic organs are present, it is certain that some intra-thoracic tumor exists, and the probability will be that the pleurisy is a mere accidental complication, excited by the irritation of the tumor.

The causes and modes of death in these cases of intra-thoracic tumors are various. In most instances the patient lingers on, the sense of oppression and the distress of breathing gradually increasing, until at length the appetite fails, he becomes wretchedly thin, and, worn out by want of sleep and oppressed with anasarca, he sinks thoroughly exhausted. But sometimes he gradually succumbs in consequence of the interference with the respiration induced by pressure on the par vagum, or on the phrenic nerve, or on the pulmonary plexus, or by obstruction of the pulmonary artery; or he sinks rapidly from hemorrhage resulting from perforation of the pulmonary artery or one of the larger vessels, or, in the case of an aneurismal tumor, from the giving way of the aneurismal sac; or he is cut off by pneumonia, sometimes followed by gangrene resulting from pressure of the morbid mass; or he dies suddenly from spasm of the glottis, induced by pressure on the recurrent nerve.

The treatment is necessarily of a palliative nature, and when the disease is far advanced our utmost efforts are of little avail to give even temporary relief; but in the earlier stages much good may be effected by remedies calculated to tranquillize the circulation and regulate the secretions, and also, as occasion may require, by dry cupping, blisters, poultices, opiate and iodine lotions, and other local applications, which have the effect of subduing pain, relieving local inflammation, and producing absorption of the inflammatory matters which have been effused.

INDEX.

www.ingramcontent.com/pod-product-compliance
Lightning Source LLC
Chambersburg PA
CBHW052340110726
47901CB00005B/1302
* 9 7 8 3 3 3 7 8 1 2 5 5 3 *